MEDIA AND COMMUNICATION – TECHNOLOGIES, POLICIES AND CHALLENGES

TAPPING THE TELEVISION WHITE SPACE SPECTRUM: A REVOLUTION IN PUBLIC AIRWAVE USE

MEDIA AND COMMUNICATION – TECHNOLOGIES, POLICIES AND CHALLENGES

Additional books in this series can be found on Nova's website under the Series tab.

Additional E-books in this series can be found on Nova's website under the E-books tab.

MEDIA AND COMMUNICATION – TECHNOLOGIES, POLICIES AND CHALLENGES

TAPPING THE TELEVISION WHITE SPACE SPECTRUM: A REVOLUTION IN PUBLIC AIRWAVE USE

JULIE E. HENDRIKS
EDITOR

Nova Science Publishers, Inc.
New York

For permission to use material from this book please contact us:
Telephone 631-231-7269; Fax 631-231-8175
Web Site: http://www.novapublishers.com

LIBRARY OF CONGRESS CATALOGING-IN-PUBLICATION DATA

Tapping the television white space spectrum : a revolution in public airwave use / editors, Julie E. Hendriks.
p. cm.
Includes index.
ISBN 978-1-60692-956-8 (hardcover)
1. Television broadcasting policy--United States. 2. Broadband communication systems--United States. I. Hendriks, Julie E.
HE8700.8.T326 2010
384.55'21--dc22
2010015633

Published by Nova Science Publishers, Inc. ✢ New York

CONTENTS

PREFACE

The Federal Communications Commission's Laboratory Division has completed a second phase of its measurement studies of the spectrum sensing and transmitting capabilities of prototype "TV white space" devices. These devices have been developed to demonstrate capabilities that might be used in unlicensed low power radio transmitting devices that would operate on frequencies in the broadcast television bands that are unused in each local area. This new book describes the tests performed during the second (Phase II) series of tests and provides a compilation of the results of measurements of the spectrum sensing and transmitting functions of various prototype "white space devices" (WSDs).

Chapter 1 - *Introduction.* The Federal Communications Commission's Laboratory Division has completed a second phase of its measurement studies of the spectrum sensing and transmitting capabilities of prototype "TV white space" devices. These devices have been developed to demonstrate capabilities that might be used in unlicensed low power radio transmitting devices that would operate on frequencies in the broadcast television bands that are unused in each local area. At this juncture, we believe that the burden of "proof of concept" has been met. We are satisfied that spectrum sensing in combination with geo-location and database access techniques can be used to authorize equipment today under appropriate technical standards and that issues regarding future development and approval of any additional devices, including devices relying on sensing alone, can be addressed.

The Commission is conducting a rulemaking proceeding to consider authorization of new, low power transmitting devices in the television broadcast spectrum at locations where channels are not being used for TV or other authorized services (ET Docket No. 04-186). This locally unused spectrum is often referred to as "TV white space." As established thus far by the Commission, white space devices (WSDs) that operate from a fixed location will be allowed into the TV spectrum simultaneous with the completion of the transition from analog to digital television broadcasting on February 17, 2009. This action will open for use a significant amount of spectrum with very desirable propagation characteristics that has heretofore lain fallow. It will also allow the development of new and innovative types of unlicensed devices that provide broadband data and other services for businesses and consumers without disrupting the incumbent television and other authorized services that operate in the TV bands. The Commission is considering whether to also allow "personal/portable" WSDs to operate in the TV spectrum.

One of the principal considerations in the white space proceeding is how to reliably determine the availability of unused frequencies in local areas. A number of parties participating in the proceeding have suggested an approach for identifying unused frequencies whereby a WSD would employ a "listen before talk" or "detect and avoid" strategy. This approach would use "spectrum sensing" techniques that listen for the signals of TV stations, wireless microphones, and other incumbent services. The Commission has requested comment on whether to require that the sensing capability of devices using this approach be able to detect signals as low as -116 dBm, or some alternative value. A second technical consideration in this matter is the potential for WSDs to interfere with TV reception and wireless microphone operations. To address these issues, the Commission announced that it would perform testing of the spectrum sensing and transmitting capabilities of the prototype WSDs.

The purpose of the testing program is to provide additional information for the record that will be considered along with other information in developing the Commission's final decision on white space devices. The tests are not intended for equipment authorization or to determine whether the devices would comply with any possible standards that the Commission might adopt. Rather, they will provide information in support of the Commission's action in this matter. Initial tests (Phase I) under this program were completed in July of 2007. This chapter describes the tests performed during the second (Phase II) series of tests and provides a compilation of the results of measurements of the spectrum sensing and transmitting functions of various prototype WSDs.

In October 2007, the Commission's Office of Engineering and Technology (OET) issued a Public Notice announcing that it would perform additional laboratory and field tests of white space devices and invited interested parties to submit such devices for testing at the FCC Laboratory in Columbia, Maryland. Subsequently, OET issued a second Public Notice announcing that the second phase testing would begin on January 24, 2008.

WSD Prototype Devices Submitted for Evaluation. Five devices were submitted for examination in the Phase II tests. These devices were provided by Adaptrum, the Institute for Infocomm Research (I2R), Microsoft Corporation, Motorola Inc., and Philips Electronics North America (Philips). These devices are not intended as actual consumer products but rather are development tools for evaluating the viability of spectrum sensing and potential interference. They do not communicate with other devices. Not all tests were performed on all devices. For example, the Microsoft device was available only for limited tests in the laboratory and the I2R device was submitted after the initial tests were completed.

Spectrum Sensing for TV Broadcast Signals. This portion of the study examined the ability of the WSDs to detect whether channels are occupied by ATSC (digital) TV signals. All of the prototype devices had capabilities for detecting TV broadcast signals on UHF channels 21-5 1, the operating range of the prototype devices. The tests were initially performed in the laboratory under various controlled conditions. Spectrum sensing sensitivity to clean digital TV signals in isolation and also in the presence of other TV signals on adjacent channels was measured. In addition, sensitivity tests were performed using recorded TV signals to simulate "real world" conditions. Spurious emissions generated by the prototype WSDs were also measured. The laboratory tests were followed by field tests at nine sites in Maryland and the District of Columbia to evaluate the DTV sensing performance.

Spectrum Sensing for Wireless Microphones. The wireless microphone portion of the testing looked at the ability of the WSDs to detect wireless microphones authorized under Part 74 of FCC rules. Microsoft, Philips and I2R indicated that their devices were capable of sensing wireless microphones. Tests of this capability were initially performed in a controlled environment in the laboratory. Those tests were followed by field tests at two sites, one in Maryland and one in New York City, to evaluate the capability of the devices to detect wireless microphones under field conditions.

Transmitter Characterization and Interference Testing. The Adaptrum device included transmitting capability. Laboratory tests were performed to characterize the transmitter's signal, which is an important element for assessing the interference potential of WSD devices. Field tests were also performed to evaluate potential interference from the Adaptrum transmitter; however, these tests were limited.

Chapter 2 - This appendix provides information regarding the measurement equipment used to perform this test/measurement program. Table A-1 provides a list of the equipment used, including the manufacturer and model number, a brief description and where appropriate, the most recent calibration date.

Calibration information resulting from measurements performed in the laboratory to determine the signal attenuation in various circuits used to support the test/measurements is also provided.

The following information is provided with respect to calibrations performed as a part of this program to determine the signal attenuation (or gain) associated with the various components used to support the test/measurements.

The input signal level used in the detection threshold was set at the input to the WSD scanner antenna input port and thus, no signal calibration was necessary.

A short length (2-feet) of RG-223/U coaxial cable was used to connect the WSD transmitter output port directly to the spectrum analyzer for performing the transmitter emissions measurements. The signal attenuation in this cable was measured and found to exhibit a maximum loss of 0.4 dB over the frequency range involved (5 12-698 MHz). The insertion loss associated with the channel-reject (notch) filter is 2.9 dB, including the additional coaxial cable used to integrate into the measurements system.

Chapter 3 - The Commission is conducting a proceeding in ET Docket No. 04-186 to consider authorizing the operation of new, low power devices in the television (TV) broadcast spectrum at locations where channels are not being used for authorized services.[1] This unused spectrum is often referred to as the "TV white spaces."

As part of this proceeding, the Office of Engineering and Technology (OET) is conducting a test program on TV white space prototype devices. The purpose of this testing program is to provide additional information for the record that will be considered in assessing the interference potential of such devices and appropriate requirements.

Initial tests (Phase I) were completed in July of 2007.[2] The Office of Engineering and Technology on October 5, 2007, issued a public notice inviting submittal of prototype devices for further tests (Phase II).[3] The Public Notice stated that further details on the testing would be released at a later time.

As part of this effort, several prototype devices were recently submitted for testing.[4] The devices are not finished consumer products. They are designed to demonstrate the feasibility of various concepts for sharing the TV broadcast spectrum without causing harmful interference. The Commission may ultimately establish requirements that the current prototype devices do not meet. However, all products would need to be certified as complying with the final rules before they may be marketed to consumers.

This document generally describes the tests that will be conducted in Phase II. The Phase II tests largely follow the procedures used for Phased I with some modifications. Comments and suggestions offered within the public record in the Commission's proceeding on the TV white spaces with respect to both the previous and current testing were considered and included where appropriate and practicable. The Phase II tests will be open to attendance and observation by any interested party.

The testing will include both laboratory (bench) tests and field tests. The laboratory tests will measure the performance capabilities of the prototype devices under controlled conditions. The field tests will be conducted at a variety of locations to provide information on the performance of the devices under real world conditions OET staff will adapt the test plan as appropriate based on the specific capabilities of each device and circumstances that may arise as the tests progress.

At the conclusion of the Phase II testing, a report will be issued and placed into the record in the Commission's TV white spaces proceeding. Any deviations from the test plan set forth herein will be described in this test report.

Chapter 4 - The following is the peer review of OET Report FCC/OET 08-TR-1005, entitled "Evaluation of the Performance of Prototype TV-Band White Spaces Devices Phase II." The report is a result of OET's second phase study of hardware development devices with capabilities for operating on an unlicensed basis in the TV bands as part of the "TV white spaces" proceeding, ET Docket No. 04-186. This second phase study examined four development devices and follows OET's initial study of TV white space development devices, which was released July 31, 2007, *see* "Initial Evaluation of the Performance of Prototype TV-Band White Space Devices," OET Report FCC/OET 07-TR-1006 (Phase I Report).[1] The devices examined in this second phase study were submitted by industry in response to an OET public notice inviting submittal of additional prototype white space devices and are product development platforms, rather than models of products that could actually be marketed. One of these devices has both transmitting and spectrum sensing capabilities; the other three have only spectrum sensing capability. Each of the devices has unique spectrum sensing capabilities and was examined in the laboratory and the field with respect to its ability to detect the presence of digital TV, analog TV, and/or Part 74 wireless microphone signals. The one device with a transmitter was also examined for its potential to cause interference to the reception of digital television signals and cable service.

Peer review of the OET report was performed as required under the OMD Information Improvement Act for influential scientific and engineering studies. The review panel was made up of five engineers, three from the Enforcement Bureau and two from the Wireless Telecommunications Bureau. The review panel analyzed and discussed various subject areas in the OET report, both independently and jointly. Specifically, as requested in the OET memo, the review panel addressed the following:

1. Whether the scope of testing in terms of spectrum sensing abilities and signal conditions examined was appropriate and sufficient;
2. Whether the measurement methodologies used in the testing of the prototype devices spectrum sensing abilities was appropriate;
3. Whether the scope of testing of the device with transmitting capability for its potential to cause interference to digital TV, analog TV, and wireless microphone signals was appropriate
4. Whether the various tests performed were properly conducted consistent with the selected methodologies.

Chapter 5 - The following information and comments are submitted in response to the October 1, 2008 memorandum of the peer review panel that provides their report on the review of the Office of Engineering and Technology's Phase II measurement study of prototype TV-band white space devices (WSDs). The study and its findings are set forth in a report entitled "Evaluation of the Performance of Prototype TV-Band White Space Devices, Phase II", OET Report FCC/OET 08-TR-1005. This study examined the performance of the "detect and avoid" (spectrum sensing) techniques used by several prototype WSDs to evaluate their ability to detect television channels occupied with incumbent signals (digital or analog TV and wireless microphones). The peer review of this study was performed by staff of the Commission's Wireless Telecommunications and Enforcement Bureaus pursuant to the Office of Management and Budget's requirement under the Information Quality Act that influential scientific assessments be subject to peer review to enhance the quality and credibility of the government's scientific information.

Chapter 6 - In its continuing efforts to promote efficient use of spectrum and to extend the benefits of such use to the public, the Federal Communications Commission (FCC) today adopted a Second Report and Order (Second R&O) that establishes rules to allow new, sophisticated wireless devices to operate in broadcast television spectrum on a secondary basis at locations where that spectrum is open. (This unused TV spectrum is now commonly referred to as television "white spaces"). The rules adopted today will allow for the use of these new and innovative types of unlicensed devices in the unused spectrum to provide broadband data and other services for consumers and businesses.

Chapter 7 – These remarks were delivered as Separate Statement of Michael J. Copps, Commissioner, Federal Communications Commission, Second Report and Order and Memorandum Opinion and Order, dated November 2008.

Chapter 8 – These remarks were delivered as Separate Statement of Deborah Taylor Tate, Commissioner, Federal Communications Commission, Second Report and Order and Memorandum Opinion and Order, dated November 2008.

Chapter 9 – These remarks were delivered as Statement of Kevin J. Martin, Chairman, Federal Communications Commission, Second Report and Order and Memorandum Opinion and Order, dated November 2008.

Chapter 10 – These remarks were delivered as Statement of Robert M. McDowell, Commissioner, Federal Communications Commission, Second Report and Order and Memorandum Opinion and Order, dated November 2008.

Chapter 11 – These remarks were delivered as Statement of Jonathan S. Adelstein, Commissioner, Federal Communications Commission, Second Report and Order and Memorandum Opinion and Order, dated November 2008.

In: Tapping the Television White Space Spectrum: A Revolution... ISBN: 978-1-60692-956-8
Editor: Julie E. Hendriks

Chapter 1

EVALUATION OF THE PERFORMANCE OF PROTOTYPE TV - BAND WHITE SPACES DEVICES PHASE II

Steven K. Jones, Thomas W. Phillips, Hugh L. Van Tuyl and Robert D. Weller

EXECUTIVE SUMMARY

Introduction. The Federal Communications Commission's Laboratory Division has completed a second phase of its measurement studies of the spectrum sensing and transmitting capabilities of prototype "TV white space" devices. These devices have been developed to demonstrate capabilities that might be used in unlicensed low power radio transmitting devices that would operate on frequencies in the broadcast television bands that are unused in each local area. At this juncture, we believe that the burden of "proof of concept" has been met. We are satisfied that spectrum sensing in combination with geo-location and database access techniques can be used to authorize equipment today under appropriate technical standards and that issues regarding future development and approval of any additional devices, including devices relying on sensing alone, can be addressed.

The Commission is conducting a rulemaking proceeding to consider authorization of new, low power transmitting devices in the television broadcast spectrum at locations where channels are not being used for TV or other authorized services (ET Docket No. 04-186). This locally unused spectrum is often referred to as "TV white space." As established thus far by the Commission, white space devices (WSDs) that operate from a fixed location will be allowed into the TV spectrum simultaneous with the completion of the transition from analog to digital television broadcasting on February 17, 2009. This action will open for use a significant amount of spectrum with very desirable propagation characteristics that has heretofore lain fallow. It will also allow the development of new and innovative types of unlicensed devices that provide broadband data and other services for businesses and consumers without disrupting the incumbent television and other authorized services that operate in the TV bands. The Commission is considering whether to also allow "personal/portable" WSDs to operate in the TV spectrum.

One of the principal considerations in the white space proceeding is how to reliably determine the availability of unused frequencies in local areas. A number of parties participating in the proceeding have suggested an approach for identifying unused frequencies whereby a WSD would employ a "listen before talk" or "detect and avoid" strategy. This approach would use "spectrum sensing" techniques that listen for the signals of TV stations, wireless microphones, and other incumbent services. The Commission has requested comment on whether to require that the sensing capability of devices using this approach be able to detect signals as low as -116 dBm, or some alternative value. A second technical consideration in this matter is the potential for WSDs to interfere with TV reception and wireless microphone operations. To address these issues, the Commission announced that it would perform testing of the spectrum sensing and transmitting capabilities of the prototype WSDs.

The purpose of the testing program is to provide additional information for the record that will be considered along with other information in developing the Commission's final decision on white space devices. The tests are not intended for equipment authorization or to determine whether the devices would comply with any possible standards that the Commission might adopt. Rather, they will provide information in support of the Commission's action in this matter. Initial tests (Phase I) under this program were completed in July of 2007.[1] This chapter describes the tests performed during the second (Phase II) series of tests and provides a compilation of the results of measurements of the spectrum sensing and transmitting functions of various prototype WSDs.

In October 2007, the Commission's Office of Engineering and Technology (OET) issued a Public Notice announcing that it would perform additional laboratory and field tests of white space devices and invited interested parties to submit such devices for testing at the FCC Laboratory in Columbia, Maryland.[2] Subsequently, OET issued a second Public Notice announcing that the second phase testing would begin on January 24, 2008.[3]

WSD Prototype Devices Submitted for Evaluation. Five devices were submitted for examination in the Phase II tests. These devices were provided by Adaptrum, the Institute for Infocomm Research (I2R), Microsoft Corporation, Motorola Inc., and Philips Electronics North America (Philips). These devices are not intended as actual consumer products but rather are development tools for evaluating the viability of spectrum sensing and potential interference. They do not communicate with other devices. Not all tests were performed on all devices. For example, the Microsoft device was available only for limited tests in the laboratory and the I2R device was submitted after the initial tests were completed.

Spectrum Sensing for TV Broadcast Signals. This portion of the study examined the ability of the WSDs to detect whether channels are occupied by ATSC (digital) TV signals. All of the prototype devices had capabilities for detecting TV broadcast signals on UHF channels 21-5 1, the operating range of the prototype devices. The tests were initially performed in the laboratory under various controlled conditions. Spectrum sensing sensitivity to clean digital TV signals in isolation and also in the presence of other TV signals on adjacent channels was measured. In addition, sensitivity tests were performed using recorded TV signals to simulate "real world" conditions. Spurious emissions generated by the prototype WSDs were also measured. The laboratory tests were followed by field tests at nine sites in Maryland and the District of Columbia to evaluate the DTV sensing performance.

Spectrum Sensing for Wireless Microphones. The wireless microphone portion of the testing looked at the ability of the WSDs to detect wireless microphones authorized under Part 74 of FCC rules. Microsoft, Philips and I2R indicated that their devices were capable of sensing wireless microphones. Tests of this capability were initially performed in a controlled environment in the laboratory. Those tests were followed by field tests at two sites, one in Maryland and one in New York City, to evaluate the capability of the devices to detect wireless microphones under field conditions.

Transmitter Characterization and Interference Testing. The Adaptrum device included transmitting capability.[4] Laboratory tests were performed to characterize the transmitter's signal, which is an important element for assessing the interference potential of WSD devices. Field tests were also performed to evaluate potential interference from the Adaptrum transmitter; however, these tests were limited.

Observations

- All of the devices were able to reliably detect the presence a clean DTV signal on a single channel at low levels in the range of -116 dBm to -126 dBm; the detection ability of each device varied little relative to the channel on which the clean signal was applied. These measurements did not take into account the antenna that would be used with personal/portable devices.
- The detection threshold sensitivity of the devices varied from -106 dBm to -128 dBm when recorded off-air DTV signals, which included multi-path fading and other "real-world" distortion, were used. The impact on detection ability varied considerably among the WSDs and with the characteristics of the different recorded signals.
- Several tests were performed with DTV signals present in adjacent channels. These tests showed that in the presence of moderate-to-strong signals in a first adjacent channel, the detection threshold sensitivity of all of the devices was severely impacted. For some of the devices, the degradation in the detection sensitivity was as much as 60- 70 dB. In some cases, the degradation was such that the detection threshold could not be measured. This could impact significantly the ability of the devices to reliably detect TV signals within stations' service areas.
- The Microsoft, Philips and I2R devices were tested for their ability to sense for the presence of wireless microphones (both FM/analog and digital) operating within UHF TV channels. With no other signals present, the devices were able to detect wireless microphones at levels ranging from -103 dBm to -129 dBm depending on the type of microphone, and the device. In the presence of DTV signals in adjacent channels, the detection threshold was degraded such that it affected the ability of the devices to reliably detect the microphone signals.

- Scan time for the devices varied from 0.1 second per channel (Motorola) to 185 seconds per channel (Adaptrum). The Adaptrum device was modified during the tests and its scan time changed from 37 seconds per channel to 185 seconds per channel.
- Channel Occupancy (TV sensing) field tests were performed at nine locations for the Adaptrum, I2R, Motorola, and Philips devices.
    - In most cases, the devices correctly reported channels as occupied when the device was operated within the service contour of the stations broadcasting on those channels and viewable signals were observed on the channels.
    - In some instances, the Adaptrum, I2R, and Motorola (in sensing only mode) devices incorrectly reported channels as unoccupied (available) when the WSD was operated within a station's service contour and the signal was viewable.
    - All of the devices reported some channels as occupied when the WSD was operated outside of the service contours of stations broadcasting on those channels whether the signal was viewable or not.
    - The Philips device generally reported most channels occupied, whether the WSD was operating inside or outside any service contours whether the signal was viewable or not.
    - During the field tests, the Motorola device's geolocation/database feature was used in combination with its sensing capabilities. In those tests, the Motorola device correctly reported all occupied channels used by stations within whose contours the WSD was operated.
- Wireless microphone sensing tests were performed using the I2R and Philips devices at two field locations. The tests were conducted first with microphones off, and then turned on, in pre-determined channels to determine if the devices could sense the presence of wireless microphones. At both sites and all the test locations, the Philips device reported all the channels on which the microphones were designated to transmit as occupied whether the microphone was transmitting or not. The I2R device indicated several channels as available even when the microphones were on.
- The Adaptrum device's transmitter was characterized in the laboratory and was used to investigate interference potential to DTV signal reception. Anecdotal tests demonstrated that co-channel interference would occur at line-of-sight distances of up to 360 meters at an EIRP of approximately +7 dBm when the DTV was receiving a weak signal using a receive antenna at a height of 9.3 meters. No interference was observed when the Adaptrum device transmitted on an immediate adjacent channel even with the transmitter in close proximity to the receiver with a roof-top antenna. No other configurations were tested for interference.
- Anecdotal tests were performed at two field sites to assess the interference potential from WSD transmitters to cable television reception via direct pick-up of signals by cable system components. These tests showed that under certain circumstances, when the transmit antenna was placed in close proximity to a cable connected TV, direct pick-up interference was observed. The direct pick-up interference potential appears to be highly dependent on the interconnection among various system components (*e.g.*, cable amplifiers, splitters and set-top boxes) being used.

1. INTRODUCTION

The Federal Communications Commission's (FCC) Laboratory Division has completed the second phase of its program to measure the spectrum sensing and transmitting capabilities of devices intended to operate on frequencies in the broadcast television bands that are unused in each local area. These so-called "white space devices" (WSDs) have been developed to demonstrate features that might be used in unlicensed low power radio transmitting devices.

This measurement project was undertaken in support of the Commission's pending consideration in ET Docket No. 04-186 on whether to permit low power radio transmitting devices to operate on an unlicensed basis in the frequency bands that are currently allocated to the television broadcast and certain other licensed services.[5] Such low power operations would only be allowed on frequencies that are not used by TV stations or other licensed services in each local area. These unused, *i.e.*, vacant or "available," frequencies in local areas are often termed spectrum "white spaces." As established thus far by the Commission, fixed white space devices (WSDs) will be allowed into the TV spectrum simultaneous with the completion of the transition from analog to digital television broadcasting on February 17, 2009. This action will open for use a significant amount of spectrum with very desirable propagation characteristics that has heretofore lain fallow and will allow the development of new and innovative types of unlicensed devices that provide broadband data and other services for businesses and consumers without disrupting the incumbent television and other authorized services that operate in the TV bands. The Commission is considering whether to also allow "personal/portable" WSDs to operate in the TV spectrum.

The Commission is considering two categories of devices for unlicensed operation in the TV white spaces.[6] The first category consists of low-power "personal/portable" WSDs that will function similar to WiFi transceivers in laptop computers, and wireless in-home local area networks (LANs). The second category consists of higher-powered "fixed/access" WSDs that would typically be operated from a fixed location and might be used to provide a commercial service such as wireless broadband access. As established previously by the Commission, fixed white space devices will be allowed into the TV spectrum upon completion of the transition from analog to digital television broadcasts on February 17, 2009.[7] The Commission is now considering whether to allow non-fixed "personal/portable" WSDs to also operate in the TV spectrum.

An important consideration in the proceeding is how to ensure that unlicensed devices operate only on vacant frequencies. One approach under consideration is for the WSD to employ "smart radio" features that would use a "detect and avoid" or "spectrum sensing" strategy. An alternative approach would rely on accessing a database of licensed services to identify active services near the device's location. The device location would be determined by an integral geo-location technology, such as GPS. This technique is commonly known as a "geolocation" detection approach. A second issue is the potential for signals transmitted by a WSD to cause radio frequency interference (RFI) to incumbent users of these bands (*i.e.*, TV reception and wireless microphone operations). The Commission indicated that it would test prototype devices to collect technical information regarding both of these issues.

Consistent with the Commission's plan for white space testing, the Office of Engineering and Technology (OET), through its Laboratory Division, performed initial tests of two prototype devices and issued a report on the results of those tests in July of 2007.[8]

Subsequently, on October 5, 2007, OET issued a Public Notice inviting submittal of additional prototype devices for further tests (Phase II).[9] On January 17, 2008, OET issued a test plan for the Phase II tests to help ensure that the testing would be open to outside observers, and announced that the testing would commence on January 24, 2008.[10]

Five parties responded to the announcement of the Phase II testing and provided prototype WSDs to the Laboratory for testing: Adaptrum, the Institute for Infocomm Research (I2R), Microsoft Corporation (Microsoft), Motorola Inc., and Philips Electronics North America Corp (Philips).[11] The Microsoft devices malfunctioned during the test process and were not replaced. The I2R device was submitted after the initial laboratory tests were completed. As a result, not all the tests were performed on these two devices. All of these prototypes have a sensing capability, but only two (Microsoft and Adaptrum) have a transmitter. Since the Microsoft device malfunctioned before the transmitter tests were performed, its transmitter was not used in any of these tests.

This chapter describes the tests and measurements performed on the prototype WSDs and the results obtained, with focus on impacts to broadcast television and wireless microphone operations. In particular, this chapter provides the results of tests of the spectrum sensing capabilities of the prototype devices as a means for identifying TV channels unused by TV broadcast or wireless microphone operations and, where the devices included transmission capability, their emission characteristics and potential for causing interference to those services. While other incumbent services operate in the TV bands, those services are not specifically being examined in this test program as the Commission has proposed other methods for protecting them from WSD operations.

1.1. Test Scope and Approach

The Phase II tests first examined the devices in the laboratory under various controlled test conditions. The WSDs' ability to sense DTV signals was evaluated in several ways: using a single DTV signal, with two DTV signals (one placed on an adjacent channel), and with recorded RF captures that exhibit various signal fading conditions. Similarly, for those devices capable of sensing wireless microphones, tests were performed with wireless microphones alone and with DTV signals present on adjacent channels. The wireless microphones were operated on various frequencies within a TV channel (both near the center of the channel and near the edges).

In preparation for the field tests, the performance of the devices in sensing off-the-air signals was recorded at various locations in the laboratory. Also, controlled tests of the Adaptrum transmitter were performed at the laboratory to estimate the distances at which it might cause interference to TV reception. Finally with input from various interested parties, nine sites were selected for field tests of the device's ability to sense TV signals, and two sites were chosen for field tests of their ability to sense wireless microphones. The field sites were selected to represent a variety of environments including urban, suburban, rural, residential and business locations. At two of the field test sites, anecdotal direct pick-up interference tests were performed to assess the impact of a WSD transmitter on cable TV reception. The following sections and the appendices provide a tabulation of the data collected during the tests.

2. White Space Prototype Devices

2.1. Adaptrum Prototype White Space Device

Adaptrum indicates that its "Cognitive Radio (CR) Platform" is an integrated hardware and software development system that has been specifically designed for TV white space operation on UHF television channels 21-51. The system is capable of various forms of TV signal sensing including waveform/signature sensing, spectral identification, signal power estimation, and network-level cooperative sensing. It detects both analog and digital TV signals, but performs those functions in two separate scans. It is not designed to detect wireless microphone signals. The system is also capable of signal transmission in the TV bands with flexible waveform, modulation, and signal bandwidth construction. It incorporates transmit power control and chain linearization to reduce adjacent channel interference. The maximum transmitter output power specification is 100 mW (+20 dBm) over the selected bandwidth. Figure 2-1 provides a photograph of the prototype device/platform.

The key components of the platform are: 1) a wide-band high dynamic-range radio frequency (RF) transceiver operating over the frequency range 400 MHz to 1000 MHz, 2) a field-programmable gate array (FPGA)-based hardware development board with integrated high-speed analog-to-digital converters (ADCs) and a high-density FPGA where CR baseband and protocol-layer functions are implemented, and 3) a Matlab-based integrated development environment (IDE) where the CR hardware functions are controlled using a Matlab graphical user interface (GUI) and Matlab scripts.

The scan time associated with this prototype device was initially 37 seconds per channel (sec/ch). However, a software modification installed to improve the detection sensitivity with respect to DTV signals with offset pilot carriers slowed the scan time by a factor of five to 185 sec/ch. Additional information can be obtained from the Adaptrum *ex-parte* filing in the subject rulemaking proceeding.[12]

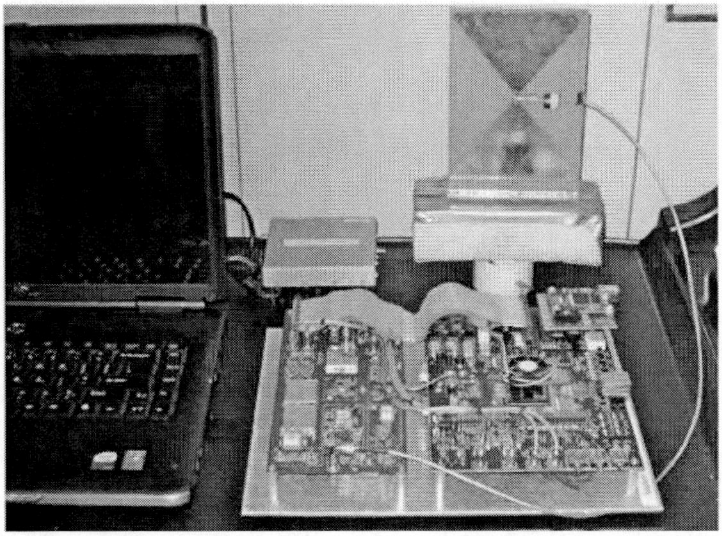

Figure 2-1. Adaptrum Prototype WSD

2.2. Institute for Infocomm Research (I2R) Prototype White Space Device

The I2R prototype WSD operates on channels 21-51 and is capable of sensing analog and digital TV signals and wireless microphone signals. The device consists of: 1) a commercial TV tuner to receive television signals and translate them to their associated intermediate frequency (IF), 2) mixed-signal digital processing boards for performing ADC, downconversion and filtering, 3) a laptop computer used to configure the hardware, provide a GUI and generate reports, and 4) a fixed-length whip receive antenna.

The (I2R) device was also able to detect wireless microphones. It provided two scan modes – "DTV" and "All" and two test modes – "Lab" and "Field". In the DTV scan mode it searched only for DTV signals and in the "All" mode it searched only for microphones. The "Lab" test mode was intended only for conducted bench testing and the "Field" mode was intended only for reception of radiated signals with an antenna. Figure 2-2 presents a photograph of the I2R prototype device/platform. Additional information can be obtained from the I2R *ex-parte* filing in the subject rulemaking proceeding.[13]

2.3. Microsoft Prototype White Space Device

The Microsoft WSD platform operates on channels 21-5 1 and is capable of sensing analog and digital TV signals and wireless microphone signals. The system consists of two core system subassemblies: 1) a wide-band spectrum scanner, a network processor and a tunable UHF half-duplex transceiver controlled by the network processor and 2) a Windows-based laptop computer that utilizes the Internet Explorer browser to establish a command and control user interface via an Ethernet connection. This device is also capable of transmitting on its operating channels.

Figure 2-2. I2R Prototype WSD

The scanner/sensor function consists of a broadband (521-698 MHz) computer-controlled frequency scanner and high-speed digitizer used to incrementally scan over UHF TV channels 21-5 1 in 6-MHz segments. The accumulated digitized time-domain information is then passed to the network analyzer where a 2048-point Fast Fourier Transform (FFT) is performed. Signal feature templates for DTV, analog TV, and wireless microphone waveforms are sequentially compared to the resulting FFT output to determine those channels occupied by DTV or analog TV signals. Channels determined not to be occupied by DTV or analog TV signals are subsequently analyzed for potential narrowband incumbent signals such as wireless microphones. Those channels determined not to be occupied by either DTV, analog TV, or wireless microphone signals are declared to be available "white space" channels. User control and scanner results are provided via the laptop computer connection.

The UHF transmitter assembly consists of three sub-components: 1) an S-Band (2.4 GHz) 802.11g OFDM modem; 2) a Half-duplex S-Band to UHF block converter; and 3) and a network processor browser to exercise control over frequency and power. Two external fixed-frequency bandpass filters were provided for use in transmitter testing. A photograph of the Microsoft WSD platform is provided in Figure 2-3.

Additional information can be obtained from the Microsoft *ex parte* filings in the subject rulemaking proceeding.[14]

2.4. Motorola Prototype White Space Device

The Motorola WSD platform operates on channels 21-51 and includes capabilities for geo-location and sensing of digital TV signals. It does not sense analog television signals or wireless microphone signals. The system consists of a Cognitive Radio Rack and a laptop computer host connected via Ethernet. The rack consists of a UHF radio and two PRO-3500 carrier boards co-located in a compact Peripheral Component Interconnect (PCI) chassis. The cognitive engine runs on the lower board.

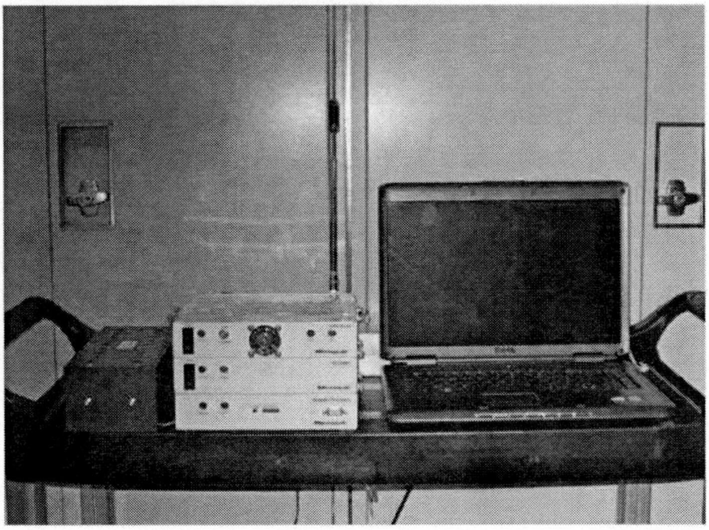

Figure 2-3. Microsoft Prototype WSD

This WSD implements a geo-location-based approach as its primary method for the determination of occupied TV channels with a spectrum-sensing capability used to refine the results of the geo-location solution and to prioritize those channels found to be available. The geo-location capability as tested requires the user to manually input the geographic coordinates and relies on a downloaded version of the FCC's TV database for a particular geographic area.

This WSD exhibited the fastest scan execution time of 0.1 sec/ch. A photograph of the Motorola WSD is provided in Figure 2-4. More information can be obtained from Motorola *ex parte* filings in the subject rulemaking proceeding.[15]

2.5. Philips Prototype White Space Device

The Philips WSD platform operates on channels 21-51 and is capable of sensing analog and digital TV signals and wireless microphone signals consists of: 1) a commercial TV tuner for tuning to a specified television channel and translating to IF, 2) a digital signal processing board for ADC and processing and 3) a desktop computer used to configure the hardware, provide a GUI and store detection results.

Philips claims that the prototype WSD will scan UHF channels 21-51 and detect ATSC (DTV), NTSC (analog TV) or wireless microphones to a level of at least -114 dBm over a 6-MHz television channel. The channel scan time for this device varies between 8 and 50 sec/ch due to the sequential application of separate ATSC, NTSC and wireless microphone detection algorithms.

Figure 2-5 presents a photograph of the Philips WSD platform. Additional information can be obtained from Philips *ex-parte* filing in the subject rulemaking proceeding.[16]

Figure 2-4. Motorola Prototype WSD

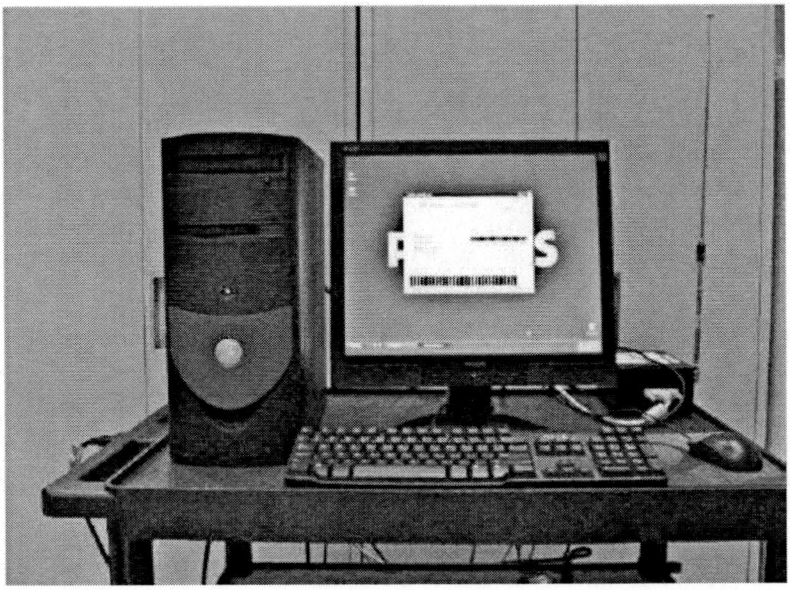

Figure 2-5. Philips Prototype WSD

3. DTV Scanning/Spectrum Sensing Capability Tests

This section describes the laboratory measurements performed on each of the prototype WSDs submitted for testing. For the sensing measurements performed in the laboratory, a simulated or recorded signal was applied to the antenna input of the WSD prototype utilizing a coaxial cable connection (*i.e.*, conducted measurements). As such, the effects of the WSD receive antenna were ignored.

Ignoring the WSD receive antenna contribution is analogous to assuming an isotropic receive antenna (*i.e.,* a theoretical antenna having uniform gain and directivity). An isotropic receive antenna for ultra-high frequency (UHF) operation represents an ideal that cannot be physically realized. Rather, the receive antenna of WSDs will likely have low and variable (or negative gain) and thus the results reported herein would have to be adjusted accordingly to describe practical devices. These measurements also make no effort to quantify other potential contributors to television signal loss such as might be experienced from shielding/absorption effects introduced by proximate human bodies.

The general approach to these tests was to provide a DTV input signal at progressively lower amplitude levels until the prototype could no longer detect it with complete (100%) reliability.

It is recognized that some of the WSD detection algorithms interpret detection reliability as low as 50% as indicative of an occupied channel, but the design intentions of the algorithms in other devices were not readily apparent.

3.1. Clean DTV Signal Measurements

The objective of the clean DTV signal measurements was to determine the detection sensitivity of the scanning/sensing capability of the WSD prototype devices submitted for evaluation with respect to a "clean" (undistorted) DTV signal.

For these measurements the input was a single, simulated DTV signal produced by a Rhode and Schwarz DTV Test Broadcast System (SFU). The generated DTV signal was provided to the antenna input of the WSD prototype under test via a coaxial cable. An external variable attenuator bank was used to adjust the signal power. Block diagrams showing the measurement equipment and related interconnections in addition to detailed equipment information are provided in Appendix A. The detailed procedure used to perform these measurements follows:

- The SFU was used to produce a single, clean DTV signal that was provided via coaxial cable through a variable attenuator and into the RF input of a spectrum analyzer.
- With the variable attenuator set to zero, the SFU power control was adjusted to produce an average power level of -60 dBm/6 MHz, referenced to the input connection of the spectrum analyzer. This level provides a displayed signal-to-noise ratio (S/N) of at least 20 dB. The signal power was measured using a rootmean-square (rms) detector and averaged over 100 sweeps.
- The cable from the SFU was then removed from the analyzer input and connected to the antenna input of the WSD prototype.
- The SFU-generated DTV signal was further attenuated using the external variable attenuator bank from the -60 dBm level to a level slightly higher than the expected detection threshold (*e.g.*, the variable attenuator was typically set to provide an additional 54 dB attenuation so as to produce an initial input signal level of -114 dBm)
- Thirty (30) independent detection trials were performed with the WSD tuned to channel 21 (lowest channel in tuning range). The number of successful detections of an occupied channel was recorded.
- If the successful detection rate over the thirty trials was 100%, the input signal power was reduced by increasing the variable attenuator by 1 dB and the procedure repeated.
- If the successful detection rate over the thirty trials was less than 100%, then the input signal power was increased by decreasing the variable attenuator by 1 dB and the procedure repeated.
- This iterative process was repeated until the DTV power level was found where the WSD prototype could no longer detect the occupying signal in the channel with 100% reliability.
- Additional data points were then collected above and below this breakpoint.
- The process was repeated in its entirety on two additional test channels, 36 and 51, representing the middle and upper channels within the tuning range, respectively.[17]

The figures that follow represent the results obtained from the clean signal DTV sensing measurements for each of the WSD prototype devices. Two separate plots are provided for the I2R device, representing results obtained in the two available modes of operation ("laboratory" test mode and the "field" test mode). The field test mode utilizes separate circuitry to suppress spurious reception.

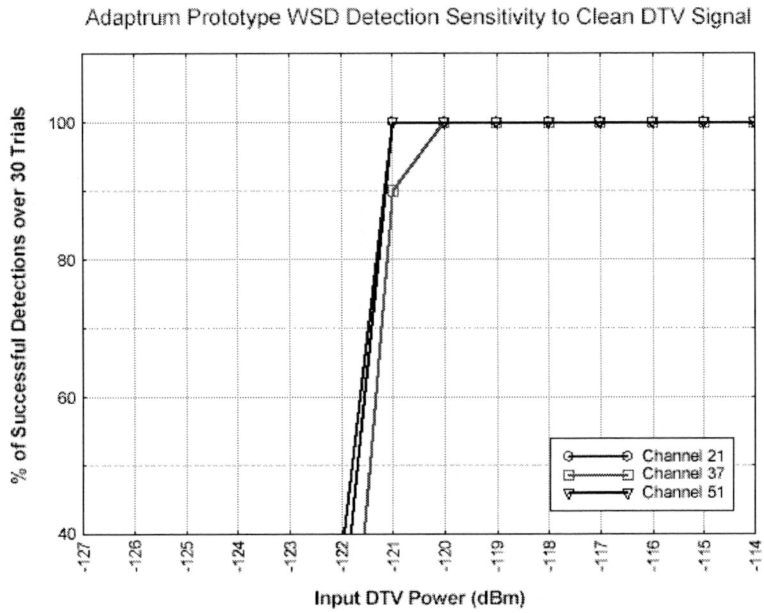

Figure 3-6. Clean Signal Sensitivity Results for Adaptrum WSD Prototype

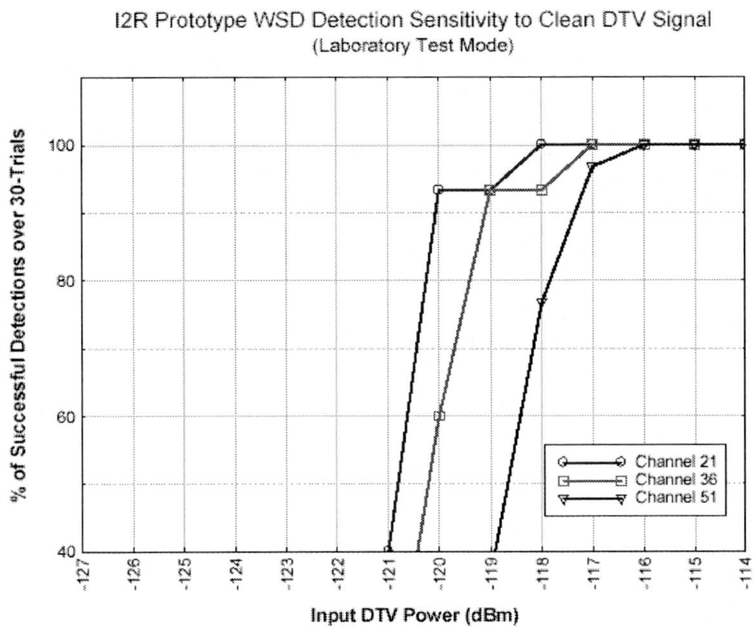

Figure 3-7. Clean Signal Sensitivity Results for I2R WSD Prototype in Laboratory Test Mode

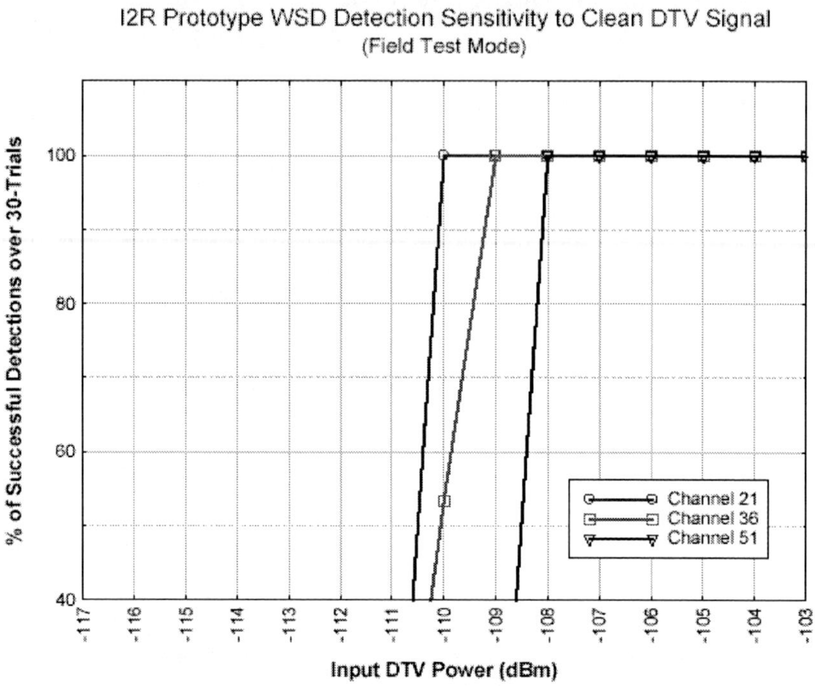

Figure 3-8. Clean Signal Sensitivity Results for I2R WSD Prototype in Field Test Mode

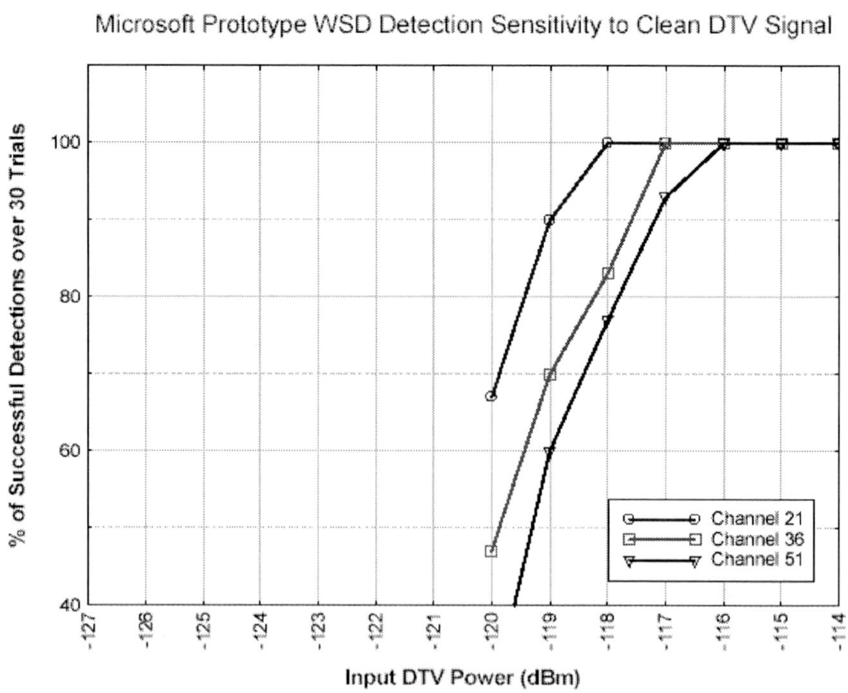

Figure 3-9. Clean Signal Sensitivity Results for Microsoft WSD Prototype

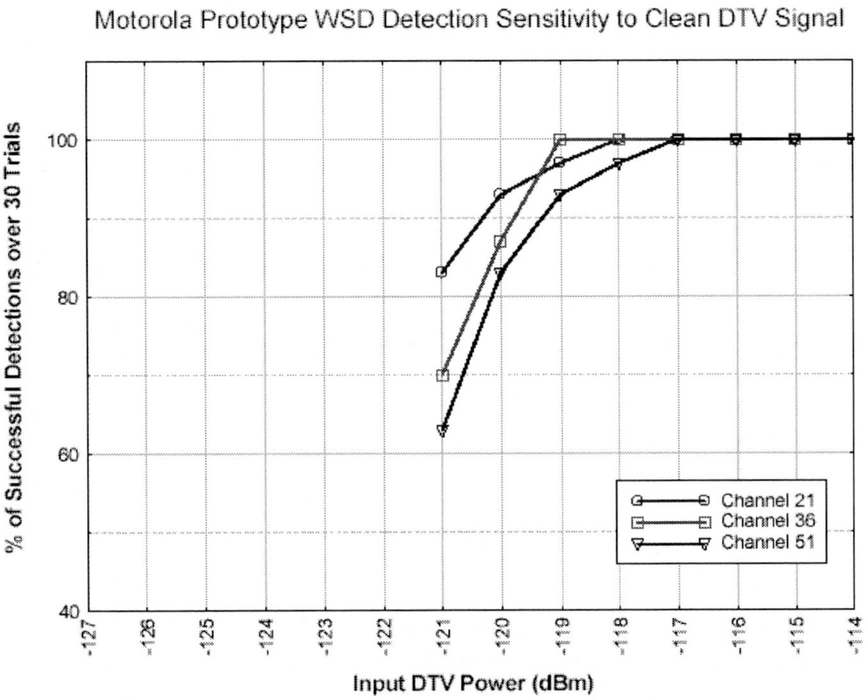

Figure 3-10. Clean Signal Sensitivity Results for Motorola WSD Prototype in Test Mode

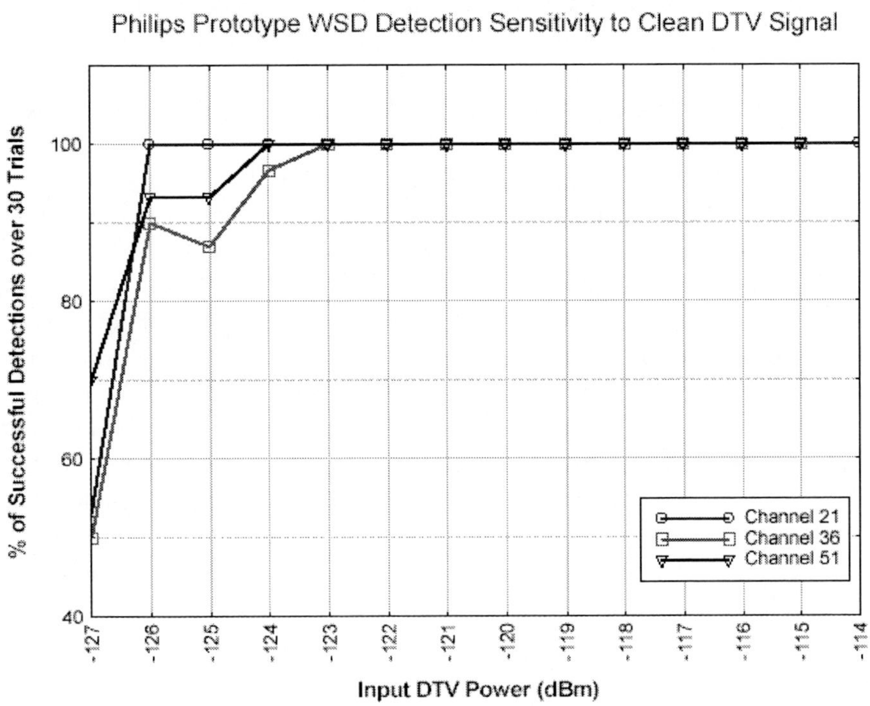

Figure 3-11. Clean Signal Sensitivity Results for Philips WSD Prototype

3.2. Recorded (Captured) DTV Signal Measurements

The objective of the recorded DTV signal measurements was to determine the detection sensitivity of the scanning/sensing capability of the WSD prototype devices with respect to recorded over-the-air DTV signals that incorporate "real world" signal distortions resulting from reflections, multi-path fading, and other factors.

The signal input for these measurements consisted of a set of twelve recorded DTV waveforms ("captures") selected by the IEEE 802.22 subcommittee.[18] Detailed information, including frequency domain and temporal representations of each of these twelve waveforms, is provided in Appendix B of this chapter.

A Wavetech Services WS-2 100 RF player® was used to play back the captures on the desired RF channel and to provide them to the antenna input of each of the WSD prototypes via coaxial cable. Two external variable attenuator banks were utilized to adjust the power level of the input signal. A 75-50 Ω transformer was utilized at the Wavetech player output to match system impedance.

Although different algorithms are utilized among the prototype devices to detect the presence of a DTV signal, they all appear to share a similarity in that they all sample the spectrum in proximity of an anticipated DTV pilot signal. However, the amount of time dedicated to sampling and signal processing differs significantly among the prototype devices. Since there are significant instantaneous temporal variations in many of the RF captures over their 25-second playback period (see Appendix B), the resulting detection decisions can also vary depending on when the sampling actually occurs. Efforts were made to minimize this variable (*e.g.*, the waveform playback was restarted for each independent trial); however its effect is evident in the results obtained from some of the prototype devices. Another method for reducing this variability is to increase the number of trials beyond the stated number of thirty. For some of the devices, the additional time required to execute more than thirty trials was excessive and was thus deemed impractical. However, for others (*e.g.*, the Motorola WSD), it was practical to perform additional trials. Therefore, for this device, in addition to performing the measurement over thirty independent trials, a second measurement was also performed utilizing one thousand independent trials. The results of both measurements are provided herein.

After the laboratory measurements were completed but before the field trials commenced, the prototype manufacturers were offered an opportunity to make minor adjustments to their devices based on the laboratory results. For example, one issue observed by the manufacturer of the Adaptrum prototype was difficulty in detecting a DTV signal having a frequency offset pilot tone. As a result, Adaptrum utilized this opportunity to improve this with a software modification.[19] A second set of data is presented for the Adaptrum device showing the effect of this modification.

Block diagrams showing the measurement equipment and related interconnections in addition to detailed equipment information are provided in Appendix A. The measurement procedure was as follows:

- A Wavetech Services WS-2 100 RF player® was used to play back each of twelve recorded DTV signals and provide the RF output via coaxial cable through an external variable attenuator bank and into the RF input of a spectrum analyzer.

- With the external variable attenuators set to zero, the average output signal power in the channel bandwidth was measured using the spectrum analyzer (over 100- sweeps utilizing an rms detector).
- The coaxial cable was then removed from the RF input of the analyzer and connected to the receive antenna input of the prototype device.
- The signal was further attenuated using the variable attenuator bank to achieve a power level near the threshold level recorded in Section 3.1.
- Thirty (30) independent trials were performed and the number of successful detections recorded.[20] Since the data from the Section 3.1 indicated relative consistency over the three channels examined, these tests were performed only on a single channel (typically channel 36, mid-channel in the WSD tuning range).[21]
- If the successful detection rate over the thirty scans was 100%, then the input signal power was reduced by increasing the attenuation by 1 dB and the procedure repeated.
- If the successful detection rate over the thirty scans was less than 100%, then the input signal power was increased by reducing the attenuation by 1 dB and the procedure repeated.
- The process was repeated until the DTV power level was found where the WSD prototype could no longer detect the signal in the channel with 100% reliability.
- Additional data points were then collected above and below this breakpoint.

The results of these measurements are provided in the figures that follow.

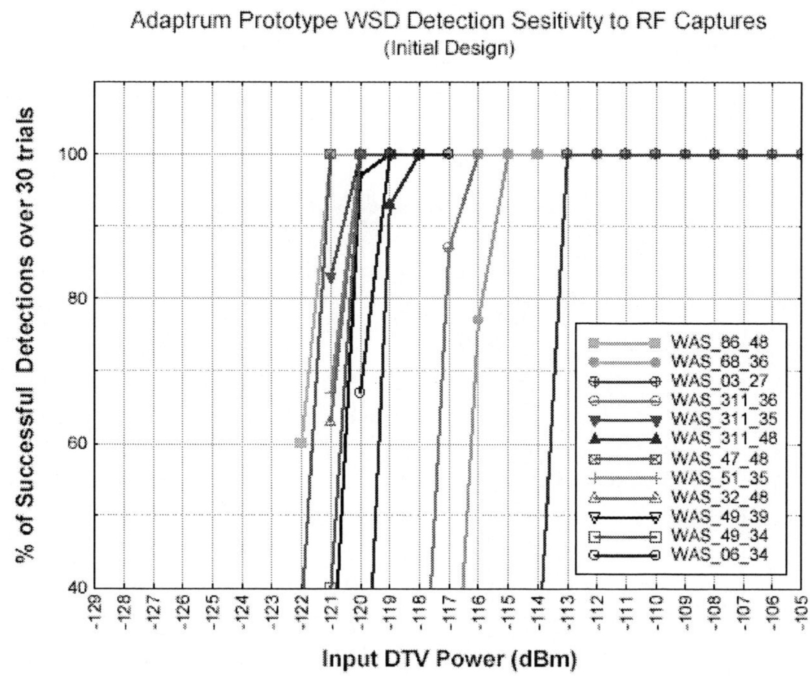

Figure 3-12. DTV Capture Results for Adaptrum WSD Prototype (initial configuration with limited freqency offset pilot capability)

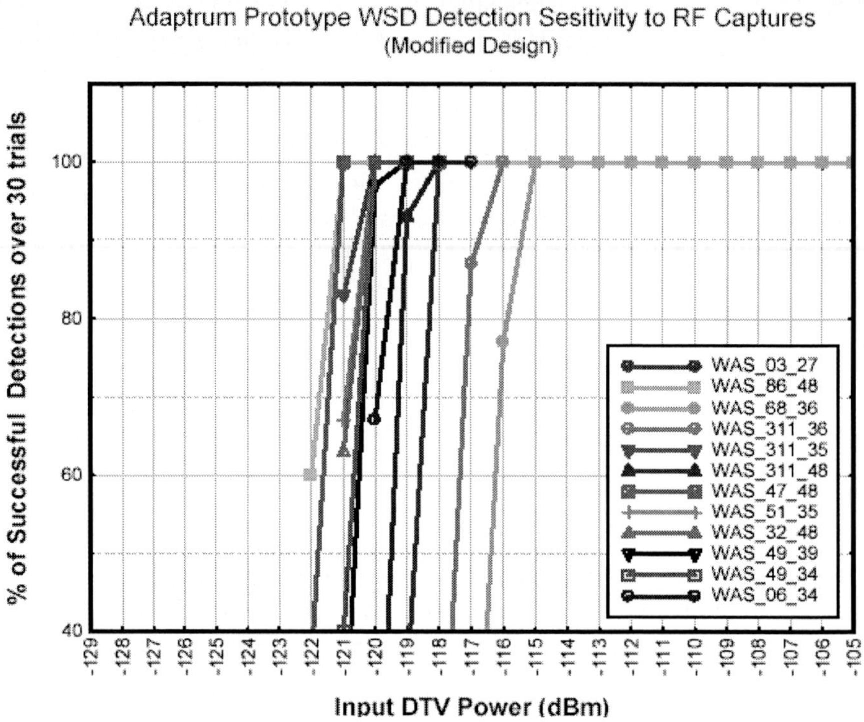

Figure 3-13. DTV Capture Results for Adaptrum WSD Prototype (modified configuration with improved frequency offset pilot capability)

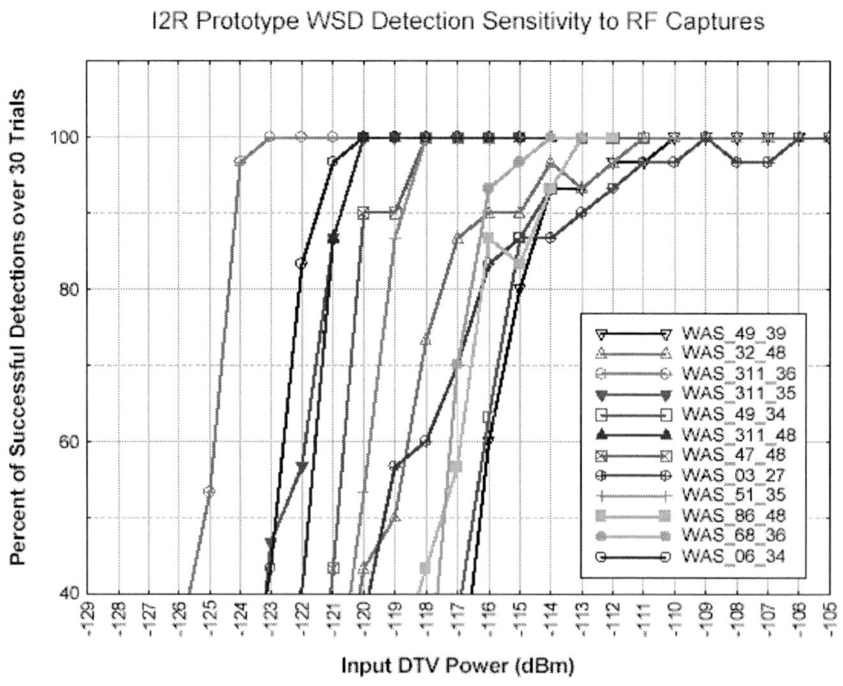

Figure 3-14. DTV Capture Results for I2R WSD Prototype (Laboratory Mode)

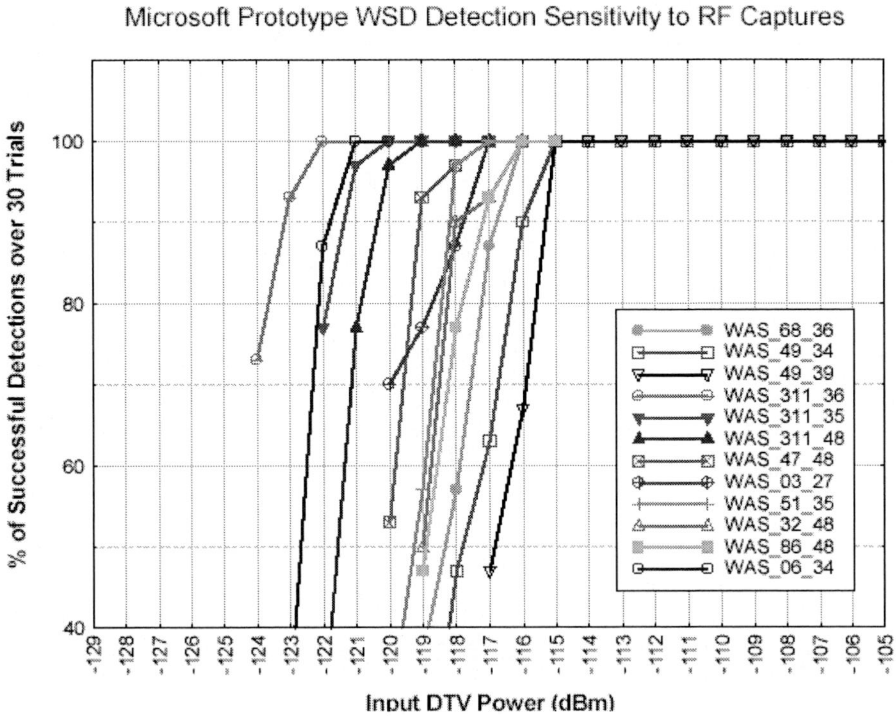

Figure 3-15. DTV Capture Results for Microsoft WSD Prototype

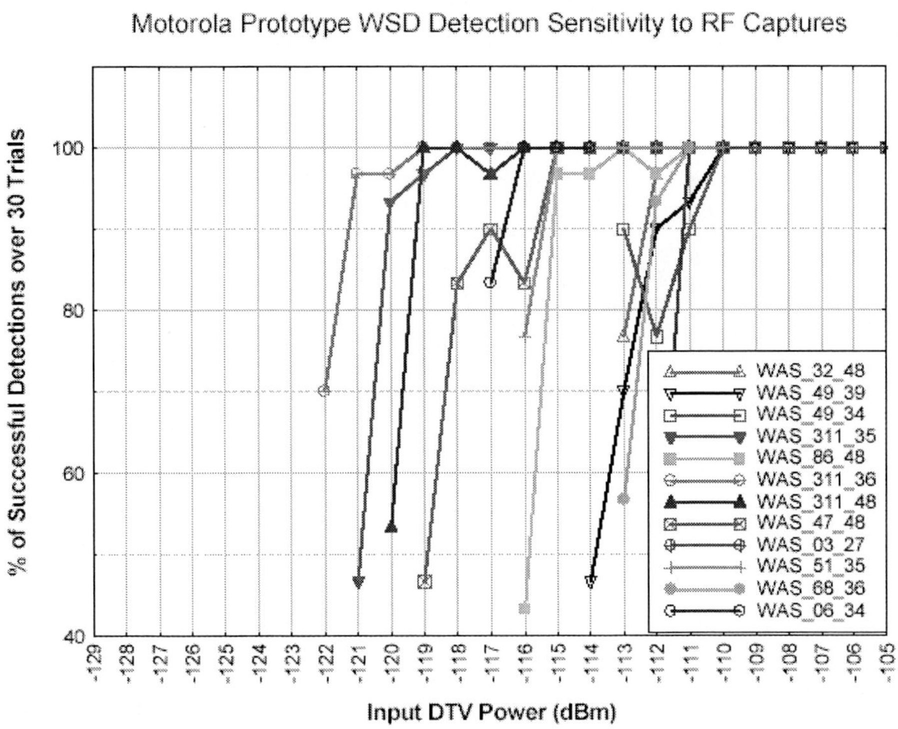

Figure 3-16. DTV Capture Results for Motorola WSD Prototype (30-trials)

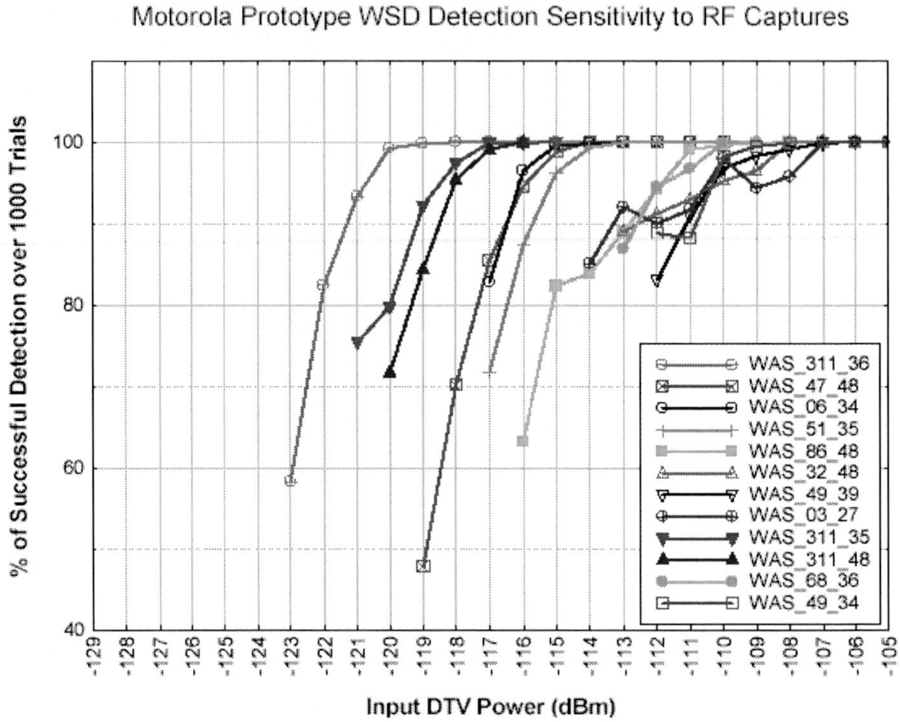

Figure 3-17. DTV Capture Results for Motorola WSD Prototype (1000 Trials)

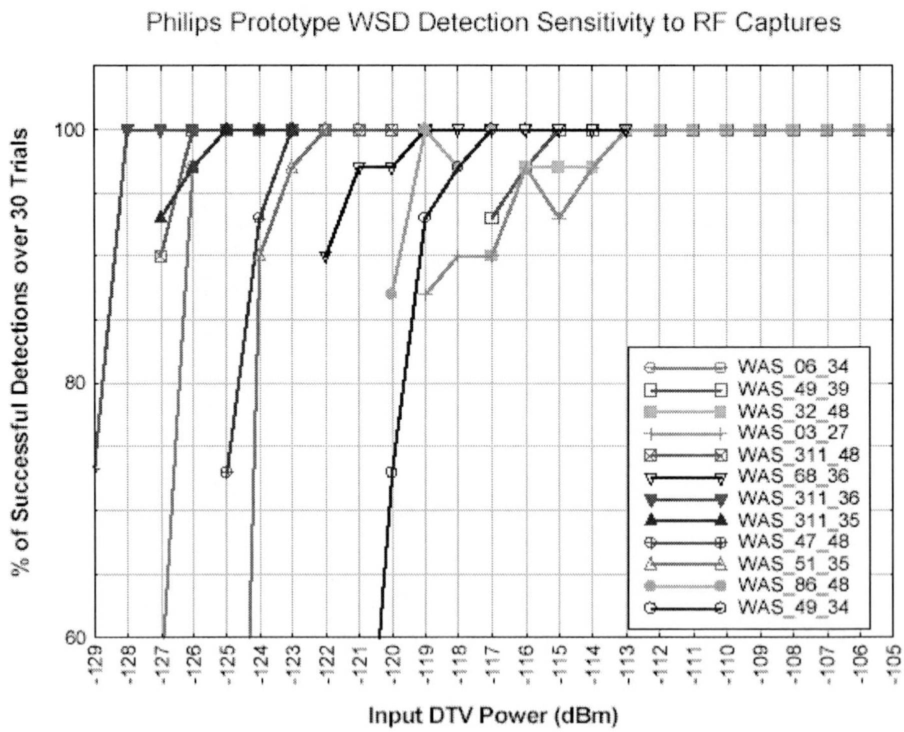

Figure 3-18. DTV Capture Results for Philips WSD Prototype

3.3. Two-Signal (Adjacent Channel Interference) Measurements

The objective of the two-signal adjacent-channel interference measurements was to assess the potential for degradation to the DTV signal detection capability when another DTV signal is present on an adjacent channel. Because of equipment limitations, these measurements were limited to assessing the effect of one adjacent channel at a time (*i.e.*, two-signals, one on the detection channel and one on an adjacent channel). The effects of DTV signals on both first- and second-adjacent channels were tested.

Detection threshold measurements were made in a manner similar to that described in Section 3.1, but included one additional DTV signal placed sequentially on the first (N±1) and then second (N±2) adjacent channels relative to the detection channel. Detection threshold measurements were performed with the adjacent-channel signal power set to the -28 dBm (high), -53 dBm (moderate) and -68 dBm (low) levels. It should be noted that DTV signal levels in "real world" environments can exceed -28 dBm and actual examples of this were observed in the field test component of this effort (see Section 5.2).

For these measurements, the R&S SFU was used to generate the adjacent-channel DTV signal. The Wavetech player was used to play back a "clean" DTV signal (Zone Plate test signal) in the detection channel. The power level of the adjacent channel DTV signal was set using the SFU's internal power controls and verified with a spectrum analyzer. The power level of the detection-channel signal was set utilizing a bank of external variable attenuators and verified with a spectrum analyzer.

The SFU provides a good representation of a DTV signal within the 6 MHz TV channel passband but it does not precisely match the out-of-band characteristics of operating DTV stations (*i.e.,* the "DTV Mask" for full service stations). Therefore, in order to preclude the possibility of excessive signal leakage ("splatter") from the adjacent channel into the detection channel, the DTV signal was filtered through a 7-pole RF bandpass filter in order to more accurately simulate the out-of-band characteristics of a "real world" DTV broadcast signal.

The filter employed was a Micro Communications, Inc. fixed-tuned Interdigital Bandpass (IDBP) filter.[22] Passing the SFU-generated DTV waveform through this filter provides out-of-band characteristics that adhere to the DTV emissions mask (see Appendix A).[23] However, since the filter is fixed-tuned to channel 29, the adjacent- channel signal had to be maintained on channel 29 while the detection channel was varied to represent the N±1 and N±2 relationships with respect to the detection channel. For example, in order to examine the effect of an N+1 adjacent channel interaction, the SFU DTV signal (representing the adjacent-channel signal) was generated on channel 29 and the Wavetech DTV signal (representing the detection-channel signal) was generated on channel 28. Likewise, the N-1 adjacent channel scenario was simulated by transmitting the detection signal on channel 30 and the N+2 and N-2 scenarios were simulated by transmitting the detection signals on channels 27 and 3 1, respectively, while the adjacent- channel DTV signal remained on channel 29.

During the course of these measurements, the Microsoft prototype sample device began to malfunction and eventually ceased to operate, necessitating the abandonment of further measurements utilizing this device.

As mentioned above, before the field trials were initiated, the prototype manufacturers were offered an opportunity to make minor adjustments to modify their devices based on the bench-test results. One issue observed by the manufacturer of the Motorola prototype was degradation to the detection capability in the presence of an adjacent-channel DTV signal (particularly one at moderate and higher power levels). Motorola utilized this opportunity to add an automatic gain control (AGC) circuit to the device.[24] This necessitated that a second set of laboratory data be collected from this prototype to demonstrate and verify the effect of the modification.

Block diagrams showing the measurement equipment and related interconnections in addition to detailed equipment information are provided in Appendix A.

The measurement procedure was as follows:

- The SFU was used to generate a DTV waveform on channel 29 that was delivered via coaxial cable through a fixed-tuned bandpass filter to provide the adjacent-channel signal. The Wavetech player was used to playback another DTV waveform on the pertinent detection channel. The detection-channel signal was provided via coaxial cable through a bank of external variable attenuators and combined with the adjacent channel signal. The combined signals were split with one output connected with coaxial cable to the antenna input of the WSD prototype device and the other output connected with identical coaxial to a spectrum analyzer. A block diagram of this set-up is provided in Appendix A.
- The SFU-generated adjacent-channel signal was set to produce a signal with amplitude of -28 dBm at the spectrum analyzer input. For subsequent measurements, this same procedure was used to set the adjacent-channel signal power to -53 dBm (representing a moderate level) and -68 dBm (representing a low level), respectively.
- The Wavetech RF player was set-up to provide the detection signal on channel 28, simulating an N+1 adjacent-channel scenario (*i.e.,* channel 28 detection channel with a second DTV signal in the upper adjacent channel 29). In subsequent measurements, the output channel of the Wavetech player was set to channel 30 (N- 1), channel 27 (N+2), and channel 31 (N-2) to represent the remaining three adjacent-channel scenarios under examination.
- With the external variable attenuator set to zero, the power in the detection channel was measured with the spectrum analyzer. Using this measurement as a baseline, the variable attenuator banks were adjusted to reduce the signal power in the detection channel to a level near the previously observed detection threshold.
- Thirty (30) independent trials of the WSD detector were performed in the detection channel and the number of successful detections was recorded.
- If the successful detection rate over the thirty independent trials was 100%, then the input signal power was reduced (by increasing the signal attenuation) and the procedure repeated.
- If the successful detection rate over the thirty trials was less than 100%, then the input signal power was increased (by decreasing the signal attenuation) and the procedure repeated.
- The process was repeated until the DTV power level was found where the WSD prototype could no longer detect the signal with 100% reliability.
- Additional data points were collected at levels above and below this breakpoint.

Table 3-1. Adjacent Channel Test Results

Prototype WSD	Detection Thresholds in the Presence of DTV-Occupied Adjacent Channels (dBm)											
	N+1			N-1			N+2			N-2		
	H	M	L	H	M	L	H	M	L	H	M	L
Adaptrum	A	-108	-120	-100	A	A	-108	-120	-120	-105	-119	-120
I2R	-97	-117	-119	A	A	-117	-118	-118	-118	-118	-118	-117
Microsoft	-67	-110	A	B	B	B	-76	-118	-120	B	B	B
Motorola	-52	-97	-114	-47	-68	-109	-79	-115	-114	-76	-115	-114
Motorola (w/modifi-cation)	-102	-111	-110	-73	-98	-110	-96	-112	-112	-96	-112	-112
Philips	A	A	-122	A	A	-122	A	-122	-124	A	A	-123

Notes:

H = High adjacent signal level (-28 dBm);

M = Moderate adjacent signal level (-53 dBm);

L = Low adjacent signal level (-68 dBm).

A = Insufficient receiver selectivity and/or receiver desensitization prevented collection of meaningful data.

B = Device malfunction prevented collection of data.

The data resulting from these measurements are presented below in Table 3-1. Only the final signal level where 100% correct detection was observed is reported in this table.

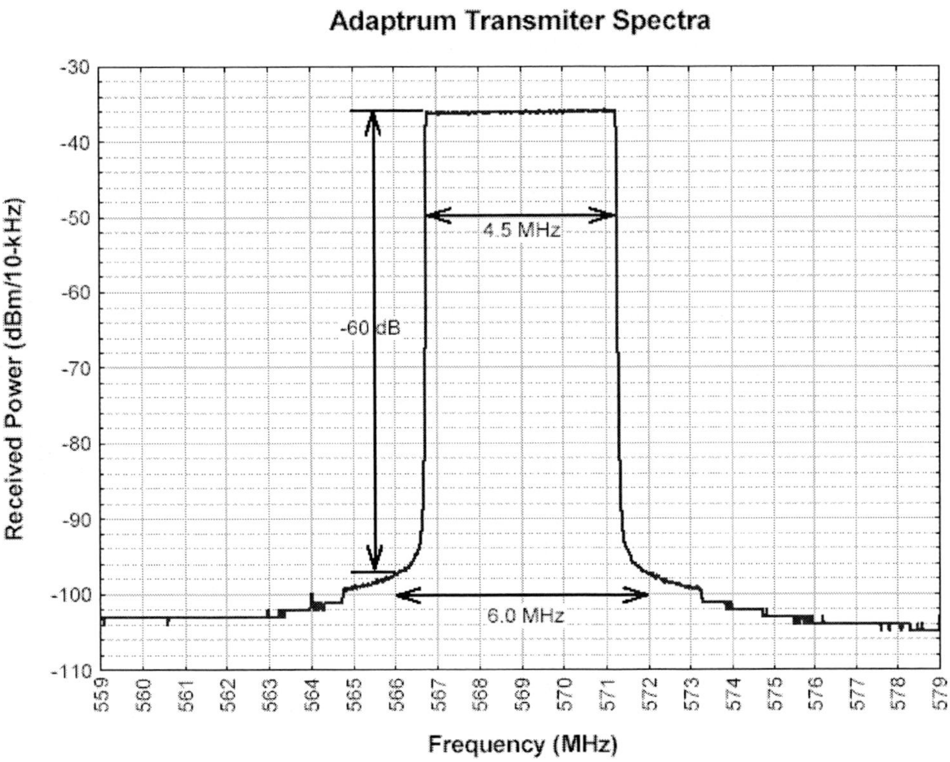

Figure 3-19. Adaptrum Transmitter Emission Spectrum

3.4. Transmitter Measurements

3.4.1. Conducted Measurement

The objective of the conducted transmitter measurements was to examine the emitted spectral characteristics of those WSD prototype devices having a transmit function. Two of the originally-provided WSD prototypes include active transmitters, the Adaptrum and Microsoft devices. Prior to these measurements, the Microsoft device malfunctioned. However, the spectral characteristics associated with the Microsoft WSD transmitter are expected to be unchanged from those measured and reported in the Phase I measurements.[25]

The Adaptrum WSD was used with an external wide-band amplifier to boost the power applied to the transmit antenna. The emission spectrum was measured by connecting the transmitter/amplifier output through a 30 dB attenuator to a spectrum analyzer. Figure 3-13 presents the resulting spectral plot.

The following information can be determined from this data:

- The bandwidth of the transmitted signal is approximately 4.5 MHz.
- The out-of-band (channel) emission (OOBE) suppression is slightly more than 86.5 dB (60 dB + 10 log(4.5 MHz/10 kHz)) relative to the total power in the carrier.
- The output power is shown as -36 dBm as measured in a 10-kHz resolution bandwidth. This converts to a level of -9.5 dBm (-36 + 10log (4.5 MHz/10 kHz)) in the 4.5 MHz emission bandwidth. A 30-dB external attenuator was utilized to protect the spectrum analyzer from overload and the coaxial cable loss was measured at 1.2 dB. Therefore, the final conducted output power from the WSD is (-9.5 + 30 + 1.2) = 21.7 dBm or approximately 150 milliwatts.

3.4.2. Radiated Measurement

Measurements were made to derive the equivalent isotropically radiated power (EIRP) based on measurements of the radiated power of the transmitter in the Adaptrum WSD prototype. In order to determine the EIRP of this device's transmitter, a set of radiated measurements were performed utilizing the laboratory's semi-anechoic chamber. For these measurements, the WSD was placed atop a wooden test bench that is mounted on a turntable and activated in the transmit mode of operation. A calibrated laboratory- grade receive antenna was affixed to a variable-height mast at a separation distance of 3- meters relative to the transmit antenna. The WSD transmit antenna and the calibrated receive antenna were varied in relative orientation (*e.g.*, with respect to polarization, height and angular position) until a maximum signal level was obtained on a spectrum analyzer. This measurement process was repeated on three distinct frequencies representing the lower, middle and upper tuning range of the transmitter. The maximum received signal level was recorded at each frequency.

The maximum EIRP of the transmitter can be determined from the measured maximum received power by applying the following equation:

$$EIRP = P_R - G_R + L_C + L_P \qquad (1)$$

where;
P_R = received power measured with the spectrum analyzer, in dBm,
G_R = receive antenna gain, in dBi,
L_C = signal attenuation in coaxial connecting cable, in dB,
L_P = free space propagation path loss, in dB.

The receive antenna gain was determined from manufacturer-provided calibration data, the signal attenuation in the connecting coaxial cable was measured and the free- space propagation path loss[26] was calculated for each measurement frequency. The parameters used in Equation (1) along with the results are provided in Table 3-2 below.

3.5. Miscellaneous Laboratory Measurements

This section describes some miscellaneous measurements performed on the WSDs that were not a part of the original test plan. These measurements were typically undertaken to help explain unusual and/or unexpected observations.

In preparation for the field trials, the detection capabilities of the prototype devices were first exercised in the over-the-air environment at the laboratory facility (sensing the real world environment, not laboratory test signals). Several of the devices indicated that all (or most) channels were occupied even though it was known that unoccupied (available) channels existed at the laboratory.

These miscellaneous measurements represent an attempt to investigate working hypotheses regarding these unanticipated observations. Since these measurements were solely exploratory, they were not necessarily performed consistently over the set of all prototype devices. In fact, because of the late submission of the I2R device, none of these measurements were performed with it. In other cases, only a partial suite was performed, particularly if preliminary results did not raise concerns.

The first of these measurements consisted of exercising the detection capability of the WSD device using its associated antenna at locations within the laboratory (indoors), in the laboratory's parking lot (outdoors), and inside of the laboratory's semi-anechoic chamber. When some of the devices indicated significant channel occupancy inside the anechoic chamber (where all environmental UHF signals are significantly attenuated), a set of measurements were performed to examine possible explanations. Radiated measurements were performed to determine the spurious energy emitted by the prototypes under the assumption that at sufficient intensity they may trigger false detections. No effort was made to isolate the source of the spurious emissions among the individual system components.

Table 3-2. EIRP Measurement Results

Frequency (MHz)	P_R (dBm)	G_R(dBi)	L_C (dB)	L_P(dB)	EIRP (dBm)
515	-10.9	5.8	2.9	36.2	22.4
605	-12.6	6.3	3.1	37.7	21.9
695	-16.2	4.8	3.3	38.9	21.2

The results of these measurements are presented in Appendix H. The tables provided in Section H.1 provide the results obtained from exercising the WSD sensing/detection capability within the laboratory. Section H.2 presents spectral plots representing the data collected from the radiated spurious emissions measurements. These tables and plots show that many of the prototype platforms generate emissions internally at levels that create a potential for degrading a device's detection capability (in the form of high missed detection and/or false alarm rates).

3.6. Summary of Laboratory Measurement Results

The following table summarizes the results obtained from the laboratory measurements performed with each of the WSD prototype devices.

The clean signal results indicate that the measured detection threshold ranges between -116 and -126 dBm (ignoring the I2R results in field test mode) over the complete set of prototype devices. The results also indicate that the detection threshold is relatively consistent over the tuning range of the device.

The measured detection thresholds with respect to the twelve DTV waveform captures range between -106 and -128 dBm over the total set of prototype devices. Variability observed in the results of the waveform capture tests when compared with the results of the clean waveform tests is explained partly by temporal amplitude variations in the recorded waveform.

The adjacent-channel sensitivity measurements indicate that moderate-to-strong DTV signals occupying channels adjacent to the detection channel can significantly degrade detection capability, thereby affecting the ability of a device to reliably detect DTV signals.

Transmitter measurements on the one WSD with transmit capability demonstrated a clean output spectra with good out-of-band emission (OOBE) suppression. The maximum EIRP of the device was determined to nominally be +22 dBm (~150 mW) over the tuning range and the OOBE suppression was found to be at least -86.5 dB.

Table 3-3. Summary of Bench Test Results

WSD Prototype	Clean Signal Threshold Range (dBm/6-MHz)	Capture Signal Threshold Range (dBm-6-MHz)	Clean Signal Threshold Range with Adjacent Channel Signal Present (dBm/6-MHz)
Adaptrum	-120 to -121	-113 to -121	-100 to -120
I2R	-116 to -118 (lab mode) -108 to -100 (field mode)	-106 to -123	-97 to -119
Microsoft	-116 to -118	-115 to -122	-67 to -120
Motorola	-120 to -121	-107 to -119	-47 to -115 -73 to -112 (w/mod)
Philips	-123 to -126	-113 to -128	Unknown to -123

Note: Thresholds are referenced to device antenna input (receive antenna characteristics ignored).

The miscellaneous measurements showed that many of the prototype platforms generate emissions internally, at levels that create a potential for degrading the devices' detection capabilities (in the form of high missed detection and/or false alarm rates).

4. WSD Transmitter Interference Tests and Demonstrations

This section describes transmitter interference tests performed as a part of the Phase II measurement effort. These tests are considered anecdotal for a number of reasons. For example, the use of over-the-air signals implies that little control could be exercised upon significant test variables. In addition, since the tests were performed at a limited number of locations and at particular times of day, variations in signal level occurred which could not be characterized. Also, since the Adaptrum WSD prototype was the only available working device with a transmitter, the interference effects to DTV reception were examined for only the waveform of this particular device, which may or may not be representative of a final TV white space product.

These tests all utilized the Adaptrum WSD transmitter to radiate an undesired signal while the potential victim receive system was monitored visually for interference. Attempts were made to operate the victim receiver in a minimum desired signal condition to represent a worst-case interference scenario.

Section 4.1 describes a test performed to assess potential adjacent-channel and co-channel interference interactions when a proximate DTV receive system is tuned to a station near the recognized threshold of visibility (TOV) of -84 dBm. This test was performed within the FCC laboratory facility.

Section 4.2 describes tests performed at two of the residential field trial sites to assess the potential for interference to the reception of cable television programming via the direct pick-up of an undesired WSD-transmitted signal either by the coaxial cable/connections, the television tuner and/or a set-top converter box. In these tests, the precise interference entry point was not identified.

4.1. Channel Interference Potential to DTV

The objective of this test was to examine interference distances between the prototype WSD transmitter and a typical consumer-installed DTV receiving system, with WSD transmissions either on the same or on an adjacent channel relative to a weak-signal DTV signal. The co-channel scenario is one that should be avoided by the WSD channel detection scheme, but is investigated as a part of this effort in order to provide an estimate of the interference distance in the case where the detection scheme fails to identify a weak-signal occupied channel.

4.1.1. Undesired Signal (U) Transmission System

The undesired signal was produced by the Adaptrum transmitter coupled with an external 21 dB gain wideband amplifier. The WSD is capable of transmitting an OFDM signal with a signal bandwidth adjustable from 0.5 to 18 MHz. A vertically-polarized planar bi-conical transmit antenna at a height of 2 meters (6.6 feet) above ground level (AGL) was used to radiate the signal generated by the transmitter. The EIRP of the WSD prototype transmitter was measured in the semi-anechoic chamber and determined to nominally be +22 dBm (~150 mW). See Section 3.4 for details. Figure 4-1 presents a photograph of the WSD signal transmission system on the laboratory grounds.

4.1.2. Desired Signal (D) Receive System

The receive system used in these tests consisted of a consumer-grade horizontally-polarized television antenna (Radio Shack VU-90XR) mounted at a height of 9.3 meters (30.5 feet) on the rotor-driven extendable mast of an FCC measurement truck and connected via coaxial cable sequentially to 1) an Agilent E7405A spectrum analyzer with internal pre-amplifier and 2) a commercial set-top box with a late- generation DTV tuner (Magnavox) feeding a consumer DTV receiver (JVC). See Figures 4-2 and 4-3 for photographs of the desired signal receive and measurement system.

Figure 4-1. Adaptrum WSD Prototype Transmit System

Figure 4-2. FCC Measurement Truck

Figure 4-3. DTV Receive System

4.1.3. Test Methodology

The following test procedure was used for the transmitter field tests:

- The DTV receive antenna was mounted on the extendable mast of the FCC measurement truck and elevated to 9.3 meters (30.5 feet) above ground level (AGL) to simulate a consumer-installed roof-top mounted antenna. The truck was parked on one end of the FCC laboratory property.
- A 75-50 Ω transformer was used to connect the antenna through the antenna feed (coaxial cable) to a spectrum analyzer.
- The antenna was rotated and the spectrum analyzer used to examine the received power levels associated with all available DTV channels.
- Power measurements were compared to determine the weakest receivable signal. This signal was used as the desired signal in these tests. By this method, channel 30 (WNVT-DT in Goldvein, VA, approximately 50-miles distant) was selected as the desired channel for this test. The measured channel power of WNVT-DT was found to vary between -75 and -77 dBm at the test location.
- The antenna feed was then switched from the spectrum analyzer to the set-top box and the signal content displayed on the DTV receiver.
- The signal was observed to produce a stable picture without obvious artifacts.
- A step attenuator was inserted into the signal path and the desired signal was attenuated in 1 dB steps until the picture was lost. By this method it was determined that a 5-6 dB margin existed between the received and threshold signal levels. This confirmed the TOV of the receiver to be within the range -80 to -83 dBm.
- The Adaptrum WSD transmit system (transmitter and supplied antenna) was initially placed at a distance of 12.2 meters (40 feet) from the base of the receive antenna mast along the same radial as the receive antenna boresight and oriented to maximize mainbeam-to-mainbeam coupling (not considering cross-polarization losses). The transmitter was tuned to each of the immediately adjacent channels (29 and 31) and the OFDM signal was transmitted at full power in a 4.5 MHz bandwidth. No

interference was observed to the set-top box/DTV receiver while receiving WNVT programming on channel 30.

- The distance was decremented in 3.1 meters (10 feet) steps and the adjacent- channel tests repeated. No adjacent-channel interference was observed on the DTV at any distance; however, alternative test configurations were not explored.
- This test was repeated with a transmit bandwidth of 6 MHz. Again, no interference could be produced with the WSD transmitter tuned to either of the immediately adjacent channels.
- The WSD transmitter was then tuned to channel 30 (co-channel) with an emission bandwidth of 4.5 MHz.
- The transmit antenna was initially placed at a distance of 45.7 meters (150 feet) from the base of the receive antenna along the same radial as before.
- The transmitter was activated at full power. Interference was immediately observed in the form of complete loss of the television picture.
- The transmit channel was changed to 31 in order to isolate from the desired signal and the power through the receiver chain measured. It was determined that at this distance, the undesired power level was 28 dB above the desired power (D/U = - 28 dB) which suggested that considerable additional distance would likely have to be realized to mitigate the interference (*i.e.*, to achieve a D/U of 15 dB requires 43 dB additional attenuation of the undesired signal).
- In order to achieve as much attenuation as possible from distance separation (while still maintaining a line-of-sight propagation path), the transmitter and receiver systems were separated as far as practical within in the confines of the laboratory property.
- The maximum distance that could be practically obtained between the transmitter system and the receiver system while staying within the receive antenna mainbeam was 360 meters (1180 feet). The transmitter was again activated at full power within a 4.5-MHz bandwidth and co-tuned with the desired signal (on channel 30). Interference was once again observed as a complete loss of picture. The desired signal level was measured on channel 30 and determined to be -76.9 dBm and the undesired signal level was measured on channel 31 and determined to be -77.4 dBm (D/U = 0.5 dB)
- Since no greater distance separation could be practically realized, a step attenuator was placed between the transmitter amplifier and antenna and used to further attenuate the undesired signal until the interference was eliminated. It was found that an additional 15 dB of attenuation was required to resolve the interference (D/U = 15.5 dB). The results of this test are summarized in Table 4-1.
- The test was repeated with the WSD transmitter located to the side and behind the DTV receive antenna, in order to observe the variation in interference level as the WSD was moved off boresight of the receive antenna (away from the axis of symmetry of the mainbeam). The resulting data are summarized in Table 4-2. The approximate locations associated with the side- and back-lobe tests are shown in the aerial image provided as Figure 4-4.

Table 4-1. Co-Channel Interference Test Results in Rx Antenna Mainbeam

Tx Location (WGS-84)	Rx Location (WGS-84)	Path Length (m)	Bearing w.r.t. Rx MB axis (degrees)	Ix ?	Required Attenuation (dB)
39°10.098'N 076°49.445'W	39° 10.098'N 076° 49.445'W	360	0	Y	15
Receive antenna oriented at a bearing of 239° relative to North.					

Figure 4-4. Site Orientation for Co-Channel Interference Tests in Rx Back- and Side-Lobes

Table 4-2. Co-Channel Interference Test Results in Rx Back-Lobe and Side-Lobes

Tx Site	Coordinates (WGS-84)	Path Length (m)	Bearing w.r.t. Rx MB axis (degrees)	IX	Required Attenuation (dB)
1	39° 10.100'N 076° 49.441'W	370	180.0	Y	3
2	39° 10.161'N 076° 49.554'W	340	147.8	Y	14
3	39° 10.096'N 076° 49.801'W	270	72.7	Y	27
4	39° 10.024'N 076° 49.808'W	220	43.7	Y	24
Receive system located at 39°10.098'N; 076°49.445'W with antenna oriented at a bearing of 239° relative to North.					

4.2. Direct Pick-Up Interference to Cable Service

Anecdotal checks for direct pick-up interference to cable service were performed at the two residential test sites with cable television subscriptions (the third residential site did not subscribe to cable service). The intent was to observe the susceptibility of cable TV reception to interference caused by the direct pick-up of emissions from a WSD prototype transmitter. This interference mechanism has been previously examined and documented by the FCC laboratory.[27]

Consumer cable service networks in the United States typically utilize quadrature-amplitude-modulated (QAM) signals to provide digital television programming. This requires a dedicated QAM demodulator, distinct from that used to demodulate ATSC 8- VSB broadcast signals. Many DTV receivers include both QAM and ATSC 8-VSB demodulators to permit reception of digital cable signals without the need for a separate set-top box. A DTV equipped with a QAM tuner can thus receive unencrypted digital cable TV programming (and also encrypted cable programming if equipped with an active CableCard).

The frequency plan of cable channels differ from the frequency plan of broadcast television channels. Cable channels are offset in frequency relative to over-the-air (broadcast) channel allocations.[28] Cable channels also use different numeric designations and include channels on frequencies not used by broadcast television. At each test site, it was necessary to find an analog and an unencrypted (clear QAM) digital cable channel whose frequencies overlapped the spectrum of a broadcast channel within the tuning range of the Adaptrum prototype WSD transmitter.

4.2.1. Field Test Site 3 (Residential)

Cable channel 73, (519.0 ± 3.0 MHz) was identified as an analog cable channel available at this test site. The Adaptrum WSD prototype can tune its transmitter to channel 21 (515.0 ± 3.0 MHz), which overlaps cable channel 73. Therefore, this channel combination was used to observe the interference potential to analog CATV reception at test site 3.

First, a spectrum analyzer was used to measure the channel power at the cable service outlet, utilizing a 75-50Ω transformer. The measured channel power from the cable TV outlet at site 3 was -7 dBmV. The applicable standard states that the carrier level at the "input terminals of the first device located on the subscriber premises" should be -12 to +15 dBmV.[29] A step attenuator was inserted into the cable signal path and used to reduce the power on channel 73 at the DTV "cable" input to -11 dBmV (1 dB above the minimum specified level).

The cable service was then attached to the "cable" input of a digital widescreen television, bypassing the existing cable routing (including a distribution amplifier, splitters, various cables, and a set-top converter box). Figure 4-5 presents a line-diagram of the cable receive system at test site 3. The DTV receiver was tuned to cable channel 73 and a good quality analog picture was observed.

Interference to Analog Cable Service

The Adaptrum antenna was placed on a tripod at a height of approximately 1.5 meters and approximately 1 meter horizontally distant from the TV. The antenna was connected to the WSD transmitter output port via coaxial cable through a step attenuator and mounted in its normal polarization (vertical). The prototype transmitter was tuned to channel 21 and activated at full power (no external attenuation). The picture quality on the cable channel was observed to be significantly degraded. The Adaptrum prototype provides for a variable transmit bandwidth. Using that feature, the transmit bandwidth was reset from 4.5 MHz to 6.0 MHz and the demonstration repeated. Interference to the TV became noticeably worse.

The cable service signal path was changed to include the existing cable routing (including the distribution amplifier, etc., but without the converter box) and the test was repeated. Again, interference was observed on the TV receiver and was also noted on an analog TV in an adjacent room, which was tuned to channel 73, in the form of a complete loss of picture.

The step attenuator in the transmitter signal path was used to decrease the transmitted signal level from the maximum power (+22 dBm EIRP based on data presented in Section 3.4) in 1 dB steps while observing the DTV for interference. With 15 dB of signal attenuation (7 dBm EIRP), the interference was barely perceptible (on either TV).

Interference to Digital Cable Service

Next, an attempt was made to observe the interference potential from the WSD transmitter to a digital cable (QAM) channel. Cable channel 77 (543.0 ± 3.0 MHz) was identified as a clear QAM Cable TV channel available at this test site. The Adaptrum WSD prototype transmitter can tune to broadcast TV channel 26 (545.0 ± 3.0 MHz), which overlaps CATV channel 77. Therefore, this channel combination was used in the demonstration.

The digital widescreen TV was tuned to channel 77 but could not demodulate the signal. Thus, a demonstration of the interference potential to the TV's QAM tuner could not be performed at this site.

The cable converter (set-top) box was inserted into the signal path and connected to the DTV's tuner input. The set-top box was tuned to virtual channel 220 (cable channel 77) and a good quality digital picture was observed. The Adaptrum transmitter was tuned to channel 26 and activated at full power (no external attenuation). Interference was immediately observed in the form of a complete loss of picture.

The variable attenuator was incrementally increased to reduce the radiated power in 1 dB steps. At a reduction of 18 dB (4 dBm EIRP) the picture reappeared. The attenuation in the cable signal path was removed and the test repeated. Picture loss was again observed at a transmit attenuation setting of 12 dB (10 dBm EIRP).

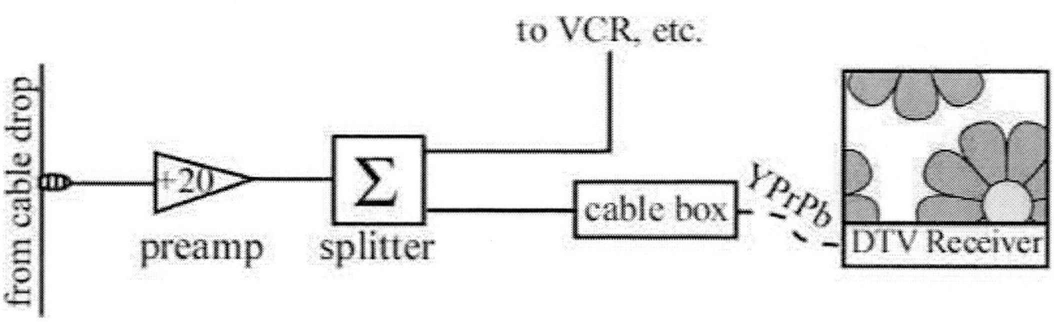

Figure 4-5. Line Diagram of Cable TV Receive System at Test Site #3

Finally, all of the existing routing from the cable service wall outlet to the converter box (including amplifier, splitters and cables) was removed and replaced with a laboratory-grade patch connecting the wall outlet and the converter box input. The Adaptrum transmitter was reset to its maximum power (all external attenuation removed) and turned on. No interference was observed until the transmit antenna was moved very close to the converter box (within 0.3 meters).

Tentative Conclusions at Test Site #3

- The residence's wiring, preamps, and television receivers were susceptible to direct pick-up (DPU) interference in analog mode.
- The cable converter set-top box operating in QAM mode was almost immune to interference from the WSD transmitter (this set-top box was not tested in the analog mode).

4.2.2. Field Test Site 6 (Residential)

A line diagram of the cable service receive system at test site #6 is presented as Figure 4-6. Digital cable signals were processed by the digital set top box; analog cable signals passed through the digital set top box and were processed by the tuner/demodulator set top box.

The peak visual carrier level on cable channel 72 (511.25 MHz)[30] from the connector at the wall was measured at +1 dBmV.[31] No additional attenuation was used for the tests at this site.

Interference to Analog Cable Service

The Adaptrum transmitter was set to operate with a 6 MHz transmit bandwidth on TV channel 21 (512–518) MHz, and its antenna was located about two meters from the analog cable converter box (which fed via HDMI a video monitor). The picture showed perceptible interference (TASO Grade 3) when the transmitter was energized. The antenna was then relocated about one foot from the set top box and the picture degraded further and showed definitely objectionable interference (TASO Grade 5).

With the WSD antenna located about two meters from the analog cable converter box the transmit power output from the WSD was reduced in one decibel steps until the interference became barely visible on the video monitor. This threshold of visibility was found at +13 dBm (9 dB below full transmitter output power).

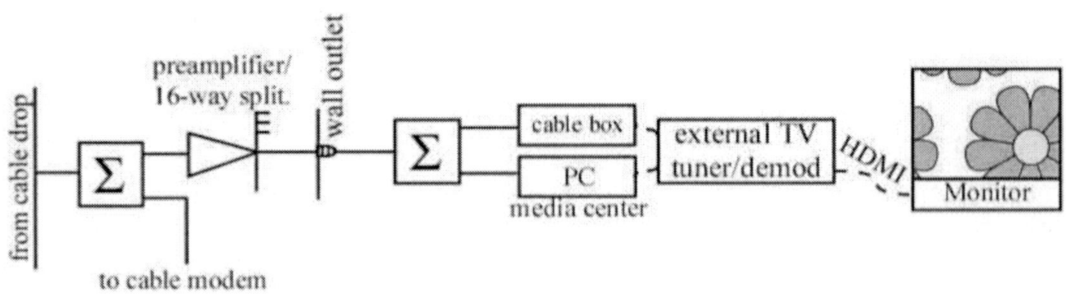

Figure 4-6. Line Diagram of Cable TV Receive System at Test Site #6

Interference to Digital Cable Service

The Adaptrum transmitter was set to operate on TV Channel 30 (566–572 MHz). The cable signal path ran from the wall outlet through a splitter to a digital cable converter set top box.[32] Cable channel 81 (564–570 MHz) was used for the tests; the measured channel power in 6 MHz was -7 dBmV.[33] The 256 QAM cable channel provided six program streams, corresponding to virtual channels 2, 4, 5, 6, 9, and 12. The set-top box was tuned to (virtual) channel 2 for the tests.

The WSD transmitter's antenna was located about four meters from the set-top box. When the transmitter was energized at +22 dBm, the picture experienced occasional blocking, indicating "weak" interference. When the transmitter antenna was relocated to a position two meters from the set-top box, the blocking became frequent, and at a distance of one meter, the picture froze and then disappeared. The same severity of interference was observed at each distance with the transmitter set to either 4.5 or 6 MHz transmit bandwidth.

With the WSD transmitter's antenna at a distance of two meters from the set-top box, 3 dB of attenuation was introduced between the WSD transmitter and its antenna, and the interference ceased. This represents a transmitter power output of +19 dBm.

The additional 3 dB of signal attenuation was removed and the WSD transmit antenna was maintained at a distance of 2 meters from the set-top box, and rotated from vertical to horizontal. The severity of the interference increased when the antenna was rotated. Additional attenuation of 11 dB was required to eliminate the interference, representing a power output of +11 dBm.

Tentative Conclusions from Location #6 Tests

- The residence's cable receiving system was susceptible to direct pick-up (DPU) interference in analog mode when the WSD transmit antenna was within two meters of the RF components.
- The residence's digital cable set-top box was susceptible to interference from the WSD in QAM mode (analog mode not tested) at a distance of about two meters.
- The digital set-top box was more susceptible to interference from the WSD transmitter when the transmit antenna was horizontally polarized.

4.3. Summary of Results

The results of the tests described in Section 4.1 demonstrate that a DTV receive system tuned to a weak DTV channel can experience interference at significant separation distances (data extrapolation indicates to up to 1.2 km) from the WSD transmitter when it is radiating a signal at ~1 50 mW EIRP.

With the transmitter operating at full power on a channel adjacent to a weak channel being received by the DTV receiver system, interference was not observed in a simulated rooftop-mounted receive antenna scenario.

These tests served to informally validate currently accepted D/U thresholds for co-channel interference (15 dB).

Direct pick-up interference to cable television reception was demonstrated at transmit power levels consistent with those reported in previous FCC measurement efforts. The interference appears to be dependent on the configuration of the cable installation and whether or not a set-top box was used and if so, the susceptibility of the box to signal ingress.

5. TV-Sensing (Channel Occupancy) Field Tests

This section describes the field measurement component of the Phase II measurement effort and provides a summary of the data collected. This component involved field trials performed with each of the WSD prototypes in an effort to assess their ability to detect incumbent television stations in "real world" environments. Measurements of over-the-air TV signals were also performed at each site to determine the occupancy of each television channel over the range 21-51.

Four of the originally submitted WSD prototypes were used for these tests. As previously discussed, prior to the commencement of the field tests, Adaptrum modified its device to improve the detection of DTV signals with frequency offset pilot carriers.[34] Unfortunately, this software patch resulted in the device taking considerably longer to perform a full 31-channel scan.[35] This limited the number of trials that could be practically performed with this device at each test location. Motorola also modified its device prior to the field tests to mitigate overload problems resulting from strong adjacent channel signals.[36]

Nine field sites were selected for these tests, representing a mix of urban, suburban and rural locales. The sites included three private residences, an office building and five outdoor locations. Table 5-1 lists these nine test sites. More information is provided on a site-by-site basis in the following subsections.

Trials of the detection capabilities of each of the WSD prototypes were performed from two distinct locations at each test site (the Adaptrum device was often limited to one full-scan trial due to time limitations). The rationale for performing trials from two separate locations at each site was to attempt to avoid the possibility of testing in a signal null zone. This practice also provided an opportunity to examine the consistency in the results between proximate locations. Except for the Adaptrum WSD, two trials were performed at each of two locations, for a total of four separate trials at each site.

The WSDs were generally operated separately and sequentially in order to prevent spurious energy generated by one prototype from affecting the results of another (exceptions were at test sites 5 and 8, where an adequate distance between the WSDs could be achieved). Although only four WSDs were tested, the Motorola prototype was tested in two configurations at each site ("normal or geolocation" and "test or sensing" modes), and at several sites the Philips prototype was also tested in a second configuration (with external attenuation added between the antenna and the receiver input).

At each test site, measurements were performed to determine the occupancy of each of the channels in the 21-51 range. These measurements involved using the FCC truck, equipped with a pneumatic mast that was used to extend a consumer television receive antenna to a height (relative to ground level) of approximately 9.1 meters (30 feet) in an effort to simulate a typical rooftop antenna installation. The antenna was connected first to a spectrum analyzer via coaxial cable and rotated to optimize reception on each channel. A spectrum analyzer was used to identify the signal type and to measure the associated signal power level. For digital TV (ATSC) signals, the power level was measured as the integrated power over the 6-MHz channel. For analog TV (NTSC) signals, the power level was measured in terms of the peak-of-sync power of the visual carrier. After completing these measurements, the antenna lead was connected to an appropriate receiver to determine whether a "viewable" picture could be displayed. For those channels occupied by DTV signals, a digital-to-analog set-top box was used as the tuner, paired with the display portion of a digital television receiver. For channels with analog signals, the antenna lead was connected directly to the TV receiver. Additional details regarding these on-site measurements are presented in Appendix C.

Several anecdotal interference tests were also performed as a part of these field measurements. As discussed above, at sites 3 and 6, tests were performed to aid in assessing the potential for direct pick-up interference to cable television reception (see Section 4.2). Additionally, anecdotal tests of the wireless microphone detection capability implemented in two of the prototype devices (Philips and I2R) were performed at test sites 3 and 5. This data is described elsewhere in this chapter (see Section 6).

Table 5-1. Test Site Information

Site Number	Site Location	GPS Coordinates (NAD 27)	Site Description
1	Patapsco State Park Avalon Area Rear Parking Lot Elkridge, MD	39° 14' 33" N 076° 45' 06" W	Public state park aside Patapsco riverbed in Baltimore suburbs
2	Thomas A. Dixon, Jr Aircraft Observation Area Dorsey Road, Glen Burnie, MD	39° 09' 44" N 076° 39' 51" W	Public park in Baltimore suburbs; Under approach path to BWI/ Thurgood Marshall Airport runway 15R/33L
3	Private Residence Ellicott City, MD	39° 14' N 076° 46' W	Single family home on hilltop in Baltimore suburbs
4	Private Residence College Park, MD	38° 59' N 076° 54' W	Single family home in Washington DC suburbs
5	FCC Headquarters (Portals) 445 12th St., SW Washington, DC	38° 52' 59" N 077° 01' 43" W	High rise office building in downtown Washington DC (urban)
6	Private Residence Galesville, MD	38° 50' N 076° 32' W	Single family home on waterfront in rural southern Maryland
7	Doub's Park Wolfsville Rd (Rt 17) @ Rt 40 Myersville, MD	39° 30' 33" N 077° 33' 37" W	Public baseball park in rural western Maryland
8	Horse Farm Rt 17 @ Middle Point Road North of Myersville, MD	39° 33' 48" N 077° 31' 30" W	Driveway entrance to horse farm in rural western Maryland
9	Grossnickle Church of the Brethren Rt 17 @ Meeting House Road North of Myersville, MD	39° 32' 36" N 077° 31' 46" W	Parish Hall parking lot in rural western Maryland

The following subsections describe the test sites in greater detail and present the information and data collected at each. For each test site, a table is presented that summarizes information associated with each channel in the 21-51 range (excluding channel 37). This table provides six columns of data obtained from the FCC database, augmented with a column of data obtained from on-site observations. The data under each column header are as follows:

- **CH** – lists each channel from 21-51, excluding channel 37 (which is reserved for Radio Astronomy and in-hospital wireless medical telemetry operations).
- **Station Call Sign** – lists the station identifier (call sign) for each TV station (transmitter location) within a 100-km radius of the test site. The 100-km radius was chosen because it is larger than the service areas of most analog and digital TV broadcast stations.
- **Station Location** – lists the community associated with each licensed station.
- **Signal Type** – identifies whether the channel is anticipated to be occupied with an analog or digital television broadcast signal based on information extracted from the FCC consolidated database system (CDBS).
- **Separation Distance** – provides the distance (in kilometers) between the station's broadcast antenna site and the test site location.
- **Within Service Contour** – identifies whether or not the test site is located within the calculated service contour of the associated station.
- **Viewable** – identifies whether or not a television signal could be demodulated and viewed with the test receiver set-up utilized in this effort.[37] In declaring an NTSC channel as viewable, no consideration was given to received picture quality.

Subsequent tables are presented for each test site that provides the raw data from the field trials for each WSD prototype. The unique column headings used in tables are described as follows:

- **Measured Signal Type** – identifies whether the actual measured signal on each channel was digital or analog. In general, this data was consistent with the anticipated signal type; however exceptions were observed. These exceptions can be explained. Often, the anticipated signal was associated with a low-power station located some distance from the test site. In such cases, it is plausible that a full-power station located beyond 100-km will overpower the low-power station. Additionally, because of differences in the methodology for measuring and expressing analog and digital signal levels, it's possible to measure NTSC signals at a lower receive carrier level than ATSC signals. Thus, it's often possible to measure an analog signal originating beyond the 100-km radius, whereas a digital signal originating inside, but near the edge of the radius, could not be measured. In most cases where a discrepancy exists between the anticipated signal type versus the measured signal type, it involves the detection of an analog signal when a digital signal was anticipated.

- **Measured Power Level** – This column presents the measured power level on each channel for each test site. Information regarding how these measurements were performed is provided in Appendix C.

- **Occupied (O) & Available (A) Channels Reported by Each WSD** – This column indicates whether the WSD prototype reported the channel as occupied (indicated by an "O"). The status of those channels not reported as "occupied" were determined based on the rationale discussed below and indicated in the table by either an "A" to represent available or a "-" to represent indeterminate. This column may be further subdivided to indicate the results observed at each particular location where the device was tested. For example, L1/S1 denotes those results obtained from the first scan performed at location 1 and similarly, L2/S2 denotes those results obtained from the second scan performed at location 2.

All of the prototype devices, except for the Motorola device when operated in "Normal" (geo-location) mode, explicitly report only those channels determined to be occupied. When operated in this mode, the Motorola device reports those channels determined to be occupied and also provides a list of the channels that it determines to be "available".

When a prototype WSD scans for both ATSC and NTSC signals (even when it requires separate scans to accomplish this) upon completion of a scan, those channels not reported as occupied can reasonably be presumed to be vacant or "available". However, if the scan is performed to detect only channels occupied with one signal type (NTSC or ATSC), then the status of the channels not reported as occupied is not readily apparent. Examples of such scans performed in the field trials were: 1) those scans performed with the Adaptrum prototype device for NTSC-occupied channels only, 2) those scans performed with the Motorola prototype device in "test" (sensing only) mode for ATSCoccupied channels only, and 3) those scans performed with the I2R prototype device at test sites 1 and 2, where the device was operated in the incorrect mode for detecting ATSC-occupied channels (in effect, resulting in NTSC-only scans).

In populating the data tables reported for each test site, the following approach was applied to determine the status of each channel based on the scan results:

- For all devices, those detection results indicated in the table by an "O" represent exactly what was reported by the device at the conclusion of each scan.
- For those devices used to perform scans for both signal types (ATSC and NTSC) at the same location, the detection results indicated in the table by an "A" represent the channels that were not reported by the device as occupied under scans for both signals and thus were presumed to be vacant or "available". This condition is applicable to the following devices under the described conditions:
  - The Adaptrum device in those locations where both ATSC and NTSC scans were performed,

    - The I2R device at all sites except 1 and 2,
    - The Motorola device when operated in the "Normal" mode (effectively),
    - The Philips device in all configurations.
- For those scans performed for only one signal type (ATSC or NTSC), determining whether a channel not reported as occupied would have been considered vacant (available) was often ambiguous and in some cases required that some judgment be applied. The following criteria was applied in making such channel status decisions:
    - If the measured over-the-air signal type anticipated via the TV contours matched the signal type and was of the same type as that being detected by the device, then a channel not reported by the device as occupied was presumed to be available (indicated by "A" in the tables).
    - If the measured signal type was of the same type as that being detected by the device (i.e., the measured signal was NTSC and the device was scanning for an NTSC signal), then regardless of the expected signal type, a channel not reported by the device as occupied was presumed to be available (indicated by "A" in the tables).
    - If the expected signal type matched the measured signal type but was of the opposite type from that being detected by the device (i.e., the expected and measured signal were ATSC but the device was scanning for NTSC), then no presumption was made for those channels not reported by the device as occupied (indicated by "-" in the tables).
    - If there was no expected signal type reported for the channel (no station assigned to the channel) and a signal type was not measured at the site, then no presumption was made regarding those channels not reported by the device as occupied (indicated in the tables by "-").[38]
    - For any other conditions for which ambiguity existed, no presumption was made regarding the channel status for those channels not reported by the device as occupied (indicated in the tables by "-").

Finally, there are some entries to the table where more than one station within the 100-km search radius is assigned to a single channel. In these cases, the signal type measured at the site took precedence when assessing channel status.

5.1. Test Site #1 – Patapsco Valley State Park, Elkridge, MD

This site offered public access (Maryland State Park) and is representative of a setting with difficult television reception characteristics. The actual test site was a parking lot in the rear of the park, adjacent to the Patapsco river bed, in a valley with rapidly rising, wooded terrain to the Northeast (towards Baltimore) and Southwest (towards Washington). The two test locations at this site were in the same parking lot, separated by approximately 45.7 meters (150 feet).

Table 5-2. Licensed TV Station Assignments within 100-km radius of Test Site 1

<table>
<tr><th>CH</th><th>Station Call Sign</th><th>Station Location</th><th>Signal Type</th><th>Separation Distance (km)</th><th>Within Service Contour? (Y/N)</th><th>Viewable? (Y/N)</th></tr>
<tr><td>21</td><td>-</td><td>-</td><td>-</td><td>-</td><td>N</td><td>N</td></tr>
<tr><td>22</td><td>WMPT</td><td>Annapolis, MD</td><td>NTSC</td><td>28.47</td><td>Y</td><td>Y</td></tr>
<tr><td>23</td><td>WDDN-LP</td><td>Washington, DC</td><td>NTSC</td><td>37.68</td><td>N</td><td>N</td></tr>
<tr><td>24</td><td>WUTB</td><td>Baltimore, MD</td><td>NTSC</td><td>5.18</td><td>Y</td><td>Y</td></tr>
<tr><td>25</td><td>WZDC-LP</td><td>Washington, DC</td><td>NTSC</td><td>44.04</td><td>N</td><td>N</td></tr>
<tr><td>26</td><td>WETA</td><td>Washington, DC</td><td>NTSC</td><td>43.41</td><td>Y</td><td>Y</td></tr>
<tr><td>27</td><td>WETA</td><td>Washington, DC</td><td>ATSC</td><td>43.05</td><td>Y</td><td>N</td></tr>
<tr><td>28</td><td>WFPT</td><td>Frederick, MD</td><td>ATSC</td><td>48.49</td><td>Y</td><td>N</td></tr>
<tr><td>29</td><td>WMPB</td><td>Baltimore, MD</td><td>ATSC</td><td>22.98</td><td>Y</td><td>N</td></tr>
<tr><td rowspan="2">30</td><td>WGCB</td><td>Red Lion, PA</td><td>ATSC</td><td>75.07</td><td>N</td><td rowspan="2">N</td></tr>
<tr><td>WNVT</td><td>Goldvein, VA</td><td>ATSC</td><td>90.49</td><td>N</td></tr>
<tr><td>31</td><td>WRZB-LP</td><td>Annapolis, MD</td><td>NTSC</td><td>33.02</td><td>N</td><td>N</td></tr>
<tr><td>32</td><td>WHUT</td><td>Washington, DC</td><td>NTSC</td><td>43.46</td><td>Y</td><td>N</td></tr>
<tr><td>33</td><td>WHUT</td><td>Washington, DC</td><td>ATSC</td><td>43.05</td><td>Y</td><td>N</td></tr>
<tr><td>34</td><td>WUSA</td><td>Washington, DC</td><td>ATSC</td><td>43.05</td><td>Y</td><td>N</td></tr>
<tr><td>35</td><td>WDCA</td><td>Washington, DC</td><td>ATSC</td><td>42.76</td><td>Y</td><td>N</td></tr>
<tr><td>36</td><td>WTTG</td><td>Washington, DC</td><td>ATSC</td><td>42.76</td><td>Y</td><td>N</td></tr>
<tr><td>38</td><td>WJZ</td><td>Baltimore, MD</td><td>ATSC</td><td>13.49</td><td>Y</td><td>Y</td></tr>
<tr><td>39</td><td>WJLA</td><td>Washington, DC</td><td>ATSC</td><td>43.05</td><td>Y</td><td>N</td></tr>
<tr><td>40</td><td>WNUV</td><td>Baltimore, MD</td><td>ATSC</td><td>13.67</td><td>Y</td><td>Y</td></tr>
<tr><td>41</td><td>WUTB</td><td>Baltimore, MD</td><td>ATSC</td><td>5.18</td><td>Y</td><td>Y</td></tr>
<tr><td>42</td><td>WMPT</td><td>Annapolis, MD</td><td>ATSC</td><td>28.47</td><td>Y</td><td>N</td></tr>
<tr><td rowspan="2">43</td><td>WPMT</td><td>York, PA</td><td>NTSC</td><td>88.29</td><td>N</td><td rowspan="2">N</td></tr>
<tr><td>WPXW</td><td>Manassas, VA</td><td>ATSC</td><td>71.06</td><td>N</td></tr>
<tr><td>44</td><td>-</td><td>-</td><td>-</td><td>-</td><td>N</td><td>N</td></tr>
<tr><td>45</td><td>WBFF</td><td>Baltimore, MD</td><td>NTSC</td><td>13.67</td><td>Y</td><td>Y</td></tr>
<tr><td>46</td><td>WBFF</td><td>Baltimore, MD</td><td>ATSC</td><td>13.67</td><td>Y</td><td>Y</td></tr>
<tr><td>47</td><td>WPMT</td><td>York, PA</td><td>ATSC</td><td>88.29</td><td>N</td><td>N</td></tr>
<tr><td>48</td><td>WRC</td><td>Washington, DC</td><td>ATSC</td><td>44.03</td><td>Y</td><td>N</td></tr>
<tr><td>49</td><td>WGCB</td><td>Red Lion, PA</td><td>NTSC</td><td>75.07</td><td>N</td><td>N</td></tr>
<tr><td>50</td><td>WDCW</td><td>Washington, DC</td><td>NTSC</td><td>39.11</td><td>Y</td><td>Y</td></tr>
<tr><td>51</td><td>WDCW</td><td>Washington, DC</td><td>ATSC</td><td>39.11</td><td>Y</td><td>N</td></tr>
</table>

Table 5-3. Adaptrum WSD Prototype Results at Test Site 1

CH	Signal Type (Database)	Within Service Contour? (Y/N)	Viewable? (Y/N)	Measured Signal Type	Measured Power Level (dBm)	Occupied (O) & Available (A) Channels Reported by WSD			
						L1/ S1	L1/ S2[1]	L2/ S1	L2/ S2 [2]
21	-	N	N	-	-	A	-	A	-
22	NTSC	Y	Y	NTSC	-74.0	O	O	O	-
23	NTSC	N	N	-	-	O	-	A	-
24	NTSC	Y	Y	NTSC	-55.0	O	O	O	-
25	NTSC	N	N	-	-	A	-	A	-
26	NTSC	Y	Y	NTSC	-96.6	O	O	A	-
27	ATSC	Y	N	-	-	A	-	A	-
28	ATSC	Y	N	ATSC	-96.3	A	-	A	-
29	ATSC	Y	N	ATSC	-90.9	O	-	O	-
30	ATSC ATSC	N N	N	-	-	O	-	A	-
31	NTSC	N	N	-	-	O	-	A	-
32	NTSC	Y	N	NTSC	-99.3	A	A	A	-
33	ATSC	Y	N	ATSC	-86.0	A	-	A	-
34	ATSC	Y	N	-	-	O	-	O	-
35	ATSC	Y	N	ATSC	-85.5	A	-	A	-
36	ATSC	Y	N	ATSC	-85.0	O	-	O	-
38	ATSC	Y	Y	ATSC	-72.5	O	-	O	-
39	ATSC	Y	N	-	-	O	-	A	-
40	ATSC	Y	Y	ATSC	-77.0	O	-	O	-
41	ATSC	Y	Y	ATSC	-69.6	O	-	O	-
42	ATSC	Y	N	ATSC	-87.0	O	-	O	
43	NTSC ATSC	N N	N	NTSC	-97.9	A	A	A	-
44	-	N	N	-	-	A	-	A	-
45	NTSC	Y	Y	NTSC	-77.3	O	O	O	-
46	ATSC	Y	Y	ATSC	-74.0	O	-	O	-
47	ATSC	N	N	NTSC	-88.3	A	A	A	-
48	ATSC	Y	N	ATSC	-87.8	A	-	A	-
49	NTSC	N	N	NTSC	-103.2	A	A	A	-
50	NTSC	Y	Y	NTSC	-95.3	O	A	A	-
51	ATSC	Y	N	NTSC	-93.9	O	A	O	-

[1]Second scan at location 1 for NTSC signals only.

[2] No second scan performed at location 2.

Table 5-4. I2R WSD Prototype Results at Test Site 1

CH	Signal Type (Database)	Within Service Contour? (Y/N)	Viewable? (Y/N)	Measured Signal Type	Measured Power Level (dBm)	Occupied (O) & Available (A) Channels Reported by WSD			
						L1/S1	L1/S2	L2/S1	L2/S2
21	-	N	N	-	-	-	-	-	-
22	NTSC	Y	Y	NTSC	-74.0	O	A	A	A
23	NTSC	N	N	-	-	-	-	-	-
24	NTSC	Y	Y	NTSC	-55.0	A	A	A	A
25	NTSC	N	N	-	-	-	-	-	-
26	NTSC	Y	Y	NTSC	-96.6	A	A	A	A
27	ATSC	Y	N	-	-	-	-	-	-
28	ATSC	Y	N	ATSC	-96.3	-	-	-	-
29	ATSC	Y	N	ATSC	-90.9	O	O	O	O
30	ATSC ATSC	N N	N	-	-	O	O	O	O
31	NTSC	N	N	-	-	-	-	-	-
32	NTSC	Y	N	NTSC	-99.3	A	A	A	A
33	ATSC	Y	N	ATSC	-86.0	-	-	-	-
34	ATSC	Y	N	-	-	-	-	-	-
35	ATSC	Y	N	ATSC	-85.5	-	-	O	O
36	ATSC	Y	N	ATSC	-85.0	O	O	-	-
38	ATSC	Y	Y	ATSC	-72.5	O	O	O	O
39	ATSC	Y	N	-	-	-	-	-	-
40	ATSC	Y	Y	ATSC	-77.0	O	O	O	O
41	ATSC	Y	Y	ATSC	-69.6	O	O	-	O
42	ATSC	Y	N	ATSC	-87.0	-	-	O	O
43	NTSC ATSC	N N	N	NTSC	-97.9	A	A	A	A
44	-	N	N	-	-	-	-	-	-
45	NTSC	Y	Y	NTSC	-77.3	O	A	A	A
46	ATSC	Y	Y	ATSC	-74.0	O	O	O	O
47	ATSC	N	N	NTSC	-88.3	A	A	A	A
48	ATSC	Y	N	ATSC	-87.8	-	-	-	-
49	NTSC	N	N	NTSC	-103.2	A	A	A	A
50	NTSC	Y	Y	NTSC	-95.3	A	A	A	A
51	ATSC	Y	N	NTSC	-93.9	A	A	O	A

NOTE: Device was operated in "all signal" mode which was later discovered to be incorrect mode for detecting ATSC signals.

Table 5-5. Motorola WSD Prototype (Test Mode) Results at Test Site 1

CH	Signal Type (Database)	Within Service Contour? (Y/N)	Viewable? (Y/N)	Measured Signal Type	Measured Power Level (dBm)	Occupied (O) & Available (A) Channels Reported by WSD			
						L1/S1	L1/S2	L2/S1	L2/S2
21	-	N	N	-	-	O	O	O	-
22	NTSC	Y	Y	NTSC	-74.0	-	-	-	-
23	NTSC	N	N	-	-	-	-	-	-
24	NTSC	Y	Y	NTSC	-55.0	-	-	-	-
25	NTSC	N	N	-	-	-	-	-	-
26	NTSC	Y	Y	NTSC	-96.6	-	-	-	-
27	ATSC	Y	N	-	-	O	-	-	-
28	ATSC	Y	N	ATSC	-96.3	A	A	A	A
29	ATSC	Y	N	ATSC	-90.9	O	O	O	O
30	ATSC ATSC	N N	N	-	-	-	-	-	-
31	NTSC	N	N	-	-	-	-	-	-
32	NTSC	Y	N	NTSC	-99.3	-	-	-	-
33	ATSC	Y	N	ATSC	-86.0	A	A	A	A
34	ATSC	Y	N	-	-	O	O	O	O
35	ATSC	Y	N	ATSC	-85.5	O	O	O	O
36	ATSC	Y	N	ATSC	-85.0	O	O	A	A
38	ATSC	Y	Y	ATSC	-72.5	O	O	O	O
39	ATSC	Y	N	-	-	O	O	O	O
40	ATSC	Y	Y	ATSC	-77.0	O	O	O	O
41	ATSC	Y	Y	ATSC	-69.6	O	O	O	O
42	ATSC	Y	N	ATSC	-87.0	O	O	O	O
43	NTSC ATSC	N N	N	NTSC	-97.9	O	O	O	O
44	-	N	N	-	-	-	-	-	-
45	NTSC	Y	Y	NTSC	-77.3	-	-	-	-
46	ATSC	Y	Y	ATSC	-74.0	O	O	O	O
47	ATSC	N	N	NTSC	-88.3	-	-	-	-
48	ATSC	Y	N	ATSC	-87.8	O	A	O	O
49	NTSC	N	N	NTSC	-103.2	-	-	-	-
50	NTSC	Y	Y	NTSC	-95.3	-	-	-	-
51	ATSC	Y	N	NTSC	-93.9	-	-	O	O

NOTE: In the test mode this device detects only ATSC signals (*i.e.*, no NTSC detection capability).

Table 5-6. Motorola WSD Prototype (Normal Mode) Results at Test Site 1

CH	Signal Type (Database)	Within Service Contour? (Y/N)	Viewable? (Y/N)	Measured Signal Type	Measured Power Level (dBm)	Occupied (O) & Available (A) Channels Reported by WSD	
						L1/S1	L2/S1
21	-	N	N	-	-	A	A
22	NTSC	Y	Y	NTSC	-74.0	O	O
23	NTSC	N	N	-	-	A	A
24	NTSC	Y	Y	NTSC	-55.0	O	O
25	NTSC	N	N	-	-	A	A
26	NTSC	Y	Y	NTSC	-96.6	O	O
27	ATSC	Y	N	-	-	O	O
28	ATSC	Y	N	ATSC	-96.3	O	O
29	ATSC	Y	N	ATSC	-90.9	O	O
30	ATSC	N	N	-	-	O	O
	ATSC	N					
31	NTSC	N	N	-	-	A	A
32	NTSC	Y	N	NTSC	-99.3	O	O
33	ATSC	Y	N	ATSC	-86.0	O	O
34	ATSC	Y	N	-	-	O	O
35	ATSC	Y	N	ATSC	-85.5	O	O
36	ATSC	Y	N	ATSC	-85.0	O	O
38	ATSC	Y	Y	ATSC	-72.5	O	O
39	ATSC	Y	N	-	-	O	O
40	ATSC	Y	Y	ATSC	-77.0	O	O
41	ATSC	Y	Y	ATSC	-69.6	O	O
42	ATSC	Y	N	ATSC	-87.0	O	O
43	NTSC	N	N	NTSC	-97.9	A	A
	ATSC	N					
44	-	N	N	-	-	A	A
45	NTSC	Y	Y	NTSC	-77.3	O	O
46	ATSC	Y	Y	ATSC	-74.0	O	O
47	ATSC	N	N	NTSC	-88.3	O	O
48	ATSC	Y	N	ATSC	-87.8	O	O
49	NTSC	N	N	NTSC	-103.2	A	A
50	NTSC	Y	Y	NTSC	-95.3	O	O
51	ATSC	Y	N	NTSC	-93.9	O	O

Table 5-7. Philips WSD Prototype Results at Test Site 1

CH	Signal Type (Database)	Within Service Contour? (Y/N)	Viewable? (Y/N)	Measured Signal Type	Measured Power Level (dBm)	Occupied (O) & Available (A) Channels Reported by WSD			
						L1/S1	L1/S2	L2/S1	L2/S2
21	-	N	N	-	-	O	O	O	O
22	NTSC	Y	Y	NTSC	-74.0	O	O	O	O
23	NTSC	N	N	-	-	O	O	O	A
24	NTSC	Y	Y	NTSC	-55.0	O	O	O	O
25	NTSC	N	N	-	-	O	O	O	O
26	NTSC	Y	Y	NTSC	-96.6	O	O	O	O
27	ATSC	Y	N	-	-	O	O	O	O
28	ATSC	Y	N	ATSC	-96.3	O	O	A	O
29	ATSC	Y	N	ATSC	-90.9	O	O	O	O
30	ATSC ATSC	N N	N	-	-	O	A	O	O
31	NTSC	N	N	-	-	A	A	O	A
32	NTSC	Y	N	NTSC	-99.3	O	O	O	O
33	ATSC	Y	N	ATSC	-86.0	O	A	O	O
34	ATSC	Y	N	-	-	O	O	O	O
35	ATSC	Y	N	ATSC	-85.5	O	O	O	O
36	ATSC	Y	N	ATSC	-85.0	O	O	O	O
38	ATSC	Y	Y	ATSC	-72.5	O	O	O	O
39	ATSC	Y	N	-	-	O	A	O	O
40	ATSC	Y	Y	ATSC	-77.0	O	O	O	O
41	ATSC	Y	Y	ATSC	-69.6	O	O	O	O
42	ATSC	Y	N	ATSC	-87.0	O	O	O	O
43	NTSC ATSC	N N	N	NTSC	-97.9	O	O	O	O
44	-	N	N	-	-	A	A	A	A
45	NTSC	Y	Y	NTSC	-77.3	O	O	O	O
46	ATSC	Y	Y	ATSC	-74.0	O	O	O	O
47	ATSC	N	N	NTSC	-88.3	O	O	O	O
48	ATSC	Y	N	ATSC	-87.8	O	O	O	O
49	NTSC	N	N	NTSC	-103.2	A	O	A	O
50	NTSC	Y	Y	NTSC	-95.3	O	O	O	O
51	ATSC	Y	N	NTSC	-93.9	O	A	O	O

5.2. Test Site #2 - Aircraft Observation Area, Glen Burnie, MD

This is a public accessible site in the Baltimore suburbs, just south of BWI/Thurgood Marshall Airport, under the approach path to runway 1 5R/33L. The channel occupancy measurements were performed from the west side of the parking lot and the prototype field trials were performed adjacent to the parking lot (location 1) and approximately 45.7 meters (150 feet) southwest of the first location (location 2). Several high-power TV signals were observed at this site.

Table 5-8. Licensed TV Station Assignments within 100-km radius of Test Site 2

CH	Station Call Sign	Station Location	Signal Type	Separation Distance (km)	Within Service Contour? (Y/N)	Viewable? (Y/N)
21	-	-	-	-	N	N
22	WMPT	Annapolis, MD	NTSC	17.53	Y	Y
23	WDDN-LP	Washington, DC	NTSC	38.57	N	N
24	WUTB	Baltimore, MD	NTSC	16.26	Y	Y
25	WZDC-LP	Washington, DC	NTSC	43.85	N	N
26	WETA	Washington, DC	NTSC	44.09	Y	N
27	WETA	Washington, DC	ATSC	43.06	Y	Y
28	WFPT	Frederick, MD	ATSC	57.12	N	N
29	WMPB	Baltimore, MD	ATSC	33.21	Y	Y
30	WGCB	Red Lion, PA	ATSC	82.75	N	N
	WNVT	Goldvein, VA	ATSC	89.68	N	N
31	WRZB-LP	Annapolis, MD	NTSC	21.65	N	N
32	WHUT	Washington, DC	NTSC	44.15	Y	Y
33	WHUT	Washington, DC	ATSC	43.06	Y	Y
34	WUSA	Washington, DC	ATSC	43.06	Y	Y
35	WDCA	Washington, DC	ATSC	42.96	Y	Y
36	WTTG	Washington, DC	ATSC	42.96	Y	Y
38	WJZ	Baltimore, MD	ATSC	19.18	Y	Y
39	WJLA	Washington, DC	ATSC	43.06	Y	N
40	WNUV	Baltimore, MD	ATSC	19.34	Y	Y
41	WUTB	Baltimore, MD	ATSC	16.26	Y	Y
42	WMPT	Annapolis, MD	ATSC	17.53	Y	Y
43	WPMT	York, PA	NTSC	96.28	N	Y
	WPXW	Manassas, VA	ATSC	71.17	N	N
44	-	-	-	-	N	N
45	WBFF	Baltimore, MD	NTSC	19.34	Y	Y
46	WBFF	Baltimore, MD	ATSC	19.34	Y	Y
47	WPMT	York, PA	ATSC	96.28	N	N
48	WRC	Washington, DC	ATSC	43.84	Y	N
49	WGCB	Red Lion, PA	NTSC	82.75	N	N
50	WDCW	Washington, DC	NTSC	38.52	Y	Y
51	WDCW	Washington, DC	ATSC	38.52	Y	N

Table 5-9. Adaptrum WSD Prototype Results at Test Site 2

CH	SignalType (Database)	Within Service Contour? (Y/N)	Viewable? (Y/N)	Signal Type (Measured)	Measured Power Level (dBm)	Occupied (O) & Available (A) Channels Reported by WSD	
						L1/S1	L1/S2
21	-	N	N	-	< -71.0	O	O
22	NTSC	Y	Y	NTSC	-39.4	O	O
23	NTSC	N	N	ATSC	< -67.3	A	A
24	NTSC	Y	Y	NTSC	-28.0	O	O
25	NTSC	N	N	NTSC	-88.6	A	A
26	NTSC	Y	N	NTSC	-79.5	O	O
27	ATSC	Y	Y	ATSC	-65.4	O	O
28	ATSC	N	N	NTSC	-90.0	A	A
29	ATSC	Y	Y	ATSC	-47.9	O	O
30	ATSC	N	N	ATSC	< -69.8	O	O
	ATSC	N	N				
31	NTSC	N	N	-	-	O	A
32	NTSC	Y	Y	NTSC	-73.4	O	O
33	ATSC	Y	Y	ATSC	-81.6	O	O
34	ATSC	Y	Y	ATSC	-59.1	O	O
35	ATSC	Y	Y	ATSC	-62.6	A	A
36	ATSC	Y	Y	ATSC	-62.2	O	O
38	ATSC	Y	Y	ATSC	-16.0	O	O
39	ATSC	Y	N	ATSC	-65.3	O	O
40	ATSC	Y	Y	ATSC	-20.4	O	O
41	ATSC	Y	Y	ATSC	-44.7	O	O
42	ATSC	Y	Y	ATSC	-43.9	O	O
43	NTSC	N	Y	NTSC	-72.7	A	A
	ATSC	N	N				
44	-	N	N	-	-	A	A
45	NTSC	Y	Y	NTSC	-23.0	O	O
46	ATSC	Y	Y	ATSC	-20.7	O	O
47	ATSC	N	N	NTSC	-78.7	A	A
48	ATSC	Y	N	ATSC	< -69.2	O	O
49	NTSC	N	N	NTSC	-93.1	A	A
50	NTSC	Y	Y	NTSC	-66.1	O	A
51	ATSC	Y	N	NTSC	-69.1	O	O

Table 5-10. I2R WSD Prototype Results at Test Site 2

CH	Signal Type (Database)	Within Service Contour? (Y/N)	Viewable? Y/N)	Signal Type (Measured)	Measured Power Level (dBm)	Occupied (O) & Available (A) Channels Reported by WSD			
						L1/S1	L1/S2	L2/S1	L2/S2
21	-	N	N	-	< -71.0	-	-	-	-
22	NTSC	Y	Y	NTSC	-39.4	A	A	A	A
23	NTSC	N	N	ATSC	< -67.3	-	-	-	-
24	NTSC	Y	Y	NTSC	-28.0	A	O	O	A
25	NTSC	N	N	NTSC	-88.6	A	A	A	A
26	NTSC	Y	N	NTSC	-79.5	O	A	A	A
27	ATSC	Y	Y	ATSC	-65.4	-	-	-	O
28	ATSC	N	N	NTSC	-90.0	A	A	A	A
29	ATSC	Y	Y	ATSC	-47.9	O	O	O	O
30	ATSC	N	N	ATSC	< -69.8	O	O	O	O
	ATSC	N	N						
31	NTSC	N	N	-	-	-	O	-	-
32	NTSC	Y	Y	NTSC	-73.4	A	O	A	A
33	ATSC	Y	Y	ATSC	-81.6	-	-	-	-
34	ATSC	Y	Y	ATSC	-59.1	O	-	O	O
35	ATSC	Y	Y	ATSC	-62.6	-	-	O	-
36	ATSC	Y	Y	ATSC	-62.2	-	-	-	O
38	ATSC	Y	Y	ATSC	-16.0	O	O	-	-
39	ATSC	Y	N	ATSC	-65.3	-	-	O	O
40	ATSC	Y	Y	ATSC	-20.4	O	-	-	A
41	ATSC	Y	Y	ATSC	-44.7	O	O	O	O
42	ATSC	Y	Y	ATSC	-43.9	-	O	-	-
43	NTSC	N	Y	NTSC	-72.7	A	A	A	A
	ATSC	N	N						
44	-	N	N	-	-	-	-	-	-
45	NTSC	Y	Y	NTSC	-23.0	A	A	O	O
46	ATSC	Y	Y	ATSC	-20.7	O	O	O	O
47	ATSC	N	N	NTSC	-78.7	A	A	O	O
48	ATSC	Y	N	ATSC	< -69.2	-	-	-	-
49	NTSC	N	N	NTSC	-93.1	A	A	A	A
50	NTSC	Y	Y	NTSC	-66.1	A	A	A	A
51	ATSC	Y	N	NTSC	-69.1	A	A	O	A

NOTE: Device was operated in "all signal" mode which was later discovered to be incorrect mode for detecting ATSC signals.

Table 5-11. Motorola WSD Prototype (Test Mode) Results at Test Site 2

CH	Signal Type (Database)	Within Service Contour? (Y/N)	Viewable?(Y/N)	Signal Type (Measured)	Measured Power Level (dBm)	Occupied (O) & Available (A) Channels Reported by WSD			
						L1/S1	L1/S2	L2/S1	L2/S2
21	-	N	N	-	< -71.0	-	-	-	O
22	NTSC	Y	Y	NTSC	-39.4	-	-	-	-
23	NTSC	N	N	ATSC	< -67.3	A	A	A	A
24	NTSC	Y	Y	NTSC	-28.0	-	-	-	-
25	NTSC	N	N	NTSC	-88.6	-	-	-	-
26	NTSC	Y	N	NTSC	-79.5	-	-	-	-
27	ATSC	Y	Y	ATSC	-65.4	O	O	O	O
28	ATSC	N	N	NTSC	-90.0	-	-	-	-
29	ATSC	Y	Y	ATSC	-47.9	O	O	O	O
30	ATSC	N	N	ATSC	< -69.8	A	A	A	A
	ATSC	N	N						
31	NTSC	N	N	-	-	-	-	-	-
32	NTSC	Y	Y	NTSC	-73.4	-	-	-	-
33	ATSC	Y	Y	ATSC	-81.6	A	A	A	A
34	ATSC	Y	Y	ATSC	-59.1	O	O	O	O
35	ATSC	Y	Y	ATSC	-62.6	O	O	O	O
36	ATSC	Y	Y	ATSC	-62.2	O	O	O	O
38	ATSC	Y	Y	ATSC	-16.0	O	O	O	O
39	ATSC	Y	N	ATSC	-65.3	O	O	A	A
40	ATSC	Y	Y	ATSC	-20.4	O	O	O	O
41	ATSC	Y	Y	ATSC	-44.7	O	O	O	O
42	ATSC	Y	Y	ATSC	-43.9	O	O	O	O
43	NTSC	N	Y	NTSC	-72.7	O	O	O	O
	ATSC	N	N						
44	-	N	N	-	-	O	O	-	-
45	NTSC	Y	Y	NTSC	-23.0	-	-	-	-
46	ATSC	Y	Y	ATSC	-20.7	O	O	O	O
47	ATSC	N	N	NTSC	-78.7	-	-	-	-
48	ATSC	Y	N	ATSC	< -69.2	O	O	O	-
49	NTSC	N	N	NTSC	-93.1	O	O	-	O
50	NTSC	Y	Y	NTSC	-66.1	-	-	-	-
51	ATSC	Y	N	NTSC	-69.1	O	O	O	O

NOTE: In the test mode this device detects only ATSC signals (*i.e.*, no NTSC detection capability).

Table 5-12. Motorola WSD Prototype (Normal Mode) Results at Test Site 2

CH	Signal Type (Database)	Within Service Contour? (Y/N)	Viewable? (Y/N)	Signal Type (Measured)	Measured Power Level (dBm)	Occupied (O) & Available (A) Channels Reported by WSD	
						L1/S1	L2/S1
21	-	N	N	-	< -71.0	A	A
22	NTSC	Y	Y	NTSC	-39.4	O	O
23	NTSC	N	N	ATSC	< -67.3	A	A
24	NTSC	Y	Y	NTSC	-28.0	O	O
25	NTSC	N	N	NTSC	-88.6	O	O
26	NTSC	Y	N	NTSC	-79.5	O	O
27	ATSC	Y	Y	ATSC	-65.4	O	O
28	ATSC	N	N	NTSC	-90.0	O	O
29	ATSC	Y	Y	ATSC	-47.9	O	O
30	ATSC	N	N	ATSC	< -69.8	A	A
	ATSC	N	N				
31	NTSC	N	N	-	-	A	A
32	NTSC	Y	Y	NTSC	-73.4	O	O
33	ATSC	Y	Y	ATSC	-81.6	O	O
34	ATSC	Y	Y	ATSC	-59.1	O	O
35	ATSC	Y	Y	ATSC	-62.6	O	O
36	ATSC	Y	Y	ATSC	-62.2	O	O
38	ATSC	Y	Y	ATSC	-16.0	O	O
39	ATSC	Y	N	ATSC	-65.3	O	O
40	ATSC	Y	Y	ATSC	-20.4	O	O
41	ATSC	Y	Y	ATSC	-44.7	O	O
42	ATSC	Y	Y	ATSC	-43.9	O	O
43	NTSC	N	Y	NTSC	-72.7	O	O
	ATSC	N	N				
44	-	N	N	-	-	A	A
45	NTSC	Y	Y	NTSC	-23.0	O	O
46	ATSC	Y	Y	ATSC	-20.7	O	O
47	ATSC	N	N	NTSC	-78.7	O	O
48	ATSC	Y	N	ATSC	< -69.2	O	O
49	NTSC	N	N	NTSC	-93.1	A	A
50	NTSC	Y	Y	NTSC	-66.1	O	O
51	ATSC	Y	N	NTSC	-69.1	O	O

Table 5-13. Philips WSD Prototype Results at Test Site 2

CH	Signal Type (Database)	Within Service Contour? (Y/N)	Viewable? (Y/N)	Signal Type (Measured)	Measured Power Level (dBm)	Occupied (O) & Available (A) Channels Reported by WSD			
						L1/S1	L1/S2	L2/S1	L2/S2
21	-	N	N	-	< -71.0	O	O	O	O
22	NTSC	Y	Y	NTSC	-39.4	O	O	O	O
23	NTSC	N	N	ATSC	< -67.3	O	O	O	O
24	NTSC	Y	Y	NTSC	-28.0	O	O	O	O
25	NTSC	N	N	NTSC	-88.6	O	O	O	O
26	NTSC	Y	N	NTSC	-79.5	O	O	O	O
27	ATSC	Y	Y	ATSC	-65.4	O	O	O	O
28	ATSC	N	N	NTSC	-90.0	O	O	O	O
29	ATSC	Y	Y	ATSC	-47.9	O	O	O	O
30	ATSC	N	N	ATSC	< -69.8	O	O	O	O
	ATSC	N	N						
31	NTSC	N	N	-	-	O	O	O	O
32	NTSC	Y	Y	NTSC	-73.4	O	O	O	O
33	ATSC	Y	Y	ATSC	-81.6	O	O	O	O
34	ATSC	Y	Y	ATSC	-59.1	O	O	O	O
35	ATSC	Y	Y	ATSC	-62.6	O	O	O	O
36	ATSC	Y	Y	ATSC	-62.2	O	O	O	O
37				ATSC	-16.0	O	O	O	O
38	ATSC	Y	Y	ATSC	-65.3	O	O	O	O
39	ATSC	Y	N	ATSC	-20.4	O	O	O	O
40	ATSC	Y	Y	ATSC	-44.7	O	O	O	O
41	ATSC	Y	Y	ATSC	-43.9	O	O	O	O
42	ATSC	Y	Y	ATSC	-72.7	O	O	O	O
43	NTSC	N	Y	NTSC	-	O	O	O	O
	ATSC	N	N						
44	-	-	-	-	-23.0	O	A	O	A
45	NTSC	Y	Y	NTSC	-20.7	O	O	O	O
46	ATSC	Y	Y	ATSC	-78.7	O	O	O	O
47	ATSC	N	N	NTSC	< -69.2	O	O	O	O
48	ATSC	Y	N	ATSC	-93.1	O	O	O	O
49	NTSC	N	N	NTSC	-66.1	O	O	O	O
50	NTSC	Y	Y	NTSC	-69.1	O	O	O	O
51	ATSC	Y	N	NTSC	< -71.0	O	O	O	O

5.3. Test Site #3 – Private Residence in Ellicott City, MD

This site is a two-story single-family home on a hilltop in the suburbs of Baltimore, MD. The channel occupancy measurements were performed in the driveway. The prototype trials were performed in the first floor living area (location 1) and a second floor bedroom (location 2). Because of the hilltop location, excellent television reception was observed, including some stations for which the site was beyond their service contour.

Table 5-14. Licensed TV Station Assignments within 100-km radius of Test Site 3

CH	Station Call Sign	Station Location	Signal Type	Separation Distance (km)	Within Service Contour? (Y/N)	Viewable? (Y/N)
21	-	-	-	-	N	N
22	WMPT	Annapolis, MD	NTSC	28.85	Y	Y
23	WDDN-LP	Washington, DC	NTSC	36.03	N	N
24	WUTB	Baltimore, MD	NTSC	5.73	Y	Y
25	WZDC-LP	Washington, DC	NTSC	42.46	N	N
26	WETA	Washington, DC	NTSC	41.75	Y	Y
27	WETA	Washington, DC	ATSC	41.46	Y	Y
28	WFPT	Frederick, MD	ATSC	46.69	Y	Y
29	WMPB	Baltimore, MD	ATSC	23.39	Y	Y
30	WGCB	Red Lion, PA	ATSC	75.96	N	Y
	WNVT	Goldvein, VA	ATSC	88.89	N	N
31	WRZB-LP	Annapolis, MD	NTSC	33.57	N	Y
32	WHUT	Washington, DC	NTSC	41.80	Y	Y
33	WHUT	Washington, DC	ATSC	41.46	Y	Y
34	WUSA	Washington, DC	ATSC	41.46	Y	Y
35	WDCA	Washington, DC	ATSC	41.15	Y	Y
36	WTTG	Washington, DC	ATSC	41.15	Y	Y
38	WJZ	Baltimore, MD	ATSC	15.11	Y	Y
39	WJLA	Washington, DC	ATSC	41.46	Y	Y
40	WNUV	Baltimore, MD	ATSC	15.28	Y	Y
41	WUTB	Baltimore, MD	ATSC	5.73	Y	Y
42	WMPT	Annapolis, MD	ATSC	28.85	Y	Y
43	WPMT	York, PA	NTSC	89.10	N	Y
	WPXW	Manassas, VA	ATSC	69.39	N	Y
44	-	-	-	-	N	N
45	WBFF	Baltimore, MD	NTSC	15.28	Y	Y
46	WBFF	Baltimore, MD	ATSC	15.28	Y	Y
47	WPMT	York, PA	ATSC	89.10	N	Y
48	WRC	Washington, DC	ATSC	42.44	Y	Y
49	WGCB	Red Lion, PA	NTSC	75.96	N	Y
50	WDCW	Washington, DC	NTSC	37.59	Y	Y
51	WDCW	Washington, DC	ATSC	37.59	Y	Y

Table 5-15. Adaptrum WSD Prototype Results at Test Site 3

CH	Signal Type (Database)	Within Service Contour? (Y/N)	(dBm) Viewable? (Y/N)	Signal Type (Measured)	Measured Power Level	Occupied (O) & Available (A) Channels Reported by WSD	
						L1/S1	L1/S2 [1]
21	-	N	N	NTSC	-95.7	O	A
22	NTSC	Y	Y	NTSC	-29.8	O	O
23	NTSC	N	N	NTSC	-72.0	O	O
24	NTSC	Y	Y	NTSC	-39.2	O	O
25	NTSC	N	N	NTSC	-72.8	O	O
26	NTSC	Y	Y	NTSC	-37.2	O	O
27	ATSC	Y	Y	ATSC	-50.9	O	-
28	ATSC	Y	Y	ATSC	-50.9	O	-
29	ATSC	Y	Y	ATSC	-64.4	O	-
30	ATSC	N	Y	ATSC	-68.3	O	-
	ATSC	N	N				
31	NTSC	N	Y	NTSC	-90.6	O	A
32	NTSC	Y	Y	NTSC	-40.1	O	O
33	ATSC	Y	Y	ATSC	-51.7	O	-
34	ATSC	Y	Y	ATSC	-43.0	O	-
35	ATSC	Y	Y	ATSC	-41.8	A	-
36	ATSC	Y	Y	ATSC	-38.6	O	-
38	ATSC	Y	Y	ATSC	-27.4	O	-
39	ATSC	Y	Y	ATSC	-44.2	O	-
40	ATSC	Y	Y	ATSC	-30.9	O	-
41	ATSC	Y	Y	ATSC	-42.7	O	-
42	ATSC	Y	Y	ATSC	-51.1	O	-
43	NTSC	N	Y	NTSC	-66.0	A	A
	ATSC	N	Y				
44	-	N	N	-	-	A	-
45	NTSC	Y	Y	NTSC	-28.8	O	O
46	ATSC	Y	Y	ATSC	-30.0	O	-
47	ATSC	N	Y	NTSC	-82.0	O	O
48	ATSC	Y	Y	ATSC	-43.8	O	-
49	NTSC	N	Y	NTSC	-77.8	O	O
50	NTSC	Y	Y	NTSC	-40.5	O	O
51	ATSC	Y	Y	ATSC	-49.7	O	-

[1] Second scan for NTSC signals only.

Table 5-16. I2R WSD Prototype Results at Test Site 3

CH	Signal Type (Database)	Within Service Contour? (Y/N)	Viewable? (Y/N)	Signal Type (Measured)	Measured Power Level(dBm)	Occupied (O) & Available (A) Channels Reported by WSD			
						L1/S1	L1/S2	L2/S1	L2/S2
21	-	N	N	NTSC	-95.7	A	O	O	O
22	NTSC	Y	Y	NTSC	-29.8	A	O	A	O
23	NTSC	N	N	NTSC	-72.0	A	A	A	O
24	NTSC	Y	Y	NTSC	-39.2	O	O	O	O
25	NTSC	N	N	NTSC	-72.8	O	O	O	A
26	NTSC	Y	Y	NTSC	-37.2	O	O	O	O
27	ATSC	Y	Y	ATSC	-50.9	O	O	O	O
28	ATSC	Y	Y	ATSC	-50.9	O	O	O	O
29	ATSC	Y	Y	ATSC	-64.4	O	O	O	O
30	ATSC	N	Y	ATSC	-68.3	O	O	O	O
	ATSC	N	N						
31	NTSC	N	Y	NTSC	-90.6	O	O	A	A
32	NTSC	Y	Y	NTSC	-40.1	O	A	O	O
33	ATSC	Y	Y	ATSC	-51.7	O	O	O	O
34	ATSC	Y	Y	ATSC	-43.0	O	O	O	O
35	ATSC	Y	Y	ATSC	-41.8	O	O	O	O
36	ATSC	Y	Y	ATSC	-38.6	O	O	O	O
38	ATSC	Y	Y	ATSC	-27.4	O	O	O	O
39	ATSC	Y	Y	ATSC	-44.2	O	O	O	O
40	ATSC	Y	Y	ATSC	-30.9	O	O	O	O
41	ATSC	Y	Y	ATSC	-42.7	O	O	O	O
42	ATSC	Y	Y	ATSC	-51.1	O	O	O	O
43	NTSC	N	Y	NTSC	-66.0	O	O	O	O
	ATSC	N	Y						
44	-	N	N	-	-	A	A	A	A
45	NTSC	Y	Y	NTSC	-28.8	A	A	O	A
46	ATSC	Y	Y	ATSC	-30.0	O	O	O	O
47	ATSC	N	Y	NTSC	-82.0	A	A	O	O
48	ATSC	Y	Y	ATSC	-43.8	O	O	O	O
49	NTSC	N	Y	NTSC	-77.8	A	A	A	A
50	NTSC	Y	Y	NTSC	-40.5	O	A	O	O
51	ATSC	Y	Y	ATSC	-49.7	O	O	O	O

Table 5-17. Motorola WSD Prototype (Test Mode) Results at Test Site 3

CH	Signal Type (Database)	Within Service Contour? (Y/N)	Viewable? (Y/N)	Signal Type (Measured)	Measured Power Level (dBm)	Occupied (O) & Available (A) Channels Reported by WSD			
						L1/S1	L1/S2	L2/S1	L2/S2
21	-	N	N	NTSC	-95.7	-	-	-	-
22	NTSC	Y	Y	NTSC	-29.8	-	-	-	-
23	NTSC	N	N	NTSC	-72.0	-	-	-	-
24	NTSC	Y	Y	NTSC	-39.2	-	-	-	-
25	NTSC	N	N	NTSC	-72.8	-	-	-	-
26	NTSC	Y	Y	NTSC	-37.2	-	-	-	-
27	ATSC	Y	Y	ATSC	-50.9	O	O	O	O
28	ATSC	Y	Y	ATSC	-50.9	O	O	O	O
29	ATSC	Y	Y	ATSC	-64.4	O	O	O	O
30	ATSC	N	Y	ATSC	-68.3	O	O	O	O
	ATSC	N	N						
31	NTSC	N	Y	NTSC	-90.6	-	-	-	-
32	NTSC	Y	Y	NTSC	-40.1	-	-	-	-
33	ATSC	Y	Y	ATSC	-51.7	A	A	A	A
34	ATSC	Y	Y	ATSC	-43.0	O	O	O	O
35	ATSC	Y	Y	ATSC	-41.8	O	O	O	O
36	ATSC	Y	Y	ATSC	-38.6	O	O	O	O
38	ATSC	Y	Y	ATSC	-27.4	O	O	O	O
39	ATSC	Y	Y	ATSC	-44.2	O	O	O	O
40	ATSC	Y	Y	ATSC	-30.9	O	O	O	O
41	ATSC	Y	Y	ATSC	-42.7	O	O	O	O
42	ATSC	Y	Y	ATSC	-51.1	O	O	O	O
43	NTSC	N	Y	NTSC	-66.0	O	O	O	O
	ATSC	N	Y						
44	-	N	N	-	-	-	-	O	-
45	NTSC	Y	Y	NTSC	-28.8	-	-	-	-
46	ATSC	Y	Y	ATSC	-30.0	O	O	O	O
47	ATSC	N	Y	NTSC	-82.0	-	-	-	-
48	ATSC	Y	Y	ATSC	-43.8	O	O	O	O
49	NTSC	N	Y	NTSC	-77.8	-	-	-	-
50	NTSC	Y	Y	NTSC	-40.5	O	-	-	-
51	ATSC	Y	Y	ATSC	-49.7	O	O	O	O

NOTE: In the test mode this device detects only ATSC signals (*i.e.*, no NTSC detection capability).

Table 5-18. Motorola WSD Prototype (Normal Mode) Results at Test Site 3

CH	Signal Type (Database)	Within Service Contour? (Y/N)	Viewable? (Y/N)	Signal Type (Measured)	Measured Power Level (dBm)	Occupied (O) & Available (A) Channels Reported by WSD	
						L1/S1	L2/S1
21	-	N	N	NTSC	-95.7	A	A
22	NTSC	Y	Y	NTSC	-29.8	O	O
23	NTSC	N	N	NTSC	-72.0	A	A
24	NTSC	Y	Y	NTSC	-39.2	O	O
25	NTSC	N	N	NTSC	-72.8	A	A
26	NTSC	Y	Y	NTSC	-37.2	O	O
27	ATSC	Y	Y	ATSC	-50.9	O	O
28	ATSC	Y	Y	ATSC	-50.9	O	O
29	ATSC	Y	Y	ATSC	-64.4	O	O
30	ATSC	N	Y	ATSC	-68.3	O	O
	ATSC	N	N				
31	NTSC	N	Y	NTSC	-90.6	A	A
32	NTSC	Y	Y	NTSC	-40.1	O	O
33	ATSC	Y	Y	ATSC	-51.7	O	O
34	ATSC	Y	Y	ATSC	-43.0	O	O
35	ATSC	Y	Y	ATSC	-41.8	O	O
36	ATSC	Y	Y	ATSC	-38.6	O	O
38	ATSC	Y	Y	ATSC	-27.4	O	O
39	ATSC	Y	Y	ATSC	-44.2	O	O
40	ATSC	Y	Y	ATSC	-30.9	O	O
41	ATSC	Y	Y	ATSC	-42.7	O	O
42	ATSC	Y	Y	ATSC	-51.1	O	O
43	NTSC	N	Y	NTSC	-66.0	O	O
	ATSC	N	Y				
44	-	N	N	-	-	A	A
45	NTSC	Y	Y	NTSC	-28.8	O	O
46	ATSC	Y	Y	ATSC	-30.0	O	O
47	ATSC	N	Y	NTSC	-82.0	O	O
48	ATSC	Y	Y	ATSC	-43.8	O	O
49	NTSC	N	Y	NTSC	-77.8	A	A
50	NTSC	Y	Y	NTSC	-40.5	O	O
51	ATSC	Y	Y	ATSC	-49.7	O	O

Table 5-19. Philips WSD Prototype Results at Test Site 3

CH	Signal Type (Database)	Within Service Contour? (Y/N)	Viewable? (Y/N)	Signal Type (Measured)	Measured Power Level(dBm)	Occupied (O) & Available (A) Channels Reported by WSD			
						L1/S1	L1/S2	L2/S1	L2/S2
21	-	N	N	NTSC	-95.7	O	O	O	O
22	NTSC	Y	Y	NTSC	-29.8	O	O	O	O
23	NTSC	N	N	NTSC	-72.0	O	O	O	O
24	NTSC	Y	Y	NTSC	-39.2	O	O	O	O
25	NTSC	N	N	NTSC	-72.8	O	O	O	O
26	NTSC	Y	Y	NTSC	-37.2	O	O	O	O
27	ATSC	Y	Y	ATSC	-50.9	O	O	O	O
28	ATSC	Y	Y	ATSC	-50.9	O	O	O	O
29	ATSC	Y	Y	ATSC	-64.4	O	O	O	O
30	ATSC	N	Y	ATSC	-68.3	O	O	O	O
	ATSC	N	N						
31	NTSC	N	Y	NTSC	-90.6	O	O	O	O
32	NTSC	Y	Y	NTSC	-40.1	O	O	O	O
33	ATSC	Y	Y	ATSC	-51.7	O	O	O	O
34	ATSC	Y	Y	ATSC	-43.0	O	O	O	O
35	ATSC	Y	Y	ATSC	-41.8	O	O	O	O
36	ATSC	Y	Y	ATSC	-38.6	O	O	O	O
38	ATSC	Y	Y	ATSC	-27.4	O	O	O	O
39	ATSC	Y	Y	ATSC	-44.2	O	O	O	O
40	ATSC	Y	Y	ATSC	-30.9	O	O	O	O
41	ATSC	Y	Y	ATSC	-42.7	O	O	O	O
42	ATSC	Y	Y	ATSC	-51.1	O	O	O	O
43	NTSC	N	Y	NTSC	-66.0	O	O	O	O
	ATSC	N	Y						
44	-	N	N	-	-	O	O	O	O
45	NTSC	Y	Y	NTSC	-28.8	O	O	O	O
46	ATSC	Y	Y	ATSC	-30.0	O	O	O	O
47	ATSC	N	Y	NTSC	-82.0	O	O	O	O
48	ATSC	Y	Y	ATSC	-43.8	O	O	O	O
49	NTSC	N	Y	NTSC	-77.8	O	O	O	O
50	NTSC	Y	Y	NTSC	-40.5	O	O	O	O
51	ATSC	Y	Y	ATSC	-49.7	O	O	O	O

NOTE: An additional scan was run from location 1 with a 6-dB attenuator inserted between the receive antenna and the WSD but no change in the results was observed (*i.e.,* all channels still reported as occupied).

5.4. Test Site #4 – Private Residence in College Park, MD

This site is a single-family one-story home located in the suburbs of Washington, DC. The channel occupancy measurements were performed in the driveway. The prototype trials were performed from the first floor living room (location 1) and from an entertainment room in the finished basement (location 2).

Table 5-20. Licensed TV Station Assignments within 100-km radius of Test Site 4

CH	Station Call Sign	Station Location	Signal Type	Separation Distance (km)	Within Service Contour? /N)	Viewable? (Y/N)
21	-	-	-	-	N	N
22	WMPT	Annapolis, MD	NTSC	26.72	Y	Y
23	WDDN-LP	Washington, DC	NTSC	12.43	Y	Y
24	WUTB	Baltimore, MD	NTSC	36.42	Y	Y
25	WZDC-LP	Washington, DC	NTSC	15.16	Y	Y
26	WETA	Washington, DC	NTSC	16.50	Y	Y
27	WETA	Washington, DC	ATSC	14.67	Y	Y
28	WFPT	Frederick, MD	ATSC	46.13	Y	N
29	WMPB	Baltimore, MD	ATSC	52.90	Y	Y
30	WNVT	Goldvein, VA	ATSC	60.10	Y	Y1
31	WRZB-LP	Annapolis, MD	NTSC	31.39	N	N
32	WHUT	Washington, DC	NTSC	16.55	Y	Y
33	WHUT	Washington, DC	ATSC	14.67	Y	Y
34	WUSA	Washington, DC	ATSC	14.67	Y	Y
35	WDCA	Washington, DC	ATSC	14.81	Y	Y
36	WTTG	Washington, DC	ATSC	14.81	Y	Y
38	WJZ	Baltimore, MD	ATSC	45.31	Y	Y
39	WJLA	Washington, DC	ATSC	14.67	Y	Y
40	WNUV	Baltimore, MD	ATSC	45.49	Y	Y
41	WUTB	Baltimore, MD	ATSC	36.42	Y	N
42	WMPT	Annapolis, MD	ATSC	26.72	Y	Y
43	WPXW	Manassas, VA	ATSC	41.94	Y	Y
44	-	-	-	-	N	N
45	WBFF	Baltimore, MD	NTSC	45.49	Y	Y
46	WBFF	Baltimore, MD	ATSC	45.49	Y	Y
47	WMDO-CA	Washington, DC	NTSC	15.15	Y	Y
48	WRC	Washington, DC	ATSC	15.15	Y	Y
49	WWTD-LP	Washington, DC	NTSC	15.15	Y	Y
50	WDCW	Washington, DC	NTSC	9.89	Y	Y
51	WDCW	Washington, DC	ATSC	9.89	Y	Y

Table 5-21. Adaptrum WSD Prototype Results at Test Site 4

CH	Signal Type (Database)	Within Service Contour? (Y/N)	Viewable? (Y/N)	Signal Type (Measured)	Measured Power Level (dBm)	Occupied (O) & Available (A) Channels Reported by WSD	
						L1/S1	L1/S2[1]
21	-	N	N	-	-	A	-
22	NTSC	Y	Y	NTSC	-63.1	O	O
23	NTSC	Y	Y	NTSC	-48.2	O	O
24	NTSC	Y	Y	NTSC	-67.5	O	O
25	NTSC	Y	Y	NTSC	-66.6	O	O
26	NTSC	Y	Y	NTSC	-47.5	O	O
27	ATSC	Y	Y	ATSC	-47.3	O	-
28	ATSC	Y	N	-	-	A	-
29	ATSC	Y	Y	ATSC	-77.0	O	O
30	ATSC	Y	Y	ATSC	-78.1	A	-
31	NTSC	N	N	-	-	A	-
32	NTSC	Y	Y	NTSC	-42.2	O	O
33	ATSC	Y	Y	ATSC	-46.7	O	-
34	ATSC	Y	Y	ATSC	-37.2	O	-
35	ATSC	Y	Y	ATSC	-49.7	O	-
36	ATSC	Y	Y	ATSC	-45.5	O	-
38	ATSC	Y	Y	ATSC	-70.8	O	-
39	ATSC	Y	Y	ATSC	-45.7	O	-
40	ATSC	Y	Y	ATSC	-68.7	O	-
41	ATSC	Y	N	-	-	O	-
42	ATSC	Y	Y	ATSC	-74.0	O	-
43	ATSC	Y	Y	ATSC	-75.6	A	-
44	-	N	N	ATSC	-	A	-
45	NTSC	Y	Y	NTSC	-69.4	O	O
46	ATSC	Y	Y	ATSC	-68.9	O	-
47	NTSC	Y	Y	NTSC	-84.3	O	O
48	ATSC	Y	Y	ATSC	-52.1	O	-
49	NTSC	Y	Y	NTSC	-66.7	O	O
50	NTSC	Y	Y	NTSC	-40.0	O	O
51	ATSC	Y	Y	ATSC	-53.1	O	-

[1] Second scan for NTSC signals only.

Table 5-22. I2R WSD Prototype Results at Test Site 4

CH	Signal Type (Database)	Within Service Contour? (Y/N)	Viewable? (Y/N)	Signal Type (Measured)	Measured Power Level (dBm)	Occupied (O) & Available (A) Channels Reported by WSD			
						L1/S1	L1/S2	L2/S1	L2/S2
21	-	N	N	-	-	A	A	A	A
22	NTSC	Y	Y	NTSC	-63.1	A	O	A	O
23	NTSC	Y	Y	NTSC	-48.2	A	A	A	A
24	NTSC	Y	Y	NTSC	-67.5	A	O	A	A
25	NTSC	Y	Y	NTSC	-66.6	A	A	A	A
26	NTSC	Y	Y	NTSC	-47.5	O	O	A	O
27	ATSC	Y	Y	ATSC	-47.3	O	O	O	O
28	ATSC	Y	N	-	-	A	A	A	A
29	ATSC	Y	Y	ATSC	-77.0	O	O	A	A
30	ATSC	Y	Y	ATSC	-78.1	O	O	O	A
31	NTSC	N	N	-	-	A	A	A	A
32	NTSC	Y	Y	NTSC	-42.2	O	O	A	O
33	ATSC	Y	Y	ATSC	-46.7	O	O	O	O
34	ATSC	Y	Y	ATSC	-37.2	O	O	O	O
35	ATSC	Y	Y	ATSC	-49.7	O	O	O	O
36	ATSC	Y	Y	ATSC	-45.5	O	O	O	O
38	ATSC	Y	Y	ATSC	-70.8	O	O	O	O
39	ATSC	Y	Y	ATSC	-45.7	O	O	O	O
40	ATSC	Y	Y	ATSC	-68.7	O	O	O	O
41	ATSC	Y	N	-	-	A	O	A	A
42	ATSC	Y	Y	ATSC	-74.0	O	O	O	O
43	ATSC	Y	Y	ATSC	-75.6	O	O	A	A
44	-	N	N	ATSC	-	A	A	A	A
45	NTSC	Y	Y	NTSC	-69.4	A	A	A	A
46	ATSC	Y	Y	ATSC	-68.9	O	O	O	O
47	NTSC	Y	Y	NTSC	-84.3	A	A	A	A
48	ATSC	Y	Y	ATSC	-52.1	O	O	O	O
49	NTSC	Y	Y	NTSC	-66.7	A	A	A	A
50	NTSC	Y	Y	NTSC	-40.0	O	O	A	A
51	ATSC	Y	Y	ATSC	-53.1	O	O	O	O

Table 5-23. Motorola WSD Prototype (Test Mode) Results at Test Site 4

CH	Signal Type (Database)	Within Service Contour? (Y/N)	Viewable? (Y/N)	Signal Type (Measured)	Measured Power Level (dBm)	Occupied (O) & Available (A) Channels Reported by WSD			
						L1/S1	L1/S2	L2/S1	L2/S2
21	-	N	N	-	-	-	-	-	-
22	NTSC	Y	Y	NTSC	-63.1	-	-	-	-
23	NTSC	Y	Y	NTSC	-48.2	-	-	-	-
24	NTSC	Y	Y	NTSC	-67.5	-	-	-	-
25	NTSC	Y	Y	NTSC	-66.6	-	-	-	-
26	NTSC	Y	Y	NTSC	-47.5	-	-	-	-
27	ATSC	Y	Y	ATSC	-47.3	O	O	O	O
28	ATSC	Y	N	-	-	-	-	-	-
29	ATSC	Y	Y	ATSC	-77.0	O	O	O	O
30	ATSC	Y	Y	ATSC	-78.1	O	O	A	O
31	NTSC	N	N	-	-	-	-	-	-
32	NTSC	Y	Y	NTSC	-42.2	-	-	-	-
33	ATSC	Y	Y	ATSC	-46.7	A	A	A	A
34	ATSC	Y	Y	ATSC	-37.2	O	O	O	O
35	ATSC	Y	Y	ATSC	-49.7	O	O	O	O
36	ATSC	Y	Y	ATSC	-45.5	O	O	O	O
38	ATSC	Y	Y	ATSC	-70.8	O	O	O	O
39	ATSC	Y	Y	ATSC	-45.7	O	O	O	O
40	ATSC	Y	Y	ATSC	-68.7	O	O	O	O
41	ATSC	Y	N	-	-	O	O	-	O
42	ATSC	Y	Y	ATSC	-74.0	O	O	O	O
43	ATSC	Y	Y	ATSC	-75.6	O	O	O	O
44	-	N	N	ATSC	-	A	A	A	A
45	NTSC	Y	Y	NTSC	-69.4	-	-	-	-
46	ATSC	Y	Y	ATSC	-68.9	O	O	O	O
47	NTSC	Y	Y	NTSC	-84.3	-	-	-	-
48	ATSC	Y	Y	ATSC	-52.1	O	O	O	O
49	NTSC	Y	Y	NTSC	-66.7	-	-	-	-
50	NTSC	Y	Y	NTSC	-40.0	-	-	-	-
51	ATSC	Y	Y	ATSC	-53.1	O	O	O	O

NOTE: In the test mode this device detects only ATSC signals (*i.e.*, no NTSC detection capability).

Table 5-24. Motorola WSD Prototype (Normal Mode) Results at Test Site 4

CH	Signal Type (Database)	Within Service Contour? (Y/N)	Viewable? (Y/N)	Signal Type (Measured)	Measured Power Level (dBm)	Occupied (O) & Available (A) Channels Reported by WSD	
						L1/S1	L2/S1
21	-	N	N	-	-	A	A
22	NTSC	Y	Y	NTSC	-63.1	O	O
23	NTSC	Y	Y	NTSC	-48.2	O	O
24	NTSC	Y	Y	NTSC	-67.5	O	O
25	NTSC	Y	Y	NTSC	-66.6	O	O
26	NTSC	Y	Y	NTSC	-47.5	O	O
27	ATSC	Y	Y	ATSC	-47.3	O	O
28	ATSC	Y	N	-	-	O	O
29	ATSC	Y	Y	ATSC	-77.0	O	O
30	ATSC	Y	Y	ATSC	-78.1	O	O
31	NTSC	N	N	-	-	A	A
32	NTSC	Y	Y	NTSC	-42.2	O	O
33	ATSC	Y	Y	ATSC	-46.7	O	O
34	ATSC	Y	Y	ATSC	-37.2	O	O
35	ATSC	Y	Y	ATSC	-49.7	O	O
36	ATSC	Y	Y	ATSC	-45.5	O	O
38	ATSC	Y	Y	ATSC	-70.8	O	O
39	ATSC	Y	Y	ATSC	-45.7	O	O
40	ATSC	Y	Y	ATSC	-68.7	O	O
41	ATSC	Y	N	-	-	O	O
42	ATSC	Y	Y	ATSC	-74.0	O	O
43	ATSC	Y	Y	ATSC	-75.6	O	O
44	-	N	N	ATSC	-	A	A
45	NTSC	Y	Y	NTSC	-69.4	O	O
46	ATSC	Y	Y	ATSC	-68.9	O	O
47	NTSC	Y	Y	NTSC	-84.3	O	O
48	ATSC	Y	Y	ATSC	-52.1	O	O
49	NTSC	Y	Y	NTSC	-66.7	O	O
50	NTSC	Y	Y	NTSC	-40.0	O	O
51	ATSC	Y	Y	ATSC	-53.1	O	O

Table 5-25. Philips WSD Prototype Results at Test Site 4

CH	Signal Type (Database)	Within Service Contour? (Y/N)	Viewable? (Y/N)	Signal Type (Measured)	Measured Power Level (dBm)	Occupied (O) & Available (A) Channels Reported by WSD			
						L1/S1	L1/S2	L2/S1	L2/S2
21	-	N	N	-	-	O	O	O	O
22	NTSC	Y	Y	NTSC	-63.1	O	O	O	O
23	NTSC	Y	Y	NTSC	-48.2	O	O	O	O
24	NTSC	Y	Y	NTSC	-67.5	O	O	O	O
25	NTSC	Y	Y	NTSC	-66.6	O	O	O	O
26	NTSC	Y	Y	NTSC	-47.5	O	O	O	O
27	ATSC	Y	Y	ATSC	-47.3	O	O	O	O
28	ATSC	Y	N	-	-	O	O	O	O
29	ATSC	Y	Y	ATSC	-77.0	O	O	O	O
30	ATSC	Y	Y	ATSC	-78.1	O	O	O	O
31	NTSC	N	N	-	-	O	O	O	A
32	NTSC	Y	Y	NTSC	-42.2	O	O	O	O
33	ATSC	Y	Y	ATSC	-46.7	O	O	O	O
34	ATSC	Y	Y	ATSC	-37.2	O	O	O	O
35	ATSC	Y	Y	ATSC	-49.7	O	O	O	O
36	ATSC	Y	Y	ATSC	-45.5	O	O	O	O
38	ATSC	Y	Y	ATSC	-70.8	O	O	O	O
39	ATSC	Y	Y	ATSC	-45.7	O	O	O	O
40	ATSC	Y	Y	ATSC	-68.7	O	O	O	O
41	ATSC	Y	N	-	-	O	O	O	O
42	ATSC	Y	Y	ATSC	-74.0	O	O	O	O
43	ATSC	Y	Y	ATSC	-75.6	O	O	O	O
44	-	N	N	ATSC	-	A	A	A	A
45	NTSC	Y	Y	NTSC	-69.4	O	O	O	O
46	ATSC	Y	Y	ATSC	-68.9	O	O	O	O
47	NTSC	Y	Y	NTSC	-84.3	O	O	O	O
48	ATSC	Y	Y	ATSC	-52.1	O	O	O	O
49	NTSC	Y	Y	NTSC	-66.7	O	O	O	O
50	NTSC	Y	Y	NTSC	-40.0	O	O	O	O
51	ATSC	Y	Y	ATSC	-53.1	O	O	O	O

Table 5-26. Philips WSD Prototy Results at Test Site 4 (w/ 10-dB attenuator)

CH	Signal Type (Database)	Within Service Contour? (Y/N)	Viewable? (Y/N)	Signal Type (Measured)	Measured Power Level (dBm)	Occupied (O) & Available (A) Channels Reported by WSD	
						L1/S1	L2/S1
21	-	N	N	-	-	O	A
22	NTSC	Y	Y	NTSC	-63.1	O	O
23	NTSC	Y	Y	NTSC	-48.2	O	O
24	NTSC	Y	Y	NTSC	-67.5	O	O
25	NTSC	Y	Y	NTSC	-66.6	O	O
26	NTSC	Y	Y	NTSC	-47.5	O	O
27	ATSC	Y	Y	ATSC	-47.3	O	O
28	ATSC	Y	N	-	-	O	A
29	ATSC	Y	Y	ATSC	-77.0	O	O
30	ATSC	Y	Y	ATSC	-78.1	O	O
31	NTSC	N	N	-	-	A	A
32	NTSC	Y	Y	NTSC	-42.2	O	O
33	ATSC	Y	Y	ATSC	-46.7	O	O
34	ATSC	Y	Y	ATSC	-37.2	O	O
35	ATSC	Y	Y	ATSC	-49.7	O	O
36	ATSC	Y	Y	ATSC	-45.5	O	O
38	ATSC	Y	Y	ATSC	-70.8	O	O
39	ATSC	Y	Y	ATSC	-45.7	O	O
40	ATSC	Y	Y	ATSC	-68.7	O	O
41	ATSC	Y	N	-	-	O	O
42	ATSC	Y	Y	ATSC	-74.0	O	O
43	ATSC	Y	Y	ATSC	-75.6	O	O
44	-	N	N	ATSC	-	A	A
45	NTSC	Y	Y	NTSC	-69.4	O	O
46	ATSC	Y	Y	ATSC	-68.9	O	O
47	NTSC	Y	Y	NTSC	-84.3	O	O
48	ATSC	Y	Y	ATSC	-52.1	O	O
49	NTSC	Y	Y	NTSC	-66.7	O	O
50	NTSC	Y	Y	NTSC	-40.0	O	O
51	ATSC	Y	Y	ATSC	-53.1	O	O

5.5. Test Site #5 – FCC Headquarters in Washington, DC

This site is a high-rise office building in downtown southwest Washington, DC. Field trials of the WSD prototypes were performed in a training room on the 12th street level (ground floor) and in an area on the 7th floor near a window. Channel occupancy measurements were performed at the same locations, using a bi-conical receive antenna mounted on a tripod.

Table 5-27. Licensed TV Station Assignments within 100-km radius of Test Site 5

CH	Station Call Sign	Station Location	Signal Type	Separation Distance (km)	Within Service Contour? (Y/N)	Viewable? (Y/N)	
						L1	L2
21	-	-	-	-	N	N	N
22	WMPT	Annapolis, MD	NTSC	39.00	Y	N	Y
23	WDDN-LP	Washington, DC	NTSC	13.22	Y	N	Y
24	WUTB	Baltimore, MD	NTSC	50.54	Y	N	Y
25	WZDC-LP	Washington, DC	NTSC	7.81	Y	N	Y
26	WETA	Washington, DC	NTSC	11.13	Y	Y	Y
27	WETA	Washington, DC	ATSC	8.68	Y	N	Y
28	WFPT	Frederick, MD	ATSC	48.54	Y	N	N
29	WMPB	Baltimore, MD	ATSC	66.21	Y	N	N
30	WNVT	Goldvein, VA	ATSC	45.51	Y	N	N
31	WRZB-LP	Annapolis, MD	NTSC	42.79	N	N	N
32	WHUT	Washington, DC	NTSC	11.13	Y	Y	Y
33	WHUT	Washington, DC	ATSC	8.68	Y	N	Y
34	WUSA	Washington, DC	ATSC	8.68	Y	N	Y
35	WDCA	Washington, DC	ATSC	9.38	Y	N	Y
36	WTTG	Washington, DC	ATSC	9.38	Y	Y	Y
38	WJZ	Baltimore, MD	ATSC	59.85	Y	N	Y
39	WJLA	Washington, DC	ATSC	8.68	Y	N	Y
40	WNUV	Baltimore, MD	ATSC	60.03	Y	N	Y
41	WUTB	Baltimore, MD	ATSC	50.54	Y	N	N
42	WMPT	Annapolis, MD	ATSC	39.00	Y	N	N
43	WPXW	Manassas, VA	ATSC	28.2	Y	N	N
44	-	-	-	-	N	N	N
45	WBFF	Baltimore, MD	NTSC	60.03	Y	Y	Y
46	WBFF	Baltimore, MD	ATSC	60.03	Y	N	Y
47	WMDO-CA	Washington, DC	NTSC	7.82	Y	N	N
48	WRC	Washington, DC	ATSC	7.82	Y	Y	Y
49	WWTD-LP	Washington, DC	NTSC	7.82	Y	N	Y
50	WDCW	Washington, DC	NTSC	8.79	Y	Y	Y
51	WDCW	Washington, DC	ATSC	8.79	Y	N	Y

Table 5-28. Adaptrum WSD Prototype Results at Test Site 5

CH	Signal Type (Database)	Within Service Contour? (Y/N)	Viewable? (Y/N)		Measured Signal Type[1]		Measured Power Level (dBm)		Occupied (O) & Available (A) Channels Reported by WSD			
			L1	L2	L1	L2	L1	L2	L1/S1	L1/S2 [2]	L2/S1	L2/S2[2]
21	-	N	N	N	-	NT	-	-	A	-	A	A
22	NTSC	Y	N	Y	NT	NT	-96.5	-91.7	A	A	O	O
23	NTSC	Y	N	Y	-	NT	-	-74.8	A	-	O	O
24	NTSC	Y	N	Y	-	NT	-	-87.5	A	-	O	O
25	NTSC	Y	N	Y	NT	NT	-92.4	-74.9	O	A	O	O
26	NTSC	Y	Y	Y	NT	NT	-77.0	-61.4	O	O	O	O
27	ATSC	Y	N	Y	AT	AT	-92.3	-72.2	O	-	O	-
28	ATSC	Y	N	N	-	-	-	-	A	-	A	-
29	ATSC	Y	N	N	-	-	-	-	O	-	O	O
30	ATSC	Y	N	N	-	AT	-	-92.1	A	-	O	-
31	NTSC	N	N	N	-	-	-	-	A	-	O	-
32	NTSC	Y	Y	Y	NT	NT	-80.9	-56.9	O	O	O	O
33	ATSC	Y	N	Y	AT	AT	-87.1	-75.3	O	-	O	-
34	ATSC	Y	N	Y	AT	AT	-84.1	-63.9	O	-	O	-
35	ATSC	Y	N	Y	AT	AT	-83.3	-62.6	O	-	O	-
36	ATSC	Y	Y	Y	AT	AT	-86.2	-62.4	O	-	O	-
38	ATSC	Y	N	Y	AT	AT	-93.5	-72.4	O	-	O	-
39	ATSC	Y	N	Y	AT	AT	-87.1	-70.8	O	-	O	-
40	ATSC	Y	N	Y	AT	AT	-92.7	-72.4	A	-	O	-
41	ATSC	Y	N	N	-	-	-	-	A	-	O	-
42	ATSC	Y	N	N	-	AT	-	-88.4	A	-	A	-
43	ATSC	Y	N	N	-	AT	-	-92.9	A	-	A	-
44	-	N	N	N	-	-	-	-	A	-	A	-
45	NTSC	Y	Y	Y	NT	NT	-105	-75.4	A	A	A	A
46	ATSC	Y	N	Y	AT	AT	-88.5	-73.3	O	-	O	-
47	NTSC	Y	N	N	-	-	-	-	A	-	A	-
48	ATSC	Y	Y	Y	AT	AT	-79.2	-52.2	O	-	O	-
49	NTSC	Y	N	Y	NT	NT	-101	-81.7	A	A	O	O
50	NTSC	Y	Y	Y	NT	NT	-80.0	-71.7	O	O	O	O
51	ATSC	Y	N	Y	AT	AT	-86.5	-67.4	O	-	O	-

[1] NT = NTSC; AT = ATSC.

[2] Second scan for NTSC signals only.

Table 5-29. I2R WSD Prototype Results at Test Site 5

CH	Signal Type (Database)	Within Service Contour? (Y/N)	Viewable? (Y/N)		Measured Signal Type[1]		Measured Power Level (dBm)		Occupied (O) & Available (A) Channels Reported by WSD			
			L1	L2	L1	L2	L1	L2	L1/S1	L1/S2	L2/S1	L2/S2
21	-	N	N	N	-	NT	-	-	A	A	A	A
22	NTSC	Y	N	Y	NT	NT	-96.5	-91.7	A	A	A	A
23	NTSC	Y	N	Y	-	NT	-	-74.8	A	A	A	A
24	NTSC	Y	N	Y	-	NT	-	-87.5	A	A	A	A
25	NTSC	Y	N	Y	NT	NT	-92.4	-74.9	A	A	A	A
26	NTSC	Y	Y	Y	NT	NT	-77.0	-61.4	O	A	O	O
27	ATSC	Y	N	Y	AT	AT	-92.3	-72.2	O	O	O	O
28	ATSC	Y	N	N	-	-	-	-	A	A	A	A
29	ATSC	Y	N	N	-	-	-	-	A	A	O	O
30	ATSC	Y	N	N	-	AT	-	-92.1	A	A	O	O
31	NTSC	N	N	N	-	-	-	-	A	A	A	A
32	NTSC	Y	Y	Y	NT	NT	-80.9	-56.9	A	A	A	A
33	ATSC	Y	N	Y	AT	AT	-87.1	-75.3	O	O	O	O
34	ATSC	Y	N	Y	AT	AT	-84.1	-63.9	O	O	O	O
36	ATSC	Y	Y	Y	AT	AT	-86.2	-62.4	O	O	O	O
39	ATSC	Y	N	Y	AT	AT	-87.1	-70.8	O	O	O	O
40	ATSC	Y	N	Y	AT	AT	-92.7	-72.4	A	A	O	O
41	ATSC	Y	N	N	-	-	-	-	A	A	A	A
42	ATSC	Y	N	N	-	AT	-	-88.4	A	A	O	O
43	ATSC	Y	N	N	-	AT	-	-92.9	A	A	O	O
44	-	N	N	N	-	-	-	-	A	A	A	A
45	NTSC	Y	Y	Y	NT	NT	-105	-75.4	A	A	A	A
46	ATSC	Y	N	Y	AT	AT	-88.5	-73.3	A	A	O	O
47	NTSC	Y	N	N	-	-	-	-	A	A	A	A
48	ATSC	Y	Y	Y	AT	AT	-79.2	-52.2	O	O	O	O
49	NTSC	Y	N	Y	NT	NT	-101	-81.7	A	A	O	A
50	NTSC	Y	Y	Y	NT	NT	-80.0	-71.7	A	A	A	A
51	ATSC	Y	N	Y	AT	AT	-86.5	-67.4	O	O	O	O

[1] NT = NTSC; AT = ATSC.

Table 5-30. Motorola WSD Prototype (Test Mode) Results at Test Site 5

CH	Signal Type (Database)	Within Service Contour? (Y/N)	Viewable? (Y/N)		Measured Signal Type[1]		Measured Power Level (dBm)		Occupied (O) & Available (A) Channels Reported by WSD			
			L1	L2	L1	L2	L1	L2	L1/S1	L1/S2	L2/S1	L2/S2
21	-	N	N	N	-	NT	-	-	-	-	-	-
22	NTSC	Y	N	Y	NT	NT	-96.5	-91.7	-	-	-	-
23	NTSC	Y	N	Y	-	NT	-	-74.8	-	-	-	-
24	NTSC	Y	N	Y	-	NT	-	-87.5	-	-	-	-
25	NTSC	Y	N	Y	NT	NT	-92.4	-74.9	-	-	-	-
26	NTSC	Y	Y	Y	NT	NT	-77.0	-61.4	-	-	-	-
27	ATSC	Y	N	Y	AT	AT	-92.3	-72.2	O	O	O	O
28	ATSC	Y	N	N	-	-	-	-	-	-	-	-
29	ATSC	Y	N	N	-	-	-	-	-	-	-	-
30	ATSC	Y	N	N	-	AT	-	-92.1	-	-	A	A
31	NTSC	N	N	N	-	-	-	-	-	-	-	-
32	NTSC	Y	Y	Y	NT	NT	-80.9	-56.9	-	-	-	-
33	ATSC	Y	N	Y	AT	AT	-87.1	-75.3	A	A	A	A
34	ATSC	Y	N	Y	AT	AT	-84.1	-63.9	O	O	O	O
35	ATSC	Y	N	Y	AT	AT	-83.3	-62.6	O	O	O	O
36	ATSC	Y	Y	Y	AT	AT	-86.2	-62.4	O	O	O	O
38	ATSC	Y	N	Y	AT	AT	-93.5	-72.4	O	A	O	O
39	ATSC	Y	N	Y	AT	AT	-87.1	-70.8	O	O	O	A
40	ATSC	Y	N	Y	AT	AT	-92.7	-72.4	O	A	O	O
41	ATSC	Y	N	N	-	-	-	-	-	-	O	O
42	ATSC	Y	N	N	-	AT	-	-88.4	O	O	O	O
43	ATSC	Y	N	N	-	AT	-	-92.9	-	O	O	O
44	-	N	N	N	-	-	-	-	-	-	-	-
45	NTSC	Y	Y	Y	NT	NT	-105	-75.4	-	-	-	-
46	ATSC	Y	N	Y	AT	AT	-88.5	-73.3	A	A	O	O
47	NTSC	Y	N	N	-	-	-	-	-	-	-	-
48	ATSC	Y	Y	Y	AT	AT	-79.2	-52.2	O	O	O	O
49	NTSC	Y	N	Y	NT	NT	-101	-81.7	-	-	-	-
50	NTSC	Y	Y	Y	NT	NT	-80.0	-71.7	-	-	-	-
51	ATSC	Y	N	Y	AT	AT	-86.5	-67.4	O	O	O	O

[1] NT = NTSC; AT = ATSC.

NOTE: In the test mode this device detects only ATSC signals (*i.e.*, no NTSC detection capability).

Table 5-31. Motorola WSD Prototype (Normal Mode) Results at Test Site 5

CH	Signal Type (Database)	Within Service Contour? (Y/N)	Viewable ? (Y/N)		Measured Signal Type1		Measured Power Level (dBm)		Occupied (O) & Available (A) Channels Reported by WSD	
			L1	L2	L1	L2	L1	L2	L1/S1	L2/S1
21	-	N	N	N	-	NT	-	-	A	A
22	NTSC	Y	N	Y	NT	NT	-96.5	-91.7	O	O
23	NTSC	Y	N	Y	-	NT	-	-74.8	O	O
24	NTSC	Y	N	Y	-	NT	-	-87.5	O	O
25	NTSC	Y	N	Y	NT	NT	-92.4	-74.9	O	O
26	NTSC	Y	Y	Y	NT	NT	-77.0	-61.4	O	O
27	ATSC	Y	N	Y	AT	AT	-92.3	-72.2	O	O
28	ATSC	Y	N	N	-	-	-	-	O	O
29	ATSC	Y	N	N	-	-	-	-	O	O
30	ATSC	Y	N	N	-	AT	-	-92.1	O	O
31	NTSC	N	N	N	-	-	-	-	A	A
32	NTSC	Y	Y	Y	NT	NT	-80.9	-56.9	O	O
33	ATSC	Y	N	Y	AT	AT	-87.1	-75.3	O	O
34	ATSC	Y	N	Y	AT	AT	-84.1	-63.9	O	O
35	ATSC	Y	N	Y	AT	AT	-83.3	-62.6	O	O
36	ATSC	Y	Y	Y	AT	AT	-86.2	-62.4	O	O
38	ATSC	Y	N	Y	AT	AT	-93.5	-72.4	O	O
39	ATSC	Y	N	Y	AT	AT	-87.1	-70.8	O	O
40	ATSC	Y	N	Y	AT	AT	-92.7	-72.4	O	O
41	ATSC	Y	N	N	-	-	-	-	O	O
42	ATSC	Y	N	N	-	AT	-	-88.4	O	O
43	ATSC	Y	N	N	-	AT	-	-92.9	O	O
44	-	N	N	N	-	-	-	-	A	A
45	NTSC	Y	Y	Y	NT	NT	-105	-75.4	O	O
46	ATSC	Y	N	Y	AT	AT	-88.5	-73.3	O	O
47	NTSC	Y	N	N	-	-	-	-	O	O
48	ATSC	Y	Y	Y	AT	AT	-79.2	-52.2	O	O
49	NTSC	Y	N	Y	NT	NT	-101	-81.7	O	O
50	NTSC	Y	Y	Y	NT	NT	-80.0	-71.7	O	O
51	ATSC	Y	N	Y	AT	AT	-86.5	-67.4	O	O

[1]NT = NTSC; AT = ATSC.

Table 5-32. Philips WSD Prototype Results at Test Site 5

CH	Signal Type (Database)	Within Service Contour? (Y/N)	Viewable? (Y/N)		Measured Signal Type[1]		Measured Power Level (dBm)		Occupied (O) & Available (A) Channels Reported by WSD			
			L1	L2	L1	L2	L1	L2	L1/S1	L1/S2	L2/S1	L2/S2
21	-	N	N	N	-	NT	-	-	O	O	O	O
22	NTSC	Y	N	Y	NT	NT	-96.5	-91.7	O	O	O	O
23	NTSC	Y	N	Y	-	NT	-	-74.8	O	O	O	O
24	NTSC	Y	N	Y	-	NT	-	-87.5	O	O	O	O
25	NTSC	Y	N	Y	NT	NT	-92.4	-74.9	O	O	O	O
26	NTSC	Y	Y	Y	NT	NT	-77.0	-61.4	O	O	O	O
27	ATSC	Y	N	Y	AT	AT	-92.3	-72.2	O	O	O	O
28	ATSC	Y	N	N	-	-	-	-	A	O	A	A
29	ATSC	Y	N	N	-	-	-	-	O	O	O	O
30	ATSC	Y	N	N	-	AT	-	-92.1	O	O	O	O
31	NTSC	N	N	N	-	-	-	-	O	O	O	O
32	NTSC	Y	Y	Y	NT	NT	-80.9	-56.9	O	O	O	O
33	ATSC	Y	N	Y	AT	AT	-87.1	-75.3	O	O	O	O
34	ATSC	Y	N	Y	AT	AT	-84.1	-63.9	O	O	O	O
35	ATSC	Y	N	Y	AT	AT	-83.3	-62.6	O	O	O	O
36	ATSC	Y	Y	Y	AT	AT	-86.2	-62.4	O	O	O	O
38	ATSC	Y	N	Y	AT	AT	-93.5	-72.4	O	O	O	O
39	ATSC	Y	N	Y	AT	AT	-87.1	-70.8	O	O	O	O
40	ATSC	Y	N	Y	AT	AT	-92.7	-72.4	O	O	O	O
41	ATSC	Y	N	N	-	-	-	-	O	O	O	O
42	ATSC	Y	N	N	-	AT	-	-88.4	O	A	O	O
43	ATSC	Y	N	N	-	AT	-	-92.9	O	O	O	O
44	-	N	N	N	-	-	-	-	A	A	O	A
45	NTSC	Y	Y	Y	NT	NT	-105	-75.4	O	O	O	O
46	ATSC	Y	N	Y	AT	AT	-88.5	-73.3	O	O	O	O
47	NTSC	Y	N	N	-	-	-	-	O	O	O	O
48	ATSC	Y	Y	Y	AT	AT	-79.2	-52.2	O	O	O	O
49	NTSC	Y	N	Y	NT	NT	-101	-81.7	O	O	O	O
50	NTSC	Y	Y	Y	NT	NT	-80.0	-71.7	O	O	O	O
51	ATSC	Y	N	Y	AT	AT	-86.5	-67.4	O	O	O	O

[1] NT = NTSC; AT = ATSC.

Table 5-33. Philips WSD Prototype Results at Test Site 5 (Loc1) w/ 10-dB Attenuator

CH	Signal Type (Database)	Within Service Contour? (Y/N)	Viewable? (Y/N)	Measured Signal Type	Measured Power Level (dBm)	Occupied (O) & Available (A) Channels Reported by WSD
21	-	N	N	-	-	A
22	NTSC	Y	N	NTSC	-96.5	O
23	NTSC	Y	N	-	-	O
24	NTSC	Y	N	-	-	O
25	NTSC	Y	N	NTSC	-92.4	O
26	NTSC	Y	Y	NTSC	-77.0	O
27	ATSC	Y	N	ATSC	-92.3	O
28	ATSC	Y	N	-	-	A
29	ATSC	Y	N	-	-	O
30	ATSC	Y	N	-	-	O
31	NTSC	N	N	-	-	A
32	NTSC	Y	Y	NTSC	-80.9	O
33	ATSC	Y	N	ATSC	-87.1	O
34	ATSC	Y	N	ATSC	-84.1	O
35	ATSC	Y	N	ATSC	-83.3	O
36	ATSC	Y	Y	ATSC	-86.2	O
38	ATSC	Y	N	ATSC	-93.5	O
39	ATSC	Y	N	ATSC	-87.1	O
40	ATSC	Y	N	ATSC	-92.7	O
41	ATSC	Y	N	-	-	O
42	ATSC	Y	N	-	-	A
43	ATSC	Y	N	-	-	A
44	-	N	N	-	-	A
45	NTSC	Y	Y	NTSC	-105	O
46	ATSC	Y	N	ATSC	-88.5	A
47	NTSC	Y	N	-	-	A
48	ATSC	Y	Y	ATSC	-79.2	O
49	NTSC	Y	N	NTSC	-101	O
50	NTSC	Y	Y	NTSC	-80.0	O
51	ATSC	Y	N	ATSC	-86.5	O

5.6. Test Site #6 – Private Residence in Galesville, MD

This site is a two-story single-family waterfront home with a detached garage. The channel occupancy measurements were performed from the driveway. The WSD prototype field trials were performed from the first floor great room (location 1) and from a guest room over the detached garage (location 2).

Table 5-34. Licensed TV Station Assignments within 100-km radius of Test Site 6

CH	Station Call Sign	Station Location	Signal Type	Separation Distance (km)	Within Service Contour? (Y/N)	Viewable? (Y/N)
21	WBOC	Salisbury, MD	ATSC	86.95	N	Y
22	WMPT	Annapolis, MD	NTSC	19.16	Y	Y
23	WDDN-LP	Washington, DC	NTSC	47.76	N	Y
24	WUTB	Baltimore, MD	NTSC	52.57	Y	Y
25	WZDC-LP	Washington, DC	NTSC	47.87	N	Y
26	WETA	Washington, DC	NTSC	50.42	Y	Y
27	WETA	Washington, DC	ATSC	47.97	Y	Y
28	WFPT	Frederick, MD	ATSC	80.95	N	N
	WCPB	Salisbury, MD	NTSC	97.26	N	N
29	WMPB	Baltimore, MD	ATSC	69.89	Y	Y
30	WNVT	Goldvein, VA	ATSC	81.52	N	N
31	WRZB-LP	Annapolis, MD	NTSC	15.75	N	N
32	WHUT	Washington, DC	NTSC	50.46	Y	Y
33	WHUT	Washington, DC	ATSC	47.97	Y	Y
34	WUSA	Washington, DC	ATSC	47.97	Y	Y
35	WDCA	Washington, DC	ATSC	48.41	Y	Y
36	WTTG	Washington, DC	ATSC	48.41	Y	Y
38	WJZ	Baltimore, MD	ATSC	55.13	Y	Y
39	WJLA	Washington, DC	ATSC	47.97	Y	Y
40	WNUV	Baltimore, MD	ATSC	55.26	Y	Y
41	WUTB	Baltimore, MD	ATSC	52.57	Y	Y
42	WMPT	Annapolis, MD	ATSC	19.16	Y	Y
43	WPXW	Manassas, VA	ATSC	68.60	Y	N
44	WDPB	Seaford, DE	ATSC	83.72	N	N
45	WBFF	Baltimore, MD	NTSC	55.26	Y	Y
46	WBFF	Baltimore, MD	ATSC	55.26	Y	Y
47	WMDT	Salisbury, MD	NTSC	80.11	N	N
48	WRC	Washington, DC	ATSC	47.87	Y	Y
49	WWTD-LP	Washington, DC	NTSC	47.87	N	Y
50	WDCW	Washington, DC	NTSC	43.89	Y	Y
51	WDCW	Washington, DC	ATSC	43.89	Y	Y

Table 5-35. Adaptrum WSD Prototype Results at Test Site 6

CH	Signal Type (Database)	Within Service Contour? (Y/N)	Viewable? (Y/N)	Measured Signal Type	Measured Power Level (dBm)	Occupied (O) & Available (A) Channels Reported by WSD	
						L1/S1	L1/S2 [1]
21	ATSC	N	Y	ATSC	-79.1	O	-
22	NTSC	Y	Y	NTSC	-42.4	O	O
23	NTSC	N	Y	NTSC	-72.2	A	A
24	NTSC	Y	Y	NTSC	-59.0	O	O
25	NTSC	N	Y	NTSC	-90.2	A	A
26	NTSC	Y	Y	NTSC	-68.6	O	O
27	ATSC	Y	Y	ATSC	-72.3	O	-
28	ATSC	N	N	NTSC	-99.6	A	A
	NTSC	N	N				
29	ATSC	Y	Y	ATSC	-76.9	O	O
30	ATSC	N	N	ATSC	-90.0	A	-
31	NTSC	N	N	-	-	A	-
32	NTSC	Y	Y	NTSC	-67.5	O	A
33	ATSC	Y	Y	ATSC	-70.8	O	-
34	ATSC	Y	Y	ATSC	-63.5	O	-
35	ATSC	Y	Y	ATSC	-71.4	A	-
36	ATSC	Y	Y	ATSC	-73.1	O	-
38	ATSC	Y	Y	ATSC	-56.9	O	-
39	ATSC	Y	Y	ATSC	-64.0	O	-
40	ATSC	Y	Y	ATSC	-53.7	O	-
41	ATSC	Y	Y	ATSC	-75.5	O	-
42	ATSC	Y	Y	ATSC	-44.8	O	-
43	ATSC	Y	N	NTSC	-98.0	A	A
44	ATSC	N	N	-	-	A	-
45	NTSC	Y	Y	NTSC	-61.7	O	O
46	ATSC	Y	Y	ATSC	-54.3	O	-
47	NTSC	N	N	NTSC	-96.3	A	A
48	ATSC	Y	Y	ATSC	-68.5	O	-
49	NTSC	N	Y	NTSC	-89.1	A	A
50	NTSC	Y	Y	NTSC	-70.0	A	A
51	ATSC	Y	Y	ATSC	-76.7	O	-

[1] Second scan for NTSC signals only.

Table 5-36. I2R WSD Prototype Results at Test Site 6

CH	NTSC/ATSC	Within Service Contour? (Y/N)	Viewable? (Y/N)	Measured Signal Type	Measured Power Level (dBm)	Occupied (O) & Available (A) Channels Reported by WSD			
						L1/S1	L1/S2	L2/S1	L2/S2
21	ATSC	N	Y	ATSC	-79.1	O	O	A	A
22	NTSC	Y	Y	NTSC	-42.4	A	A	O	A
23	NTSC	N	Y	NTSC	-72.2	A	A	A	A
24	NTSC	Y	Y	NTSC	-59.0	O	O	A	A
25	NTSC	N	Y	NTSC	-90.2	A	A	A	A
26	NTSC	Y	Y	NTSC	-68.6	A	A	O	O
27	ATSC	Y	Y	ATSC	-72.3	A	O	O	O
28	ATSC	N	N	NTSC	-99.6	A	A	A	A
	NTSC	N	N						
29	ATSC	Y	Y	ATSC	-76.9	O	O	O	O
30	ATSC	N	N	ATSC	-90.0	A	A	O	O
31	NTSC	N	N	-	-	O	O	A	A
32	NTSC	Y	Y	NTSC	-67.5	A	A	A	A
33	ATSC	Y	Y	ATSC	-70.8	O	O	O	O
34	ATSC	Y	Y	ATSC	-63.5	O	O	O	O
35	ATSC	Y	Y	ATSC	-71.4	O	O	O	O
36	ATSC	Y	Y	ATSC	-73.1	O	O	O	O
38	ATSC	Y	Y	ATSC	-56.9	O	O	O	O
39	ATSC	Y	Y	ATSC	-64.0	O	O	O	O
40	ATSC	Y	Y	ATSC	-53.7	O	O	O	O
41	ATSC	Y	Y	ATSC	-75.5	A	A	O	O
42	ATSC	Y	Y	ATSC	-44.8	O	O	O	O
43	ATSC	Y	N	NTSC	-98.0	A	A	O	O
44	ATSC	N	N	-	-	A	A	A	A
45	NTSC	Y	Y	NTSC	-61.7	A	A	O	A
46	ATSC	Y	Y	ATSC	-54.3	O	O	O	O
47	NTSC	N	N	NTSC	-96.3	A	A	O	O
48	ATSC	Y	Y	ATSC	-68.5	O	O	O	O
49	NTSC	N	Y	NTSC	-89.1	A	A	A	A
50	NTSC	Y	Y	NTSC	-70.0	A	A	A	O
51	ATSC	Y	Y	ATSC	-76.7	O	O	O	O

Table 5-37. Motorola WSD Prototype (Test Mode) Results at Test Site 6

CH	NTSC/ATSC	Within Service Contour? (Y/N)	Viewable?(Y/N)	Measured Signal Type	Measured Power Level (dBm)	Occupied (O) & Available (A) Channels Reported by WSD			
						L1/S1	L1/S2	L2/S1	L2/S2
21	ATSC	N	Y	ATSC	-79.1	O	O	O	O
22	NTSC	Y	Y	NTSC	-42.4	-	-	-	-
23	NTSC	N	Y	NTSC	-72.2	-	-	-	-
24	NTSC	Y	Y	NTSC	-59.0	-	-	-	-
25	NTSC	N	Y	NTSC	-90.2	-	-	-	-
26	NTSC	Y	Y	NTSC	-68.6	-	-	-	-
27	ATSC	Y	Y	ATSC	-72.3	O	O	O	O
28	ATSC	N	N	NTSC	-99.6	-	-	-	-
	NTSC	N	N						
29	ATSC	Y	Y	ATSC	-76.9	O	O	O	A
30	ATSC	N	N	ATSC	-90.0	A	A	A	A
31	NTSC	N	N	-	-	-	-	-	-
32	NTSC	Y	Y	NTSC	-67.5	-	-	-	-
33	ATSC	Y	Y	ATSC	-70.8	A	A	A	A
34	ATSC	Y	Y	ATSC	-63.5	O	O	O	O
35	ATSC	Y	Y	ATSC	-71.4	O	O	O	O
36	ATSC	Y	Y	ATSC	-73.1	A	A	O	O
38	ATSC	Y	Y	ATSC	-56.9	O	O	O	O
39	ATSC	Y	Y	ATSC	-64.0	O	O	O	O
40	ATSC	Y	Y	ATSC	-53.7	O	O	O	O
41	ATSC	Y	Y	ATSC	-75.5	O	O	O	O
42	ATSC	Y	Y	ATSC	-44.8	O	O	O	O
43	ATSC	Y	N	NTSC	-98.0	O	O	O	O
44	ATSC	N	N	-	-	-	-	-	-
45	NTSC	Y	Y	NTSC	-61.7	-	-	-	-
46	ATSC	Y	Y	ATSC	-54.3	O	O	O	O
47	NTSC	N	N	NTSC	-96.3	-	-	-	-
48	ATSC	Y	Y	ATSC	-68.5	O	O	O	O
49	NTSC	N	Y	NTSC	-89.1	-	-	-	-
50	NTSC	Y	Y	NTSC	-70.0	-	-	-	-
51	ATSC	Y	Y	ATSC	-76.7	O	O	O	O

NOTE: In the test mode this device detects ATSC signals only (*i.e.*, no NTSC detection capability).

Table 5-38. Motorola WSD Prototype (Normal Mode) Results at Test Site 6

CH	Signal Type (Database)	Within Service Contour? (Y/N)	Viewable? (Y/N)	Measured Signal Type	Measured Power Level (dBm)	Occupied (O) & Available (A) Channels Reported by WSD	
						L1/S1	L1/S2
21	ATSC	N	Y	ATSC	-79.1	A	A
22	NTSC	Y	Y	NTSC	-42.4	O	O
23	NTSC	N	Y	NTSC	-72.2	A	A
24	NTSC	Y	Y	NTSC	-59.0	O	O
25	NTSC	N	Y	NTSC	-90.2	A	A
26	NTSC	Y	Y	NTSC	-68.6	O	O
27	ATSC	Y	Y	ATSC	-72.3	O	O
28	ATSC	N	N	NTSC	-99.6	A	A
	NTSC	N	N				
29	ATSC	Y	Y	ATSC	-76.9	O	O
30	ATSC	N	N	ATSC	-90.0	A	A
31	NTSC	N	N	-	-	A	A
32	NTSC	Y	Y	NTSC	-67.5	O	O
33	ATSC	Y	Y	ATSC	-70.8	O	O
34	ATSC	Y	Y	ATSC	-63.5	O	O
35	ATSC	Y	Y	ATSC	-71.4	O	O
36	ATSC	Y	Y	ATSC	-73.1	O	O
38	ATSC	Y	Y	ATSC	-56.9	O	O
39	ATSC	Y	Y	ATSC	-64.0	O	O
40	ATSC	Y	Y	ATSC	-53.7	O	O
41	ATSC	Y	Y	ATSC	-75.5	O	O
42	ATSC	Y	Y	ATSC	-44.8	O	O
43	ATSC	Y	N	NTSC	-98.0	O	O
44	ATSC	N	N	-	-	A	A
45	NTSC	Y	Y	NTSC	-61.7	O	O
46	ATSC	Y	Y	ATSC	-54.3	O	O
47	NTSC	N	N	NTSC	-96.3	A	A
48	ATSC	Y	Y	ATSC	-68.5	O	O
49	NTSC	N	Y	NTSC	-89.1	A	A
50	NTSC	Y	Y	NTSC	-70.0	O	O
51	ATSC	Y	Y	ATSC	-76.7	O	O

Table 5-39. Philips WSD Prototype Results at Test Site 6

CH	NTSC/ATSC	Within Service Contour? (Y/N)	Viewable? (Y/N)	Measured Signal Type	Measured Power Level (dBm)	Occupied (O) & Available (A) Channels Reported by WSD			
						L1/S1	L1/S2	L2/S1	L2/S2
21	ATSC	N	Y	ATSC	-79.1	O	O	O	O
22	NTSC	Y	Y	NTSC	-42.4	O	O	O	O
23	NTSC	N	Y	NTSC	-72.2	O	O	O	O
24	NTSC	Y	Y	NTSC	-59.0	O	O	O	O
25	NTSC	N	Y	NTSC	-90.2	O	O	O	O
26	NTSC	Y	Y	NTSC	-68.6	O	O	O	O
27	ATSC	Y	Y	ATSC	-72.3	O	O	O	O
28	ATSC	N	N	NTSC	-99.6	O	O	O	O
	NTSC	N	N						
29	ATSC	Y	Y	ATSC	-76.9	O	O	O	O
30	ATSC	N	N	ATSC	-90.0	O	O	O	O
31	NTSC	N	N	-	-	O	O	O	O
32	NTSC	Y	Y	NTSC	-67.5	O	O	O	O
33	ATSC	Y	Y	ATSC	-70.8	O	O	O	O
34	ATSC	Y	Y	ATSC	-63.5	O	O	O	O
35	ATSC	Y	Y	ATSC	-71.4	O	O	O	O
36	ATSC	Y	Y	ATSC	-73.1	O	O	O	O
38	ATSC	Y	Y	ATSC	-56.9	O	O	O	O
39	ATSC	Y	Y	ATSC	-64.0	O	O	O	O
40	ATSC	Y	Y	ATSC	-53.7	O	O	O	O
41	ATSC	Y	Y	ATSC	-75.5	O	O	O	O
42	ATSC	Y	Y	ATSC	-44.8	O	O	O	O
43	ATSC	Y	N	NTSC	-98.0	O	O	O	O
44	ATSC	N	N	-	-	A	A	A	O
45	NTSC	Y	Y	NTSC	-61.7	O	O	O	O
46	ATSC	Y	Y	ATSC	-54.3	O	O	O	O
47	NTSC	N	N	NTSC	-96.3	O	O	O	O
48	ATSC	Y	Y	ATSC	-68.5	O	O	O	O
49	NTSC	N	Y	NTSC	-89.1	O	O	O	O
50	NTSC	Y	Y	NTSC	-70.0	O	O	O	O
51	ATSC	Y	Y	ATSC	-76.7	O	O	O	O

5.7. Test Site #7 – Daub's Park in Myersville, MD

This site is a public baseball park located just outside of Myersville, in rural western Maryland. This was one of four sites recommended for measurement by MSTV.[39] Channel occupancy measurements were performed at the northwest end of the parking lot and the WSD prototype field trials were performed at the northwest and southwest end of the parking lot.

Table 5-40. Licensed TV Station Assignments within 100-km radius of Test Site 7

CH	Station Call Sign	Station Location	Signal Type	Separation Distance (km)	Within Service Contour? (Y/N)	Viewable? (Y/N)
21	WVPY	Front Royal, VA	ATSC	90.16	N	N
22	WMPT	Annapolis, MD	NTSC	99.07	N	Y
23	WDDN-LP	Washington, DC	NTSC	71.27	N	Y
24	WUTB	Baltimore, MD	NTSC	73.20	N	Y
25	WHAG	Hagerstown, MD	NTSC	38.61	Y	Y
26	WETA	Washington, DC	NTSC	72.19	N	Y
27	WETA	Washington, DC	ATSC	74.63	N	Y
28	WFPT	Frederick, MD	ATSC	34.92	Y	N
29	WMPB	Baltimore, MD	ATSC	67.55	Y	N
30	WNVT	Goldvein, VA	ATSC	98.26	N	Y
	WGCB	Red Lion, PA	ATSC	94.70	N	
31	WWPB	Hagerstown, MD	NTSC	38.60	Y	Y
32	WHUT	Washington, DC	NTSC	72.19	N	N
33	WHUT	Washington, DC	ATSC	74.63	N	Y
34	WUSA	Washington, DC	ATSC	74.63	Y	Y
35	WDCA	Washington, DC	ATSC	73.93	Y	Y
36	WTTG	Washington, DC	ATSC	73.93	Y	Y
38	WJZ	Baltimore, MD	ATSC	80.73	Y	Y
39	WJLA	Washington, DC	ATSC	74.63	Y	Y
40	WNUV	Baltimore, MD	ATSC	80.78	Y	N
41	WUTB	Baltimore, MD	ATSC	73.20	Y	N
42	WMPT	Annapolis, MD	ATSC	99.07	N	N
	WVPY	Front Royal, VA	NTSC	90.16	N	
43	WPXW	Manassas, VA	ATSC	82.48	N	Y
44	WWPB	Hagerstown, MD	ATSC	38.60	Y	N
45	WBFF	Baltimore, MD	NTSC	80.78	N	Y
46	WBFF	Baltimore, MD	ATSC	80.78	Y	N
47	WMDO-CA	Washington, DC	NTSC	75.49	N	Y
48	WRC	Washington, DC	ATSC	75.49	Y	Y
49	WGCB	Red Lion, PA	NTSC	94.70	N	Y
50	WDCW	Washington, DC	NTSC	76.21	N	Y
51	WDCW	Washington, DC	ATSC	76.21	N	N

Table 5-41. Adaptrum WSD Prototype Results at Test Site 7

CH	Signal Type (Database)	Within Service Contour? (Y/N)	Viewable? (Y/N)	Signal Type (Measured)	Measured Power Level (dBm)	Occupied (O) & Available (A) Channels Reported by WSD	
						L1/S1	L1/S2[1]
21	ATSC	N	N	ATSC	-92.9	A	-
22	NTSC	N	Y	NTSC	-69.0	O	A
23	NTSC	N	Y	NTSC	-89.2	A	A
24	NTSC	N	Y	NTSC	-88.5	A	A
25	NTSC	Y	Y	NTSC	-80.1	O	O
26	NTSC	N	Y	NTSC	-81.1	O	O
27	ATSC	N	Y	ATSC	-81.5	O	-
28	ATSC	Y	N	ATSC	-88.2	A	-
29	ATSC	Y	N	ATSC	-92.9	O	-
30	ATSC ATSC	N N	Y	ATSC	-75.7	O	-
31	NTSC	Y	Y	NTSC	-81.3	O	O
32	NTSC	N	N	NTSC	-96.5	A	A
33	ATSC	N	Y	ATSC	-79.6	O	-
34	ATSC	Y	Y	ATSC	-72.0	O	-
35	ATSC	Y	Y	ATSC	-73.1	A	-
36	ATSC	Y	Y	ATSC	-74.3	O	-
38	ATSC	Y	Y	ATSC	-84.3	A	-
39	ATSC	Y	Y	ATSC	-76.3	O	-
40	ATSC	Y	N	ATSC	-87.3	A	-
41	ATSC	Y	N	ATSC	-93.8	O	-
42	ATSC NTSC	N N	N	ATSC	-89.0	A	-
43	ATSC	N	Y	NTSC	-89.9	O	A
44	ATSC	Y	N	ATSC	-93.3	O	-
45	NTSC	N	Y	NTSC	-87.7	A	A
46	ATSC	Y	N	ATSC	-87.4	A	-
47	NTSC	N	Y	ATSC	-78.4	A	-
48	ATSC	Y	Y	ATSC	-85.5	O	-
49	NTSC	N	Y	NTSC	-84.5	A	A
50	NTSC	N	Y	NTSC	-82.6	A	A
51	ATSC	N	N	ATSC	-87.6	O	-

[1] Second scan for NTSC channels only.

Table 5-42. I2R WSD Prototype Results at Test Site 7

CH	Signal Type (Database)	Within Service Contour? (Y/N)	Viewable? (Y/N)	Signal Type (Measured)	Measured Power Level (dBm)	Occupied (O) & Available (A) Channels Reported by WSD			
						L1/S1	L1/S2	L2/S1	L2/S2
21	ATSC	N	N	ATSC	-92.9	A	A	A	A
22	NTSC	N	Y	NTSC	-69.0	A	A	A	A
23	NTSC	N	Y	NTSC	-89.2	A	A	O	O
24	NTSC	N	Y	NTSC	-88.5	A	A	A	A
25	NTSC	Y	Y	NTSC	-80.1	A	A	A	A
26	NTSC	N	Y	NTSC	-81.1	A	A	A	A
27	ATSC	N	Y	ATSC	-81.5	O	O	O	A
28	ATSC	Y	N	ATSC	-88.2	A	A	A	A
29	ATSC	Y	N	ATSC	-92.9	A	A	A	A
30	ATSC	N	Y	ATSC	-75.7	O	O	O	O
	ATSC	N							
31	NTSC	Y	Y	NTSC	-81.3	A	A	A	A
32	NTSC	N	N	NTSC	-96.5	A	A	A	A
33	ATSC	N	Y	ATSC	-79.6	O	O	O	A
34	ATSC	Y	Y	ATSC	-72.0	O	O	O	O
35	ATSC	Y	Y	ATSC	-73.1	O	O	O	O
36	ATSC	Y	Y	ATSC	-74.3	A	O	O	O
38	ATSC	Y	Y	ATSC	-84.3	A	A	O	O
39	ATSC	Y	Y	ATSC	-76.3	O	O	O	O
40	ATSC	Y	N	ATSC	-87.3	O	O	O	O
41	ATSC	Y	N	ATSC	-93.8	A	A	O	O
42	ATSC	N	N	ATSC	-89.0	A	A	A	A
	NTSC	N							
43	ATSC	N	Y	NTSC	-89.9	A	O	A	O
44	ATSC	Y	N	ATSC	-93.3	A	O	O	O
45	NTSC	N	Y	NTSC	-87.7	A	A	A	A
46	ATSC	Y	N	ATSC	-87.4	A	A	A	A
47	NTSC	N	Y	ATSC	-78.4	A	O	O	O
48	ATSC	Y	Y	ATSC	-85.5	O	A	A	A
49	NTSC	N	Y	NTSC	-84.5	A	A	A	A
50	NTSC	N	Y	NTSC	-82.6	A	A	A	A
51	ATSC	N	N	ATSC	-87.6	A	A	A	A

Table 5-43. Motorola WSD Prototype (Test Mode) Results at Test Site 7

CH	Signal Type (Database)	Within Service Contour? (Y/N)	Viewable? (Y/N)	Signal Type (Measured)	Measured Power Level (dBm)	Occupied (O) & Available (A) Channels Reported by WSD			
						L1/S1	L1/S2	L2/S1	L2/S2
21	ATSC	N	N	ATS	-92.9	A	A	A	A
22	NTSC	N	Y	NTS	-69.0	-	-	-	-
23	NTSC	N	Y	NTS	-89.2	-	-	-	-
24	NTSC	N	Y	NTS	-88.5	-	-	-	-
25	NTSC	Y	Y	NTS	-80.1	-	-	-	-
26	NTSC	N	Y	NTS	-81.1	-	-	-	-
27	ATSC	N	Y	ATS	-81.5	A	A	A	O
28	ATSC	Y	N	ATS	-88.2	A	A	A	O
29	ATSC	Y	N	ATS	-92.9	A	A	A	A
30	ATSC ATSC	N N	Y	ATS C	-75.7	O	O	O	O
31	NTSC	Y	Y	NTS	-81.3	-	-	-	-
32	NTSC	N	N	NTS	-96.5	-	-	-	-
33	ATSC	N	Y	ATS	-79.6	A	A	A	A
34	ATSC	Y	Y	ATS	-72.0	O	O	O	O
35	ATSC	Y	Y	ATS	-73.1	O	O	O	O
36	ATSC	Y	Y	ATS	-74.3	A	A	A	A
38	ATSC	Y	Y	ATS	-84.3	O	O	O	O
39	ATSC	Y	Y	ATS	-76.3	O	O	O	O
40	ATSC	Y	N	ATS	-87.3	O	O	O	O
41	ATSC	Y	N	ATS	-93.8	O	O	O	O
42	ATSC NTSC	N N	N	ATS C	-89.0	A	A	O	O
43	ATSC	N	Y	NTS	-89.9	O	O	O	O
44	ATSC	Y	N	ATS	-93.3	A	A	O	O
45	NTSC	N	Y	NTS	-87.7	-	-	-	-
46	ATSC	Y	N	ATS	-87.4	A	A	O	A
47	NTSC	N	Y	ATS	-78.4	A	A	A	A
48	ATSC	Y	Y	ATS	-85.5	O	O	O	A
49	NTSC	N	Y	NTS	-84.5	-	-	-	-
50	NTSC	N	Y	NTS	-82.6	-	-	-	-
51	ATSC	N	N	ATS	-87.6	A	A	O	O

NOTE: In the test mode this device detects ATSC signals only (*i.e.*, no NTSC detection capability).

Table 5-44. Motorola WSD Prototype (Normal Mode) Results at Test Site 7

CH	Signal Type (Database)	Within Service Contour? (Y/N)	Viewable? (Y/N)	Signal Type (Measured)	Measured Power Level (dBm)	Occupied (O) & Available (A) Channels Reported by WSD	
						L1/S1	L2/S1
21	ATSC	N	N	ATSC	-92.9	A	A
22	NTSC	N	Y	NTSC	-69.0	O	O
23	NTSC	N	Y	NTSC	-89.2	A	A
24	NTSC	N	Y	NTSC	-88.5	A	A
25	NTSC	Y	Y	NTSC	-80.1	O	O
26	NTSC	N	Y	NTSC	-81.1	A	A
27	ATSC	N	Y	ATSC	-81.5	O	O
28	ATSC	Y	N	ATSC	-88.2	O	O
29	ATSC	Y	N	ATSC	-92.9	O	O
30	ATSC	N	Y	ATSC	-75.7	A	A
	ATSC	N					
31	NTSC	Y	Y	NTSC	-81.3	O	O
32	NTSC	N	N	NTSC	-96.5	A	A
33	ATSC	N	Y	ATSC	-79.6	O	O
34	ATSC	Y	Y	ATSC	-72.0	O	O
35	ATSC	Y	Y	ATSC	-73.1	O	O
36	ATSC	Y	Y	ATSC	-74.3	O	O
38	ATSC	Y	Y	ATSC	-84.3	O	O
39	ATSC	Y	Y	ATSC	-76.3	O	O
40	ATSC	Y	N	ATSC	-87.3	O	O
41	ATSC	Y	N	ATSC	-93.8	O	O
42	ATSC	N	N	ATSC	-89.0	A	A
	NTSC	N					
43	ATSC	N	Y	NTSC	-89.9	A	A
44	ATSC	Y	N	ATSC	-93.3	O	O
45	NTSC	N	Y	NTSC	-87.7	O	O
46	ATSC	Y	N	ATSC	-87.4	O	O
47	NTSC	N	Y	ATSC	-78.4	A	A
48	ATSC	Y	Y	ATSC	-85.5	O	O
49	NTSC	N	Y	NTSC	-84.5	A	A
50	NTSC	N	Y	NTSC	-82.6	O	O
51	ATSC	N	N	ATSC	-87.6	O	O

Table 5-45. Philips WSD Prototype Results at Test Site 7

CH	Signal ype (Database)	Within Service Contour? (Y/N)	Viewable? (Y/N)	Signal Type (Measured)	Measured Power Level (dBm)	Occupied (O) & Available (A) Channels Reported by WSD			
						L1/S1	L1/S2	L2/S1	L2/S2
21	ATSC	N	N	ATSC	-92.9	O	O	O	A
22	NTSC	N	Y	NTSC	-69.0	O	O	O	O
23	NTSC	N	Y	NTSC	-89.2	O	O	O	O
24	NTSC	N	Y	NTSC	-88.5	O	O	O	O
25	NTSC	Y	Y	NTSC	-80.1	O	O	O	O
26	NTSC	N	Y	NTSC	-81.1	O	O	O	O
27	ATSC	N	Y	ATSC	-81.5	O	O	O	O
28	ATSC	Y	N	ATSC	-88.2	O	O	O	O
29	ATSC	Y	N	ATSC	-92.9	O	O	O	A
30	ATSC ATSC	N N	Y	ATSC	-75.7	O	O	O	O
31	NTSC	Y	Y	NTSC	-81.3	O	O	O	O
32	NTSC	N	N	NTSC	-96.5	O	O	O	O
33	ATSC	N	Y	ATSC	-79.6	O	O	O	O
34	ATSC	Y	Y	ATSC	-72.0	O	O	O	O
35	ATSC	Y	Y	ATSC	-73.1	O	O	O	O
36	ATSC	Y	Y	ATSC	-74.3	O	O	O	O
38	ATSC	Y	Y	ATSC	-84.3	O	O	O	O
39	ATSC	Y	Y	ATSC	-76.3	O	O	O	O
40	ATSC	Y	N	ATSC	-87.3	O	O	O	O
41	ATSC	Y	N	ATSC	-93.8	O	O	O	O
42	ATSC NTSC	N N	N	ATSC	-89.0	O	O	O	O
43	ATSC	N	Y	NTSC	-89.9	O	O	O	O
44	ATSC	Y	N	ATSC	-93.3	O	O	O	O
45	NTSC	N	Y	NTSC	-87.7	O	O	O	O
46	ATSC	Y	N	ATSC	-87.4	O	O	A	O
47	NTSC	N	Y	ATSC	-78.4	O	O	O	O
48	ATSC	Y	Y	ATSC	-85.5	O	O	O	O
49	NTSC	N	Y	NTSC	-84.5	O	O	O	O
50	NTSC	N	Y	NTSC	-82.6	O	O	O	O
51	ATSC	N	N	ATSC	-87.6	O	O	O	O

Table 5-46. Philips WSD Prototype Results at Test Site 7 (w/ 10-dB attenuator)

CH	Signal Type (Database)	Within Service Contour? (Y/N)	Viewable? (Y/N)	Signal Type (Measured)	Measured Power Level (dBm)	Occupied (O) & Available (A) Channels Reported by WSD (Loc 2)
21	ATSC	N	N	ATSC	-92.9	A
22	NTSC	N	Y	NTSC	-69.0	O
23	NTSC	N	Y	NTSC	-89.2	O
24	NTSC	N	Y	NTSC	-88.5	O
25	NTSC	Y	Y	NTSC	-80.1	O
26	NTSC	N	Y	NTSC	-81.1	O
27	ATSC	N	Y	ATSC	-81.5	O
28	ATSC	Y	N	ATSC	-88.2	O
29	ATSC	Y	N	ATSC	-92.9	A
30	ATSC	N	Y	ATSC	-75.7	O
	ATSC	N				
31	NTSC	Y	Y	NTSC	-81.3	O
32	NTSC	N	N	NTSC	-96.5	O
33	ATSC	N	Y	ATSC	-79.6	A
34	ATSC	Y	Y	ATSC	-72.0	O
35	ATSC	Y	Y	ATSC	-73.1	O
36	ATSC	Y	Y	ATSC	-74.3	O
38	ATSC	Y	Y	ATSC	-84.3	O
39	ATSC	Y	Y	ATSC	-76.3	O
40	ATSC	Y	N	ATSC	-87.3	A
41	ATSC	Y	N	ATSC	-93.8	A
42	ATSC	N	N	ATSC	-89.0	O
	NTSC	N				
43	ATSC	N	Y	NTSC	-89.9	O
44	ATSC	Y	N	ATSC	-93.3	O
45	NTSC	N	Y	NTSC	-87.7	O
46	ATSC	Y	N	ATSC	-87.4	A
47	NTSC	N	Y	ATSC	-78.4	A
48	ATSC	Y	Y	ATSC	-85.5	O
49	NTSC	N	Y	NTSC	-84.5	O
50	NTSC	N	Y	NTSC	-82.6	O
51	ATSC	N	N	ATSC	-87.6	O

5.8. Test Site #8 – Intersection of Rt. 17 and Middle Point Rd., NE of Myersville

This site is northeast of Myersville, in rural western Maryland. This was one of four sites recommended for measurement by MSTV.[40] Channel occupancy measurements were performed at the entrance of a long paved driveway leading to a horse barn and the WSD prototype field trials were performed at two locations in a field adjacent to the driveway, separated by approximately 91.4 meters (300 feet).

Table 5-47. Licensed TV Station Assignments within 100-km Radius of Test Site 8

CH	Station Call Sign	Station Location	Signal Type	Separation Distance (km)	Within Service Contour? (Y/N)	Viewable? (Y/N)
21	WVPY	Front Royal, VA	ATSC	96.57	N	N
22	-	-	-	-	N	N
23	WDDN-LP	Washington, DC	NTSC	74.42	N	N
24	WUTB	Baltimore, MD	NTSC	72.59	N	Y
25	WHAG	Hagerstown, MD	NTSC	39.35	Y	Y
26	WETA	Washington, DC	NTSC	75.79	N	Y
27	WETA	Washington, DC	ATSC	78.15	N	N
	WHTM	Harrisburg, PA	NTSC	96.92	N	
28	WFPT	Frederick, MD	ATSC	38.30	Y	N
29	WMPB	Baltimore, MD	ATSC	65.35	Y	N
30	WGCB	Red Lion, PA	ATSC	89.20	N	Y
31	WWPB	Hagerstown, MD	NTSC	39.52	Y	N
32	WHUT	Washington, DC	NTSC	75.80	N	N
33	WHUT	Washington, DC	ATSC	78.15	N	N
34	WUSA	Washington, DC	ATSC	78.15	Y	N
35	WDCA	Washington, DC	ATSC	77.44	Y	N
36	WTTG	Washington, DC	ATSC	77.44	Y	N
38	WJZ	Baltimore, MD	ATSC	79.40	Y	N
39	WJLA	Washington, DC	ATSC	78.15	Y	N
40	WNUV	Baltimore, MD	ATSC	79.44	Y	N
41	WUTB	Baltimore, MD	ATSC	72.59	Y	N
42	WVPY	Front Royal, VA	NTSC	96.57	N	Y
43	WPXW	Manassas, VA	ATSC	87.72	N	Y
	WPMT	York, PA	NTSC	94.55	N	
44	WWPB	Hagerstown, MD	ATSC	39.52	Y	N
45	WBFF	Baltimore, MD	NTSC	79.44	N	N
46	WBFF	Baltimore, MD	ATSC	79.44	Y	N
47	WPMT	York, PA	ATSC	94.55	N	N
48	WRC	Washington, DC	ATSC	79.06	Y	N
49	WGCB	Red Lion, PA	NTSC	89.20	N	N
50	WDCW	Washington, DC	NTSC	79.38	N	Y
51	WDCW	Washington, DC	ATSC	79.38	N	N

Table 5-48. Adaptrum WSD Prototype Results at Test Site 8

CH	Signal Type (Database)	Within Service Contour? (Y/N)	Viewable? (Y/N)	Signal Type (Measured)	Measured Power Level (dBm)	Occupied (O) & Available (A) Channels Reported by WSD			
						L1/S1	L1/S2[1]	L2/S1	L2/S2[1]
21	ATSC	N	N	ATSC	-93.4	O	-	O	-
22	-	N	N	NTSC	-90.4	O	A	A	A
23	NTSC	N	N	ATSC	-92.8	A	-	A	-
24	NTSC	N	Y	NTSC	-94.4	A	A	A	A
25	NTSC	Y	Y	NTSC	-77.7	O	O	O	O
26	NTSC	N	Y	NTSC	-93.4	A	A	A	A
27	ATSC	N	N	ATSC	-95.8	A	-	A	-
	NTSC	N							
28	ATSC	Y	N	ATSC	-96.2	A	-	A	-
29	ATSC	Y	N	NTSC	-90.8	A	A	A	A
30	ATSC	N	Y	ATSC	-72.7	O	-	O	-
31	NTSC	Y	N	NTSC	-75.6	O	O	O	O
32	NTSC	N	N	ATSC	-88.7	O	O	A	O
33	ATSC	N	N	ATSC	-93.1	A	-	A	-
34	ATSC	Y	N	ATSC	-91.2	O	-	O	-
35	ATSC	Y	N	ATSC	-91.7	A	-	A	-
36	ATSC	Y	N	ATSC	-93.2	O	-	O	-
38	ATSC	Y	N	ATSC	-93.8	O	-	A	-
39	ATSC	Y	N	ATSC	-92.3	O	-	A	-
40	ATSC	Y	N	ATSC	-93.2	A	-	A	-
41	ATSC	Y	N	ATSC	-95.0	O	-	A	-
42	NTSC	N	Y	NTSC	-96.1	A	A	O	O
43	ATSC	N	Y	NTSC	-102.3	A	A	A	A
	NTSC	N							
44	ATSC	Y	N	ATSC	-90.1	O	-	O	-
45	NTSC	N	N	NTSC	-100.5	A	A	A	A
46	ATSC	Y	N	ATSC	-94.6	O	-	A	-
47	ATSC	N	N	-	-	A	-	A	-
48	ATSC	Y	N	ATSC	-93.2	O	O	A	-
49	NTSC	N	N	ATSC	-94.6	O	-	A	-
50	NTSC	N	Y	NTSC	-95.7	O	A	A	A
51	ATSC	N	N	ATSC	-96.7	O	-	O	-

[1] Second scan for NTSC signals only.

Table 5-49. I2R WSD Prototype Results at Test Site 8

CH	Signal Type (Database)	Within Service Contour? (Y/N)	Viewable? (Y/N)	Signal Type (Measured)	Measured Power Level (dBm)	Occupied (O) & Available (A) Channels Reported by WSD			
						L1/S1	L1/S2[1]	L2/S1	L2/S2
21	ATSC	N	N	ATSC	-93.4	A	O	O	O
22	-	N	N	NTSC	-90.4	A	A	A	A
23	NTSC	N	N	ATSC	-92.8	A	A	O	O
24	NTSC	N	Y	NTSC	-94.4	A	A	A	A
25	NTSC	Y	Y	NTSC	-77.7	A	A	A	A
26	NTSC	N	Y	NTSC	-93.4	A	A	A	A
27	ATSC	N	N	ATSC	-95.8	O	O	A	A
	NTSC	N							
28	ATSC	Y	N	ATSC	-96.2	A	A	A	A
29	ATSC	Y	N	NTSC	-90.8	A	A	A	O
30	ATSC	N	Y	ATSC	-72.7	O	O	O	O
31	NTSC	Y	N	NTSC	-75.6	A	A	A	A
32	NTSC	N	N	ATSC	-88.7	A	A	A	A
33	ATSC	N	N	ATSC	-93.1	A	A	A	A
34	ATSC	Y	N	ATSC	-91.2	A	A	O	A
35	ATSC	Y	N	ATSC	-91.7	A	A	O	O
36	ATSC	Y	N	ATSC	-93.2	A	A	O	O
38	ATSC	Y	N	ATSC	-93.8	A	A	A	O
39	ATSC	Y	N	ATSC	-92.3	O	O	A	A
40	ATSC	Y	N	ATSC	-93.2	O	O	A	A
41	ATSC	Y	N	ATSC	-95.0	O	O	A	A
42	NTSC	N	Y	NTSC	-96.1	O	O	A	A
43	ATSC	N	Y	NTSC	-102.3	A	A	A	O
	NTSC	N							
44	ATSC	Y	N	ATSC	-90.1	A	A	O	O
45	NTSC	N	N	NTSC	-100.5	A	A	A	A
46	ATSC	Y	N	ATSC	-94.6	O	O	A	A
47	ATSC	N	N	-	-	A	A	A	A
48	ATSC	Y	N	ATSC	-93.2	A	A	O	O
49	NTSC	N	N	ATSC	-94.6	A	A	A	A
50	NTSC	N	Y	NTSC	-95.7	A	A	A	A
51	ATSC	N	N	ATSC	-96.7	A	A	A	A

Table 5-50. Motorola WSD Prototype (Test Mode) Results at Test Site 8

CH	Signal Type (Database)	Within Service Contour? (Y/N)	Viewable? (Y/N)	Signal Type (Measured)	Measured Power Level (dBm)	Occupied (O) & Available (A) Channels Reported by WSD			
						L1/S1	L1/S2[1]	L2/S1	L2/S2
21	ATSC	N	N	ATSC	-93.4	O	O	O	O
22	-	N	N	NTSC	-90.4	-	O	-	-
23	NTSC	N	N	ATSC	-92.8	A	A	A	A
24	NTSC	N	Y	NTSC	-94.4	-	-	-	-
25	NTSC	Y	Y	NTSC	-77.7	-	-	-	-
26	NTSC	N	Y	NTSC	-93.4	O	-	-	-
27	ATSC NTSC	N N	N	ATSC	-95.8	A	A	A	A
28	ATSC	Y	N	ATSC	-96.2	O	A	A	A
29	ATSC	Y	N	NTSC	-90.8	-	-	-	-
30	ATSC	N	Y	ATSC	-72.7	O	O	A	O
31	NTSC	Y	N	NTSC	-75.6	-	-	-	-
32	NTSC	N	N	ATSC	-88.7	A	A	A	A
33	ATSC	N	N	ATSC	-93.1	A	A	A	A
34	ATSC	Y	N	ATSC	-91.2	O	A	O	O
35	ATSC	Y	N	ATSC	-91.7	O	O	A	O
36	ATSC	Y	N	ATSC	-93.2	A	A	A	A
38	ATSC	Y	N	ATSC	-93.8	O	O	O	A
39	ATSC	Y	N	ATSC	-92.3	O	O	A	A
40	ATSC	Y	N	ATSC	-93.2	O	O	A	A
41	ATSC	Y	N	ATSC	-95.0	O	A	A	A
42	NTSC	N	Y	NTSC	-96.1	O	O	-	-
43	ATSC NTSC	N N	Y	NTSC	-102.3	O	O	O	O
44	ATSC	Y	N	ATSC	-90.1	A	O	O	A
45	NTSC	N	N	NTSC	-100.5	-	-	-	-
46	ATSC	Y	N	ATSC	-94.6	O	O	O	A
47	ATSC	N	N	-	-	-	-	-	-
48	ATSC	Y	N	ATSC	-93.2	A	O	O	O
49	NTSC	N	N	ATSC	-94.6	A	O	O	A
50	NTSC	N	Y	NTSC	-95.7	-	-	-	-
51	ATSC	N	N	ATSC	-96.7	A	A	O	A

NOTE: In the test mode this device detects only ATSC signals (*i.e.*, no NTSC detection capability).

Table 5-51. Motorola WSD Prototype (Normal Mode) Results at Test Site 8

CH	Signal Type (Database)	Within Service Contour? (Y/N)	Viewable? (Y/N)	Signal Type (Measured)	Measured Power Level (dBm)	Occupied (O) & Available (A) ChannelsReported by WSD	
						L1/S1	L2/S1
21	ATSC	N	N	ATSC	-93.4	A	A
22	-	N	N	NTSC	-90.4	A	A
23	NTSC	N	N	ATSC	-92.8	A	A
24	NTSC	N	Y	NTSC	-94.4	A	A
25	NTSC	Y	Y	NTSC	-77.7	O	O
26	NTSC	N	Y	NTSC	-93.4	A	A
27	ATSC	N	N	ATSC	-95.8	A	A
	NTSC	N					
28	ATSC	Y	N	ATSC	-96.2	O	O
29	ATSC	Y	N	NTSC	-90.8	O	O
30	ATSC	N	Y	ATSC	-72.7	A	A
31	NTSC	Y	N	NTSC	-75.6	O	O
32	NTSC	N	N	ATSC	-88.7	A	A
33	ATSC	N	N	ATSC	-93.1	A	A
34	ATSC	Y	N	ATSC	-91.2	O	O
35	ATSC	Y	N	ATSC	-91.7	O	O
36	ATSC	Y	N	ATSC	-93.2	O	O
38	ATSC	Y	N	ATSC	-93.8	O	O
39	ATSC	Y	N	ATSC	-92.3	O	O
40	ATSC	Y	N	ATSC	-93.2	O	O
41	ATSC	Y	N	ATSC	-95.0	O	O
42	NTSC	N	Y	NTSC	-96.1	A	A
43	ATSC	N	Y	NTSC	-102.3	A	A
	NTSC	N					
44	ATSC	Y	N	ATSC	-90.1	O	O
45	NTSC	N	N	NTSC	-100.5	O	O
46	ATSC	Y	N	ATSC	-94.6	O	O
47	ATSC	N	N	-	-	O	O
48	ATSC	Y	N	ATSC	-93.2	O	O
49	NTSC	N	N	ATSC	-94.6	A	A
50	NTSC	N	Y	NTSC	-95.7	A	A
51	ATSC	N	N	ATSC	-96.7	O	O

Table 5-52. Philips WSD Prototype Results at Test Site 8

CH	Signal Type (Database)	Within Service Contour? (Y/N)	Viewable? (Y/N)	Signal Type (Measured)	Measured Power Level (dBm)	Occupied (O) & Available (A) Channels Reported by WSD			
						L1/S1	L1/S2[1]	L2/S1	L2/S2
21	ATSC	N	N	ATSC	-93.4	O	O	O	O
22	-	N	N	NTSC	-90.4	O	O	O	O
23	NTSC	N	N	ATSC	-92.8	O	O	O	O
24	NTSC	N	Y	NTSC	-94.4	O	O	O	O
25	NTSC	Y	Y	NTSC	-77.7	O	O	O	O
26	NTSC	N	Y	NTSC	-93.4	O	O	O	O
27	ATSC NTSC	N N	N	ATSC	-95.8	O	O	O	O
28	ATSC	Y	N	ATSC	-96.2	O	O	O	O
29	ATSC	Y	N	NTSC	-90.8	O	O	O	O
30	ATSC	N	Y	ATSC	-72.7	O	O	O	O
31	NTSC	Y	N	NTSC	-75.6	O	O	O	O
32	NTSC	N	N	ATSC	-88.7	O	O	O	O
33	ATSC	N	N	ATSC	-93.1	O	O	O	O
34	ATSC	Y	N	ATSC	-91.2	O	O	O	O
35	ATSC	Y	N	ATSC	-91.7	O	O	O	O
36	ATSC	Y	N	ATSC	-93.2	O	O	O	O
38	ATSC	Y	N	ATSC	-93.8	O	O	O	O
39	ATSC	Y	N	ATSC	-92.3	O	O	O	O
40	ATSC	Y	N	ATSC	-93.2	O	O	O	O
41	ATSC	Y	N	ATSC	-95.0	O	O	O	O
42	NTSC	N	Y	NTSC	-96.1	O	O	O	O
43	ATSC NTSC	N N	Y	NTSC	-102.3	O	O	O	O
44	ATSC	Y	N	ATSC	-90.1	O	O	O	O
45	NTSC	N	N	NTSC	-100.5	O	O	O	O
46	ATSC	Y	N	ATSC	-94.6	O	O	A	O
47	ATSC	N	N	-	-	A	O	A	A
48	ATSC	Y	N	ATSC	-93.2	O	O	O	O
49	NTSC	N	N	ATSC	-94.6	O	O	O	O
50	NTSC	N	Y	NTSC	-95.7	O	O	O	O
51	ATSC	N	N	ATSC	-96.7	O	O	O	O

Table 5-53. Philips WSD Prototype Results at Test Site 8 (w/ 10-dB attenuator)

CH	Signal Type (Database)	Within Service Contour? (Y/N)	Viewable? (Y/N)	Signal Type (Measured)	Measured Power Level (dBm)	Occupied (O) & Available (A) Channels Reported by WSD (Loc 1)
21	ATSC	N	N	ATSC	-93.4	O
22	-	N	N	NTSC	-90.4	O
23	NTSC	N	N	ATSC	-92.8	A
24	NTSC	N	Y	NTSC	-94.4	O
25	NTSC	Y	Y	NTSC	-77.7	O
26	NTSC	N	Y	NTSC	-93.4	O
27	ATSC	N	N	ATSC	-95.8	A
	NTSC	N				
28	ATSC	Y	N	ATSC	-96.2	O
29	ATSC	Y	N	NTSC	-90.8	O
30	ATSC	N	Y	ATSC	-72.7	A
31	NTSC	Y	N	NTSC	-75.6	O
32	NTSC	N	N	ATSC	-88.7	O
33	ATSC	N	N	ATSC	-93.1	O
34	ATSC	Y	N	ATSC	-91.2	O
35	ATSC	Y	N	ATSC	-91.7	O
36	ATSC	Y	N	ATSC	-93.2	O
38	ATSC	Y	N	ATSC	-93.8	O
39	ATSC	Y	N	ATSC	-92.3	O
40	ATSC	Y	N	ATSC	-93.2	O
41	ATSC	Y	N	ATSC	-95.0	A
42	NTSC	N	Y	NTSC	-96.1	O
43	ATSC	N	Y	NTSC	-102.3	O
	NTSC	N				
44	ATSC	Y	N	ATSC	-90.1	O
45	NTSC	N	N	NTSC	-100.5	O
46	ATSC	Y	N	ATSC	-94.6	O
47	ATSC	N	N	-	-	A
48	ATSC	Y	N	ATSC	-93.2	O
49	NTSC	N	N	ATSC	-94.6	O
50	NTSC	N	Y	NTSC	-95.7	O
51	ATSC	N	N	ATSC	-96.7	A

5.9. Test Site #9 – Grossnickle Church at Rt 17 and Meeting House Road

This site is northeast of Myersville, in rural western Maryland. This was one of four sites recommended for measurement by MSTV.[41] Channel occupancy measurements were performed in the parking lot of the Church Parish Hall and the WSD prototype field trials were performed at two locations in the parking lot, separated by approximately 30.5 meters (100 feet).

Table 5-54. Licensed TV Station Assignments within 100-km radius of Test Site 9

CH	Station Call Sign	Station Location	Signal Type	Separation Distance (km)	Within Service Contour? (Y/N)	Viewable? (Y/N)
21	WVPY	Front Royal, VA	ATSC	94.78	N	N
22	WMPT	Annapolis, MD	NTSC	99.05	N	Y
23	WDDN-LP	Washington, DC	NTSC	72.80	N	N
24	WUTB	Baltimore, MD	NTSC	72.05	N	N
25	WHAG	Hagerstown, MD	NTSC	39.65	Y	Y
26	WETA	Washington, DC	NTSC	74.06	N	Y
27	WETA	Washington, DC	ATSC	76.44	N	N
	WHTM	Harrisburg, PA	NTSC	99.02	N	N
28	WFPT	Frederick, MD	ATSC	36.58	Y	N
29	WMPB	Baltimore, MD	ATSC	65.34	Y	N
30	WGCB	Red Lion, PA	ATSC	90.51	N	N
31	WWPB	Hagerstown, MD	NTSC	39.76	Y	Y
32	WHUT	Washington, DC	NTSC	74.06	N	Y
33	WHUT	Washington, DC	ATSC	76.44	N	N
34	WUSA	Washington, DC	ATSC	76.44	Y	N
35	WDCA	Washington, DC	ATSC	75.73	Y	N
36	WTTG	Washington, DC	ATSC	75.73	Y	N
38	WJZ	Baltimore, MD	ATSC	79.10	Y	N
39	WJLA	Washington, DC	ATSC	76.44	Y	N
40	WNUV	Baltimore, MD	ATSC	79.15	Y	N
41	WUTB	Baltimore, MD	ATSC	72.05	Y	N
42	WVPY	Front Royal, VA	NTSC	94.78	N	Y
	WMPT	Annapolis, MD	ATSC	99.05	N	N
43	WPXW	Manassas, VA	ATSC	85.64	N	N
	WPMT	York, PA	NTSC	96.09	N	Y
44	WWPB	Hagerstown, MD	ATSC	39.76	Y	N
45	WBFF	Baltimore, MD	NTSC	79.15	N	Y
46	WBFF	Baltimore, MD	ATSC	79.15	Y	N
47	WPMT	York, PA	ATSC	96.09	N	N
48	WRC	Washington, DC	ATSC	77.34	Y	N
49	WGCB	Red Lion, PA	NTSC	90.51	N	N
50	WDCW	Washington, DC	NTSC	77.76	N	Y
51	WDCW	Washington, DC	ATSC	77.76	N	N

Table 5-55. Adaptrum WSD Prototype Results at Test Site 9

CH	Signal Type (Database)	Within Service Contour? (Y/N)	Viewable? (Y/N)	Measured Signal Type	Measured Power Level (dBm)	Occupied (O) & Available (A) Channels Reported by WSD	
						L2/S1	L2/S2[1]
21	ATSC	N	N	ATSC	-90.6	O	-
22	NTSC	N	Y	NTSC	-92.9	O	A
23	NTSC	N	N	ATSC	-95.2	A	-
24	NTSC	N	N	NTSC	-108.0	A	A
25	NTSC	Y	Y	NTSC	-94.5	O	A
26	NTSC	N	Y	NTSC	-96.8	A	O
27	ATSC	N	N	NTSC	-105.6	A	A
	NTSC	N	N				
28	ATSC	Y	N	NTSC	-114.9	A	A
29	ATSC	Y	N	-	-	A	-
30	ATSC	N	N	ATSC	-88.8	A	-
31	NTSC	Y	Y	NTSC	-91.7	O	A
32	NTSC	N	Y	NTSC	-97.0	A	A
33	ATSC	N	N	NTSC	-100.5	A	A
34	ATSC	Y	N	ATSC	-92.5	O	-
35	ATSC	Y	N	ATSC	-93.7	A	-
36	ATSC	Y	N	ATSC	-91.4	O	-
38	ATSC	Y	N	ATSC	-93.5	O	-
39	ATSC	Y	N	ATSC	-92.7	A	-
40	ATSC	Y	N	ATSC	-94.7	A	-
41	ATSC	Y	N	-	-	O	-
42	NTSC	N	Y	NTSC	-100.5	O	O
	ATSC	N	N				
43	ATSC	N	N	NTSC	-85.6	A	A
	NTSC	N	Y				
44	ATSC	Y	N	ATSC	-96.3	A	-
45	NTSC	N	Y	NTSC	-93.4	A	A
46	ATSC	Y	N	ATSC	-96.5	O	-
47	ATSC	N	N	ATSC	-93.1	A	-
48	ATSC	Y	N	NTSC	-105.4	A	A
49	NTSC	N	N	ATSC	-	A	-
50	NTSC	N	Y	NTSC	-95.0	A	A
51	ATSC	N	N	ATSC	-95.9	O	-

[1] Second scan for NTSC channels only.

Table 5-56. I2R WSD Prototype Results at Test Site 9

CH	Signal Type (Database)	Within Service Contour? (Y/N)	Viewable? (Y/N)	Measured Signal Type	Measured Power Level (dBm)	Occupied (O) & Available (A) Channels Reported by WSD			
						L1/S1	L1/S2	L2/S1	L2/S2
21	ATSC	N	N	ATSC	-90.6	O	O	A	O
22	NTSC	N	Y	NTSC	-92.9	A	A	A	A
23	NTSC	N	N	ATSC	-95.2	A	A	A	A
24	NTSC	N	N	NTSC	-108.0	A	A	A	A
25	NTSC	Y	Y	NTSC	-94.5	A	A	A	A
26	NTSC	N	Y	NTSC	-96.8	A	A	A	A
27	ATSC	N	N	NTSC	-105.6	A	A	A	A
	NTSC	N	N						
28	ATSC	Y	N	NTSC	-114.9	A	A	A	A
29	ATSC	Y	N	-	-	A	A	A	A
30	ATSC	N	N	ATSC	-88.8	O	O	A	A
31	NTSC	Y	Y	NTSC	-91.7	A	A	A	A
32	NTSC	N	Y	NTSC	-97.0	A	A	A	A
33	ATSC	N	N	NTSC	-100.5	A	A	A	A
34	ATSC	Y	N	ATSC	-92.5	A	A	A	A
35	ATSC	Y	N	ATSC	-93.7	A	A	A	A
36	ATSC	Y	N	ATSC	-91.4	A	A	A	A
38	ATSC	Y	N	ATSC	-93.5	O	A	O	O
39	ATSC	Y	N	ATSC	-92.7	A	A	A	A
40	ATSC	Y	N	ATSC	-94.7	A	A	A	A
41	ATSC	Y	N	-	-	A	A	A	A
42	NTSC	N	Y	NTSC	-100.5	A	A	A	A
	ATSC	N	N						
43	ATSC	N	N	NTSC	-85.6	A	A	A	A
	NTSC	N	Y						
44	ATSC	Y	N	ATSC	-96.3	A	A	A	A
45	NTSC	N	Y	NTSC	-93.4	A	A	A	A
46	ATSC	Y	N	ATSC	-96.5	A	A	A	A
47	ATSC	N	N	ATSC	-93.1	A	A	A	A
48	ATSC	Y	N	NTSC	-105.4	A	A	A	A
49	NTSC	N	N	ATSC	-	A	A	A	A
50	NTSC	N	Y	NTSC	-95.0	A	A	A	A
51	ATSC	N	N	ATSC	-95.9	A	A	A	A

Table 5-57. Motorola WSD Prototype (Test Mode) Results at Test Site 9

CH	Signal Type (Database)	Within Service Contour? (Y/N)	Viewable? (Y/N)	Measured Signal Type	Measured Power Level (dBm)	Occupied (O) & Available (A) Channels Reported by WSD			
						L1/S1	L1/S2	L2/S1	L2/S2
21	ATSC	N	N	ATSC	-90.6	O	O	O	O
22	NTSC	N	Y	NTSC	-92.9	-	-	-	-
23	NTSC	N	N	ATSC	-95.2	A	A	A	A
24	NTSC	N	N	NTSC	-108.0	-	-	-	-
25	NTSC	Y	Y	NTSC	-94.5	-	-	-	-
26	NTSC	N	Y	NTSC	-96.8	-	-	-	-
27	ATSC	N	N	NTSC	-105.6	-	-	-	-
	NTSC	N	N						
28	ATSC	Y	N	NTSC	-114.9	-	-	-	-
29	ATSC	Y	N	-	-	-	-	-	-
30	ATSC	N	N	ATSC	-88.8	O	O	O	O
31	NTSC	Y	Y	NTSC	-91.7	-	-	-	-
32	NTSC	N	Y	NTSC	-97.0	-	-	-	-
33	ATSC	N	N	NTSC	-100.5	-	-	-	-
34	ATSC	Y	N	ATSC	-92.5	A	A	A	A
35	ATSC	Y	N	ATSC	-93.7	A	A	A	A
36	ATSC	Y	N	ATSC	-91.4	A	A	A	A
38	ATSC	Y	N	ATSC	-93.5	A-	A	O	O
39	ATSC	Y	N	ATSC	-92.7	A	A	O	A
40	ATSC	Y	N	ATSC	-94.7	O	O	O	O
41	ATSC	Y	N	-	-	O	O	O	O
42	NTSC	N	Y	NTSC	-100.5	-	-	-	-
	ATSC	N	N						
43	ATSC	N	N	NTSC	-85.6	O	O	O	O
	NTSC	N	Y						
44	ATSC	Y	N	ATSC	-96.3	A	A	A	A
45	NTSC	N	Y	NTSC	-93.4	-	-	-	-
46	ATSC	Y	N	ATSC	-96.5	O	A	O	A
47	ATSC	N	N	ATSC	-93.1	A	A	A	A
48	ATSC	Y	N	NTSC	-105.4	-	-	-	-
49	NTSC	N	N	ATSC	-	A	A	A	A
50	NTSC	N	Y	NTSC	-95.0	-	-	-	-
51	ATSC	N	N	ATSC	-95.9	A	A	O	A

NOTE: In the test mode this device detects only ATSC signals (*i.e.*, no NTSC detection capability).

Table 5-58. Motorola WSD Prototype (Normal Mode) Results at Test Site 9

<table>
<tr><th rowspan="2">CH</th><th rowspan="2">Signal Type (Database)</th><th rowspan="2">Within Service Contour? (Y/N)</th><th rowspan="2">Viewable? (Y/N)</th><th rowspan="2">Measured Signal Type</th><th rowspan="2">Measured Power Level (dBm)</th><th colspan="2">Occupied (O) & Available (A) Channels Reported by WSD</th></tr>
<tr><th>L1/S1</th><th>L2/S1</th></tr>
<tr><td>21</td><td>ATSC</td><td>N</td><td>N</td><td>ATSC</td><td>-90.6</td><td>A</td><td>A</td></tr>
<tr><td>22</td><td>NTSC</td><td>N</td><td>Y</td><td>NTSC</td><td>-92.9</td><td>A</td><td>A</td></tr>
<tr><td>23</td><td>NTSC</td><td>N</td><td>N</td><td>ATSC</td><td>-95.2</td><td>A</td><td>A</td></tr>
<tr><td>24</td><td>NTSC</td><td>N</td><td>N</td><td>NTSC</td><td>-108.0</td><td>A</td><td>A</td></tr>
<tr><td>25</td><td>NTSC</td><td>Y</td><td>Y</td><td>NTSC</td><td>-94.5</td><td>O</td><td>O</td></tr>
<tr><td>26</td><td>NTSC</td><td>N</td><td>Y</td><td>NTSC</td><td>-96.8</td><td>A</td><td>A</td></tr>
<tr><td rowspan="2">27</td><td>ATSC</td><td>N</td><td>N</td><td rowspan="2">NTSC</td><td rowspan="2">-105.6</td><td rowspan="2">A</td><td rowspan="2">A</td></tr>
<tr><td>NTSC</td><td>N</td><td>N</td></tr>
<tr><td>28</td><td>ATSC</td><td>Y</td><td>N</td><td>NTSC</td><td>-114.9</td><td>O</td><td>O</td></tr>
<tr><td>29</td><td>ATSC</td><td>Y</td><td>N</td><td>-</td><td>-</td><td>O</td><td>O</td></tr>
<tr><td>30</td><td>ATSC</td><td>N</td><td>N</td><td>ATSC</td><td>-88.8</td><td>A</td><td>A</td></tr>
<tr><td>31</td><td>NTSC</td><td>Y</td><td>Y</td><td>NTSC</td><td>-91.7</td><td>O</td><td>O</td></tr>
<tr><td>32</td><td>NTSC</td><td>N</td><td>Y</td><td>NTSC</td><td>-97.0</td><td>A</td><td>A</td></tr>
<tr><td>33</td><td>ATSC</td><td>N</td><td>N</td><td>NTSC</td><td>-100.5</td><td>O</td><td>O</td></tr>
<tr><td>34</td><td>ATSC</td><td>Y</td><td>N</td><td>ATSC</td><td>-92.5</td><td>O</td><td>O</td></tr>
<tr><td>35</td><td>ATSC</td><td>Y</td><td>N</td><td>ATSC</td><td>-93.7</td><td>O</td><td>O</td></tr>
<tr><td>36</td><td>ATSC</td><td>Y</td><td>N</td><td>ATSC</td><td>-91.4</td><td>O</td><td>O</td></tr>
<tr><td>38</td><td>ATSC</td><td>Y</td><td>N</td><td>ATSC</td><td>-93.5</td><td>O</td><td>O</td></tr>
<tr><td>39</td><td>ATSC</td><td>Y</td><td>N</td><td>ATSC</td><td>-92.7</td><td>O</td><td>O</td></tr>
<tr><td>40</td><td>ATSC</td><td>Y</td><td>N</td><td>ATSC</td><td>-94.7</td><td>O</td><td>O</td></tr>
<tr><td>41</td><td>ATSC</td><td>Y</td><td>N</td><td>-</td><td>-</td><td>O</td><td>O</td></tr>
<tr><td rowspan="2">42</td><td>NTSC</td><td>N</td><td>Y</td><td rowspan="2">NTSC</td><td rowspan="2">-100.5</td><td rowspan="2">A</td><td rowspan="2">A</td></tr>
<tr><td>ATSC</td><td>N</td><td>N</td></tr>
<tr><td rowspan="2">43</td><td>ATSC</td><td>N</td><td>N</td><td rowspan="2">NTSC</td><td rowspan="2">-85.6</td><td rowspan="2">A</td><td rowspan="2">A</td></tr>
<tr><td>NTSC</td><td>N</td><td>Y</td></tr>
<tr><td>44</td><td>ATSC</td><td>Y</td><td>N</td><td>ATSC</td><td>-96.3</td><td>O</td><td>O</td></tr>
<tr><td>45</td><td>NTSC</td><td>N</td><td>Y</td><td>NTSC</td><td>-93.4</td><td>O</td><td>O</td></tr>
<tr><td>46</td><td>ATSC</td><td>Y</td><td>N</td><td>ATSC</td><td>-96.5</td><td>O</td><td>O</td></tr>
<tr><td>47</td><td>ATSC</td><td>N</td><td>N</td><td>ATSC</td><td>-93.1</td><td>O</td><td>O</td></tr>
<tr><td>48</td><td>ATSC</td><td>Y</td><td>N</td><td>NTSC</td><td>-105.4</td><td>O</td><td>O</td></tr>
<tr><td>49</td><td>NTSC</td><td>N</td><td>N</td><td>ATSC</td><td>-</td><td>A</td><td>A</td></tr>
<tr><td>50</td><td>NTSC</td><td>N</td><td>Y</td><td>NTSC</td><td>-95.0</td><td>O</td><td>A</td></tr>
<tr><td>51</td><td>ATSC</td><td>N</td><td>N</td><td>ATSC</td><td>-95.9</td><td>O</td><td>O</td></tr>
</table>

Table 5-59. Philips WSD Prototype Results at Test Site 9

CH	Signal Type (Database)	Within Service Contour? (Y/N)	Viewable? (Y/N)	Measured Signal Type	Measured Power Level (dBm)	Occupied (O) & Available (A) Channels Reported by WSD			
						L1/S1	L1/S2	L2/S1	L2/S2
21	ATSC	N	N	ATSC	-90.6	O	O	O	O
22	NTSC	N	Y	NTSC	-92.9	O	O	O	O
23	NTSC	N	N	ATSC	-95.2	O	A	O	O
24	NTSC	N	N	NTSC	-108.0	O	O	O	O
25	NTSC	Y	Y	NTSC	-94.5	O	O	O	O
26	NTSC	N	Y	NTSC	-96.8	O	O	O	O
27	ATSC	N	N	NTSC	-105.6	O	O	O	O
	NTSC	N	N						
28	ATSC	Y	N	NTSC	-114.9	O	A	A	O
29	ATSC	Y	N	-	-	O	A	O	O
30	ATSC	N	N	ATSC	-88.8	O	O	O	O
31	NTSC	Y	Y	NTSC	-91.7	O	O	O	O
32	NTSC	N	Y	NTSC	-97.0	O	O	O	O
33	ATSC	N	N	NTSC	-100.5	O	O	O	O
34	ATSC	Y	N	ATSC	-92.5	O	O	O	O
35	ATSC	Y	N	ATSC	-93.7	O	O	O	O
36	ATSC	Y	N	ATSC	-91.4	O	O	O	O
38	ATSC	Y	N	ATSC	-93.5	O	O	O	O
39	ATSC	Y	N	ATSC	-92.7	O	A	O	O
40	ATSC	Y	N	ATSC	-94.7	O	O	O	O
41	ATSC	Y	N	-	-	O	O	O	O
42	NTSC	N	Y	NTSC	-100.5	O	O	O	O
	ATSC	N	N						
43	ATSC	N	N	NTSC	-85.6	O	O	O	O
	NTSC	N	Y						
44	ATSC	Y	N	ATSC	-96.3	O	O	O	O
45	NTSC	N	Y	NTSC	-93.4	O	O	O	O
46	ATSC	Y	N	ATSC	-96.5	O	A	O	O
47	ATSC	N	N	ATSC	-93.1	O	O	O	O
48	ATSC	Y	N	NTSC	-105.4	A	O	O	O
49	NTSC	N	N	ATSC	-	O	O	O	O
50	NTSC	N	Y	NTSC	-95.0	O	O	O	O
51	ATSC	N	N	ATSC	-95.9	A	A	O	O

5.10. TV-Sensing (Channel Occupancy) Field Test Summary

This section provides a summary of the prototype TV WSD field trial results obtained at the nine test sites.

In reducing the data from the TV sensing field tests it became obvious that two available parameters can be used to define a set of conditions which apply uniquely to each channel under consideration for "white" space use. These two parameters have been labeled "Within Contour" and "Viewable" in this chapter and each has a value of either yes (1) or no (0). This binary combination yields four possible conditions that will uniquely describe each TV channel at a particular location. These four conditions are:

- Condition I: the WSD is operating *within* the service contour of a station assigned to the channel and the broadcast signal can be displayed on a representative consumer TV receiver system (viewable).
- Condition II: the WSD is operating *within* the service contour of a station assigned to the channel but the broadcast signal *cannot* be displayed on a representative consumer TV receiver system (not viewable).
- Condition III: the WSD is operating *outside* of the service contour of a station assigned to the channel but the broadcast signal *can* be displayed on a representative consumer TV receiver system (viewable).
- Condition IV: the WSD is operating *outside* of the service contour of a station assigned to the channel and the broadcast signal *cannot* be displayed on a representative consumer TV receiver system (not viewable).

Table 5-60, presented below, defines how each channel between 21 and 51 was categorized at each of the field trial sites. In this table, each channel is assigned to one of the four conditions described above and then further segmented according to the signal type utilizing the channel (ATSC or NTSC).

Tables 5-61 through 5-66 provide the results from the field trials in terms of the ability of each WSD prototype to correctly identify Condition I, II, III or IV channels. For each condition, the number of correct detections is shown as a function of the number of scans performed. While this approach presumes that Condition II and III channels represent occupied, rather than available channels, the arithmetic complement of each reported result defines the performance under the opposite presumption (*i.e.*, that they represent available channels).

Table 5-67 shows the overall percentage of accurate signal detections confined to those channels whose occupancy status is easily established (Condition I and IV channels) over all nine of the test sites. The results are presented for Condition I (occupied) ATSC (digital) and NTSC (analog) channels and for Condition IV (available) channels.

Table 5-60. Categorization of Channels at Each Field Test Site

Test Site	ATSC Channels			NTSC Channels			Condition IV
	Condition I	Condition II	Condition III	Condition I	Condition II	Condition III	
1	38,40,41,46	27,28,29,33, 34,35,36,39, 42,48,51	None	22,24,26,4 5, 50	32	None	21,23,25,30, 31,43,44,47, 49
2	27,29,33,34, 35,36,38,40, 41,42,46	39,48,51	None	22,24,32,4 5, 50	26	43	21,23,25,28, 30,31,44,47, 49
3	27,28,29,33, 34,35,36,38, 39,40,41,42, 46,48,51	None	30,47	22,24,26,3 2, 45,50	None	31,43,49	21,23,25,44
4	27,29,30,3 3, 34,35,36,38, 39,40,42,43, 46,48,51	28,41	None	22,23,24,2 5, 26, 32, 45,47, 49, 50	None	None	21,31,44
Test Site	ATSC Channels			NTSC Channels			Condition IV
	Condition I	Condition II	Condition III	Condition I	Condition II	Condition III	
5a	36,48	27,28,29,30, 33,34,35,38, 39,40,41,42, 43,46,51	None	26,32,45, 50	22,23,24,25, 47,49	None	21,31,44
5b	27,33,34,35, 36,38,39,40, 46,48,51	28,29,30,41, 42,43	None	22,23,24,2 5, 26, 32, 45,49, 50	47	None	21,31,44
6	27,29,33,34, 35,36,38,39, 40,41,42,46, 48,51	None	21	22,24,26,3 2, 45, 50	43	23,25,49	28,30,31,44, 47
7	34,35,36,38, 39,48	28,29,40,41, 44,46	27,30,33, 43	25,31	None	22,23,24,26, 45,47,49,50	21,32,42,5 1
8	None	28,29,34,35, 36,38,39,40, 41,44,46,48	30	25	31	24,26,42,43, 50	21,22,23,27, 32,33,45,47, 49,51
9	None	28,29,34,35, 36,38,39,40, 41,44,46,48	None	25,31	None	22,26,32,42, 43,45,50	21,23,24,27, 30,33,47,49, 51

5.10.1. Observations Based on Results

The TV-sensing field trial results indicate that the Motorola WSD, which employs a geo-location/database look-up approach to identify occupied and unoccupied TV channels, was the best overall performer with respect to correctly determining occupied and unoccupied (available) TV channels within TV station service contours. This prototype also employs a spectrum sensing capability that was found to perform with less accuracy than that of other devices. The combination of geo-location/database look up and spectrum sensing techniques provide an acceptable mechanism to allow for authorization of equipment today.

Table 5-61. Adaptrum WSD Prototype Field Trial Results

Test Site	ATSC			NTSC			Condition IV
	Condition I	Condition II	Condition III	Condition I	Condition II	Condition III	
1	8/8	11/22	-	12/15	0/3	-	18/21
2	20/22	6/6	-	9/10	2/2	0/2	13/18
3	14/15	-	3/3	12/12	-	3/6	2/7
4	13/15	1/2	-	20/20	-	-	3/3
5a	2/2	9/15	-	6/8	1/12	-	3/3
5b	11/11	1/6	-	16/18	0/2	-	2/3
6	13/14	-	1/1	9/12	0/2	0/6	7/7
7	4/6	3/6	4/4	4/4	-	1/16	4/5
8	-	11/24	2/2	4/4	4/4	3/20	17/26
9	-	5/12	-	2/4	-	4/14	10/12
Totals[1]	91%	51%	100%	89%	30%	17%	75%

Notes: A significant spurious emission was observed at 600 MHz (see Appendix H) that may have prevented accurate detection on channel 35.

[1]Percentage of successful identifications of the channel occupancy under assumed conditions.

Table 5-62. I2R WSD Prototype Field Trial Results

Test Site	ATSC Channels			NTSC Channels			Condition IV
	Condition I	Condition II	Condition III	Condition I	Condition II	Condition III	
1	-	-	-	2/20	0/4	-	12/16
2	-	-	-	5/20	1/4	0/4	14/21
3	60/60	-	6/8	17/24	-	6/12	9/16
4	55/60	1/8	-	11/40	-	-	12/12
5a	4/4	12/30	-	1/8	0/12	-	6/6
5b	22/22	8/12	-	3/18	0/2	-	6/6
6	53/56	-	2/4	7/24	2/4	0/12	14/20
7	18/24	9/24	12/16	0/8	-	5/32	16/16
8	-	18/48	4/4	0/4	0/4	3/20	33/40
9	-	3/48	-	0/8	-	0/28	31/36
Totals[1]	94%	30%	75%	25%	10%	13%	81%

Notes: This device was inadvertently operated in "all signal" mode at test sites 1 and 2, which was the wrong mode for DTV detection. Therefore, no ATSC results are presented for these two test sites.

[1] Percentage of successful identifications of the channel occupancy under assumed conditions.

Table 5-63. Motorola WSD Prototype (Sensing Mode) Field Trial Results

Test Site	ATSC Channels			NTSC Channels			Condition[1] IV
	Condition I	Condition II	Condition III	Condition I	Condition II	Condition III	
1	16/16	28/44	-	-	-	-	0/7
2	40/44	9/12	-	-	-	-	8/13
3	56/60	-	4/8	-	-	-	0/1
4	55/60	3/8	-	-	-	-	4/4
5a	4/4	15/30	-	-	-	-	-
5b	19/22	6/12	-	-	-	-	-
6	49/56	-	-	-	-	-	4/4
7	19/24	12/24	9/16	-	-	-	8/12
8	-	23/48	3/4	-	-	-	21/29
9	-	13/48	-	-	-	-	15/24
Totals[2]	90%	48%	57%	-	-	-	64%

Notes:

[1] Since in the test mode the prototype device detects only ATSC signals, those channels not identified as occupied could .not be determined with certainty to be available or to be occupied by NTSC signals.

[2] Percentage of successful identifications of the channel occupancy under assumed conditions.

Table 5-64. Motorola WSD Prototype (Geolocation Mode) Field Trial Results

Test Site	ATSC Channels			NTSC Channels			Condition IV
	Condition I	Condition II	Condition III	Condition I	Condition II	Condition III	
1	8/8	22/22	-	10/10	2/2	-	14/18
2	22/22	6/6	-	10/10	2/2	2/2	12/18
3	30/30	-	4/4	12/12	-	2/6	8/8
4	30/30	4/4	-	20/20	-	-	6/6
5a	2/2	15/15	-	4/4	6/6	-	3/3
5b	11/11	6/6	-	9/9	1/1	-	3/3
6	28/28	-	0/2	12/12	2/2	0/6	10/10
7	12/12	12/12	4/8	4/4	-	6/16	6/8
8	-	24/24	0/2	2/2	2/2	0/10	14/20
9	-	24/24	-	4/4	-	3/14	12/18
Totals[1]	100%	100%	50%	100%	100%	24%	71%

Notes: [1] Percentage of successful identifications of the channel occupancy under assumed conditions.

It should be noted that the reported NTSC-detection performance may be somewhat misleading since no consideration was given to the quality of the analog picture when identifying "viewable" channels.

Table 5-65. Philips WSD Prototype Field Trial Results

Test Site	ATSC Channels			NTSC Channels			Condition IV
	Condition I	Condition II	Condition III	Condition I	Condition II	Condition III	
1	16/16	40/44	-	20/20	4/4	-	11/36
2	44/44	12/12	-	20/20	4/4	4/4	2/36
3	60/60	-	8/8	24/24	-	12/12	0/16
4	60/60	8/8	-	40/40	-	-	5/12
5a	4/4	28/30	-	8/8	12/12	-	2/6
5b	22/22	10/12	-	18/18	2/2	-	1/6
6	56/56	-	4/4	24/24	4/4	12/12	3/20
7	24/24	22/24	16/16	8/8	-	32/32	3/16
8	-	47/48	4/4	4/4	4/4	20/20	3/40
9	-	42/48	-	8/8	-	28/28	3/36
Totals[1]	100%	92%	100%	100%	100%	100%	15%

Notes: [1] Percentage of successful identifications of the channel occupancy under assumed conditions.

Table 5-66. Philips WSD Prototype (w/external attenuator) Field Trial Results

Test Site	ATSC			NTSC			Condition IV
	Condition I	Condition II	Condition III	Condition I	Condition II	Condition III	
3 (6dB)	15/15	-	2/2	6/6	-	3/3	0/4
4 (10dB)	30/30	3/4	-	20/20	-	-	5/6
5a (10dB)	2/2	11/15	-	4/4	5/6	-	3/3
7 (10dB)	6/6	2/6	3/4	2/2	-	7/8	1/4
8 (10dB)	-	11/12	0/1	1/1	1/1	5/5	4/10
Totals[1]	100%	73%	71%	100%	86%	92%	48%

Notes:
[1] Percentage of successful identifications of the channel occupancy under assumed conditions.

Table 5-67. WSD Signal Detection Performance

Prototype WSD	Condition I		Condition IV
	ATSC Occupied Channels	NTSC-Occupied Channels	Unoccupied (Available) Channels
Adaptrum	91 %	89%	75%
I2R	94%	25%	81%
Motorola (geo-loc)	100%	100%	71%
Motorola (sensing)	90%	-	64%
Philips	100%	100%	15%

The results obtained from those WSD prototypes that rely solely on spectrum sensing to identify occupied and unoccupied (available) channels demonstrate the importance of correctly specifying a detection sensitivity requirement. The Philips device, which demonstrated the most sensitivity in the laboratory tests, also performed best with respect to detecting occupied channels; however, it reported a very high percentage of channels occupied that were potentially available. The Adaptrum and I2R devices, which demonstrated less sensitivity in the laboratory measurements, did much better at correctly determining available channels but did not detect occupied channels with complete reliability. Spectrum sensing worked to some degree and it may be possible to authorize products that rely on spectrum sensing, in the future, if it can be demonstrated that they will not interfere.

The comparative received power measurements performed at each test site indicate that the difference in signal level between TV signals received by a roof-top television antenna and those received by ground-proximate antenna can vary between 0 and 34 dB, depending on the specific site conditions and whether the WSD antenna is located indoors or outdoors relative to the TV antenna.

6. Wireless Microphone Measurements

6.1. Wireless Microphone Laboratory Tests

6.1.1. Introduction

The purpose of this part of the test project was to observe the capabilities of the current white space device (WSD) prototypes to sense Part 74 wireless microphones (WM) operating on broadcast television channels at known signal levels with and without digital television (DTV) signals on adjacent channels. Tests were performed with two different types of microphones – FM modulated and digitally modulated.[42] All of the tested microphones operate in a maximum 200 kHz authorized bandwidth segment within a TV channel. Each microphone was limited to operation in a different subset of television channels 21 to 51; individual microphones were tested on frequencies within their subset as shown in the test results. The test modulation was a 1000 Hz tone at 24 kHz deviation. The two FM microphones had a 32 kHz pilot tone and the digital microphone was encrypted. The measured modulated signal for the FM microphones is shown in Figure 6-1 and the measured signal of the digital microphone, which appears the same with or without modulation, is shown in Figure 6-2.

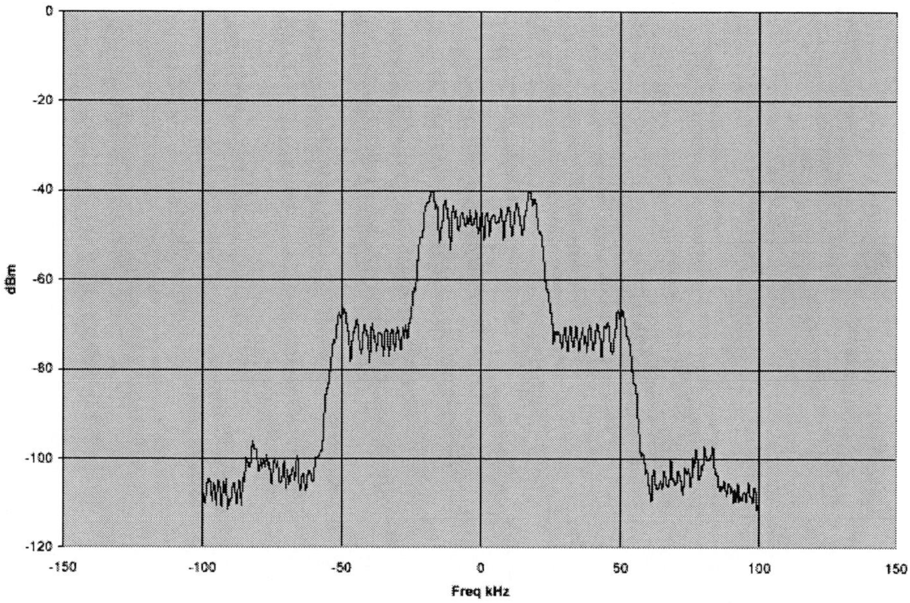

Figure 6-1. FM Microphone Signal

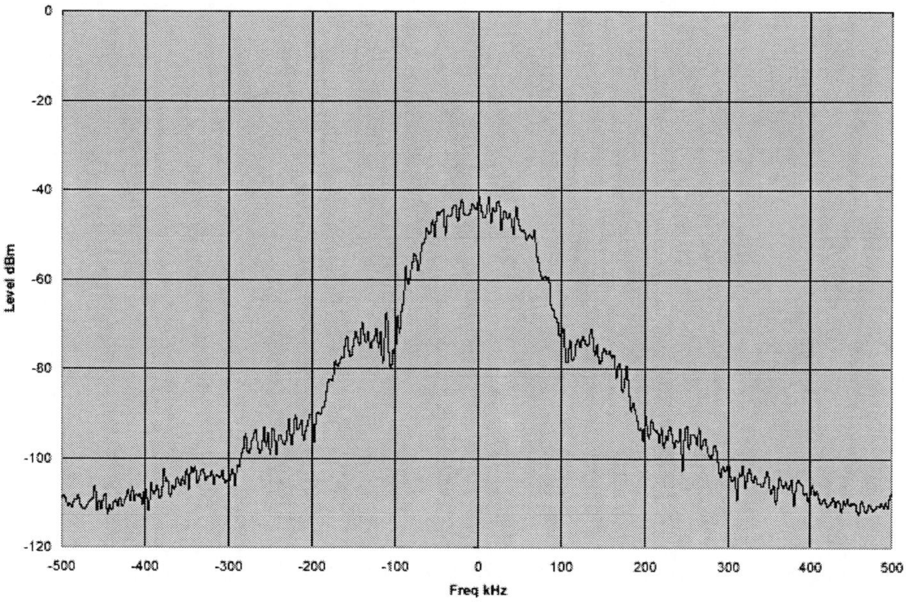

Figure 6-2. Digital Microphone Signal

The WSDs submitted by Microsoft and Philips were similar to those submitted in the previous round of testing (Phase I) with respect to their ability to detect wireless microphones with some modifications. The changes eliminate the ability to select the type of signal

(ATSC, NTSC or wireless microphone) to search for and the identification of the type of signal sensed. The versions of both of these WSDs examined in the Phase II tests searched for all three types of signals on each scan of a channel and indicated only whether the channel was occupied or available on each scan or the probability of detection for the number of scans selected. The Institute for Infocomm Research (I2R) device had the capability to search for either DTV or wireless microphone signals. It provided two scan modes – "DTV" and "All" and two test modes – "Lab" and "Field". In the DTV scan mode it searched only for DTV signals and in the "All" mode it searched only for microphones. The "Lab" test mode was intended only for conducted bench testing and the "Field" mode was intended only for reception of radiated signals with an antenna. The test results were displayed as the probability of detection for the number of scans selected.

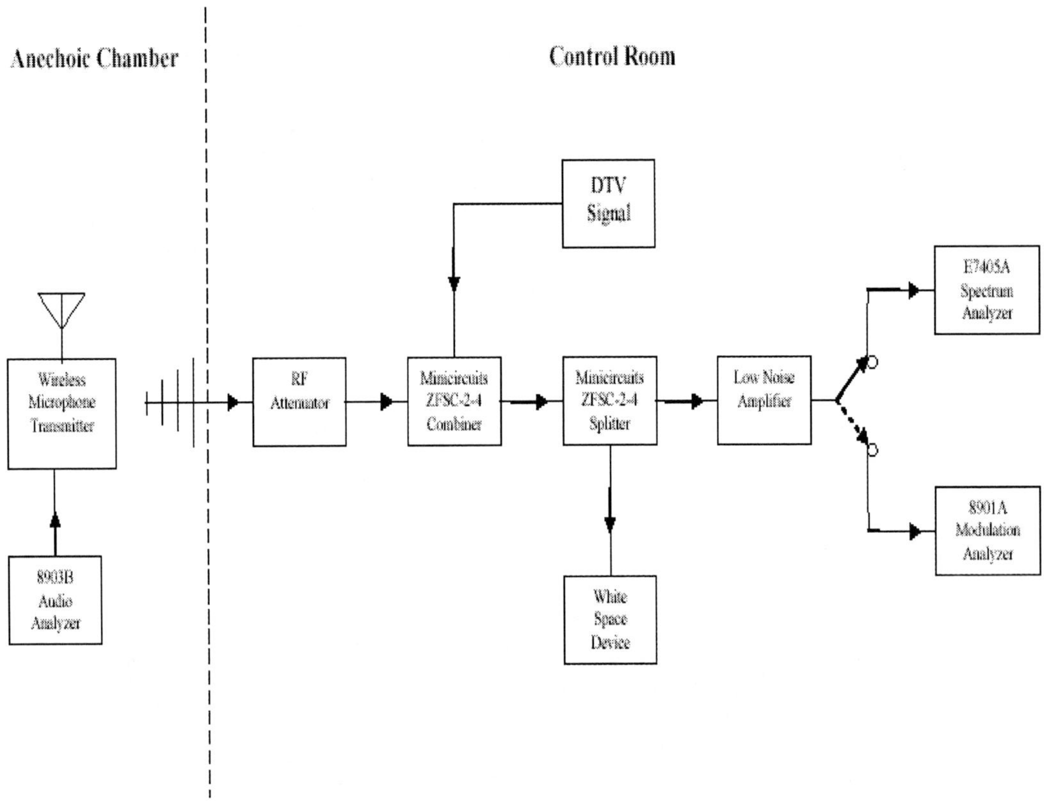

Figure 6-3. Wireless Microphone Sensing Test Setup

6.1.2. Test Procedure

Because of ambient emissions, direct pick-up and spurious radiated emissions from the WSD systems, it was not possible to derive a clean test signal on the bench by sampling the radiated signal from a wireless microphone with a receive antenna and applying it to the antenna input connector of the WSD. It was thus determined that the testing needed to be performed with the wireless microphone in an anechoic chamber. This approach avoided interference to the test from undesired signals, and allowed a sample of the radiated signal from the microphone to be received with an antenna in the chamber. This received signal was then conducted by coaxial cable to the control room for connection to the antenna terminal of the WSD, as shown in Figure 6-3.

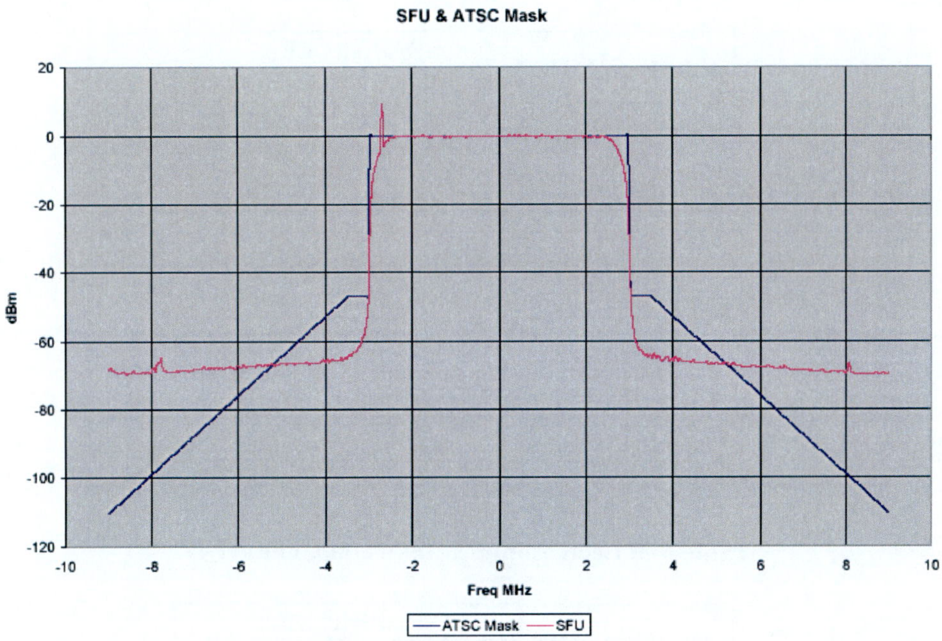

Figure 6-4. Simulated DTV Signal & ATSC Mask

The WSDs were first tested for sensitivity to a wireless microphone signal at a low, middle and high frequency within a TV channel with no other signals present. The sensitivity threshold of a WSD was defined as the lowest bandpower at which the microphone is detected on 10 of 10 successive scans. Each WSD was tested with the FM microphone modulated and, in some cases, unmodulated. The digital microphone was tested with or without modulation since there was no discernible difference between the modulated and unmodulated signal or in the response of the WSDs to the different signals. Testing was then performed with simulated DTV signals located in various combinations of first and second adjacent channels over a range of power levels. The simulated DTV signal was obtained from a Rohde& Schwarz SFU signal generator. The spectral characteristics of the simulated DTV signal with the emissions mask specified in Section 73.622(h) of the Commission's rules are shown in Figure 6.4.

6.1.3. Test Results with Microphones and No DTV Signals

The sensitivity of the WSDs with no other signals present is shown in tables 6-1 through 6-9. In most cases the FM wireless microphones were tested at 50 kHz above the low frequency edge of the channel, in the middle of the channel and 50 kHz below the high edge of the channel and, because of its wider occupied bandwidth, the digital microphone was tested 100 kHz above the low edge of the channel, in the middle of the channel and 100 kHz below the high edge of the channel. It was not always possible or considered necessary to test at all three frequencies within a channel. The Shure UR1 H4 is the same basic model as the UR1 L3 tuned to a different frequency range.

The Microsoft 205 WSD

Table 6-1. Shure UR1 L3 on Channel 44

Mic. Freq MHz	Sensitivity dBm	
	Modulated	Unmodulated
650.05	-119	-125
653	-125	-122
655.95	-124	-129

Table 6-2. Shure UR1 H4 on Channel 23

Mic. Freq MHz	Sensitivity dBm	
	Modulated	Unmodulated
524.05	-116	-123
527	-123	-120
529.95	-123	-128

Table 6-3. Lectrosonics UM 700 on Channel 44

Mic. Freq MHz	Sensitivity dBm	
	Modulated	Unmodulated
650.1	-117	---
653	-122	---
655.9	-122	---

The Philips #1 WSD

Table 6-4. Shure UR1 L3 on Channel 44

Mic. Freq MHz	Sensitivity dBm	
	Modulated	Unmodulated
650.05	-117	-115
653	-116	-103
655.95	-117	-116

Table 6-5. Shure UR1 H4 on Channel 23

Mic. Freq MHz	Sensitivity dBm	
	Modulated	Unmodulated
524.05	-117	-114
527	-116	-108
529.95	-118	-106

Table 6-6. Lectrosonics UM 700 on Channel 44

Mic. Freq MHz	Sensitivity dBm	
	Modulated	Unmodulated
650.1	-112	---
653	-115	---
655.9	-113	---

The I2R (Test Mode: Lab; Scan Mode: All) WSD

Table 6-7. Shure UR1 L3 on Channel 44

Mic. Freq MHz	Sensitivity dBm	
	Modulated	Unmodulated
650.05	-105	---
653	-127	-126
655.95	-94	---

Table 6-8. Lectrosonics UM 700 on Channel 29

Mic. Freq MHz	Sensitivity dBm	
	Modulated	Unmodulated
---	---	---
563.2	---	-121
565.9	---	-110

Table 6-9. Shure UR1 H4 on Channel 30

Mic. Freq MHz	Sensitivity dBm	
	Modulated	Unmodulated
---	---	---
569	-123	---
---	---	---

6.1.4. Test Results with Microphones and DTV Signals

The WSDs were tested to determine their ability to sense a wireless microphone on a channel and with simulated DTV signals on various combinations of first and second adjacent channels. The DTV signals were provided at power levels -28, -53, -68 and -84 dBm, which represent a high, medium, low and very low signal level. In most cases, it was found that at DTV power levels of -68 dBm or higher, the WSDs would indicate that the wireless microphone channel was occupied when no microphone signal was present (false positive) and the Microsoft WSD would not sense a wireless microphone signal at -80 dBm when the DTV signal power was -28 dBm. In a few cases when the DTV signal was on the lower adjacent channel, the WSDs would only indicate a false positive if the DTV power was increased to a power level of -53 dBm or higher. Because of test scheduling problems, the availability of test equipment or the malfunctioning of a device, formal measurements for some test conditions were not performed in all cases. Sensitivity was measured with DTV signal levels below the power level that produced false positive indications. The results are given in the tables below. For comparison, the sensitivity without DTV signals is given in parentheses.

The Microsoft 205 WSD

Table 6-10. Shure UR1 L3 (modulated) on Channel 44

DTV Channels	Microphone Freq MHz	Sensitivity dBm
43	650.05	-106 (-119)
	653	-125 (-125)
	655.95	-124 (-124)

Table 6-11. Lectrosonics UM 700 on Channel 44

DTV Channels	Microphone Freq MHz	Sensitivity dBm
43	650.1	-116 (-117)
	653	-122 (-122)
	655.9	-120 (-122)

Table 6-12. Lectrosonics UM 700 on Channel 44

DTV Channels	Microphone Freq MHz	Sensitivity dBm
45	650.1	-117 (-117)
	653	-123 (-122)
	655.9	-121 (-122)

Table 6-13. Lectrosonics UM 700 on Channel 44

DTV Channels	Microphone Freq MHz	Sensitivity dBm
43,45	650.1	-115 (-117)
	653	-122 (-122)
	655.9	-121 (-122)

Table 6-14. Lectrosonics UM 700 on Channel 44

DTV Channels	Microphone Freq MHz	Sensitivity dBm
42,43	650.1	-117 (-117)
	653	-120 (-122)
	655.9	-120 (-122)

Table 6-15. Lectrosonics UM 700 on Channel 44

DTV Channels	Microphone Freq MHz	Sensitivity dBm
45,46	650.1	-117 (-117)
	653	-121 (-122)
	655.9	-121 (-122)

The Philips #1 WSD

Table 6-16. Shure UR1 L3 on Channel 44

DTV Channels	Microphone Freq MHz	Sensitivity dBm	
		Modulated	Unmodulated
43	650.05	-114 (-117)	-112 (-115)
	653	-116 (-116)	-93 (-103)
	655.95	-116 (-117)	-116 (-116)

Table 6-17. Shure UR1 L3 on Channel 44

DTV Channels	Microphone Freq MHz	Sensitivity dBm	
		Modulated	Unmodulated
45	650.05	-116 (-117)	-110 (-115)
	653	-115 (-116)	-115 (-103)
	655.95	-117 (-117)	-114 (-116)

Table 6-18. Shure UR1 L3 on Channel 44

DTV Channels	Microphone Freq MHz	Sensitivity dBm	
		Modulated	Unmodulated
43,45	650.05	-116 (-117)	-114 (-115)
	653	-116 (-116)	-101 (-103)
	655.95	-117 (-117)	-116 (-116)

Table 6-19. Shure UR1 L3 on Channel 44

DTV Channels	Microphone Freq MHz	Sensitivity dBm	
		Modulated	Unmodulated
42,43	650.05	-115 (-117)	---
	653	-116 (-116)	---
	655.95	-118 (-117)	---

Table 6-20. Shure UR1 L3 on Channel 44

DTV Channels	Microphone Freq MHz	Sensitivity dBm	
		Modulated	Unmodulated
45,46	650.05	-116 (-117)	---
	653	-116 (-116)	---
	655.95	-117 (-117)	---

Table 6-21. Lectrosonics UM 700 on Channel 44

DTV Channels	Microphone Freq MHz	Sensitivity dBm
43	650.1	-111 (-112)
	653	-116 (-115)
	655.9	-113 (-113)

Table 6-22. Lectrosonics UM 700 on Channel 44

DTV Channels	Microphone Freq MHz	Sensitivity dBm
45	650.1	-111 (-112)
	653	-115 (-115)
	655.9	-113 (-113)

Table 6-23. Lectrosonics UM 700 on Channel 44

DTV Channels	Microphone Freq MHz	Sensitivity dBm
43,45	650.1	-112 (-112)
	653	-115 (-115)
	655.9	-113 (-113)

Table 6-24. Lectrosonics UM 700 on Channel 44

DTV Channels	Microphone Freq MHz	Sensitivity dBm
42,43	650.1	-111 (-112)
	653	-115 (-115)
	655.9	-112 (-113)

Table 6-25. Lectrosonics UM 700 on Channel 44

DTV Channels	Microphone Freq MHz	Sensitivity dBm
45,46	650.1	-112 (-112)
	653	-114 (-114)
	655.9	-112 (-113)

I2R (Test Mode: Lab; Scan Mode: All)

Table 6-26. Shure UR1 L3 (modulated) on Channel 44

DTV Channels	Microphone Freq MHz	Sensitivity dBm
43	650.05	*(-105)
45	653	-124 (-127)
	655.95	-94 (-94)

* False positive at all DTV power levels

Table 6-27. Shure UR1 H4 (Modulated)

DTV Channels	Microphone Freq MHz	Sensitivity dBm
30	565.95 (Ch 29)	-99
	572.05 (Ch 31)	-98

Table 6-28. Lectrosonics UM 700

DTV Channels	Microphone Freq MHz	Sensitivity dBm
30	565.9 (Ch 29)	-110 (-110)
	572.1 (Ch 31)	-108

6.1.5. Summary

All of the WSDs gave false positive indications of microphone detection with DTV signal levels as low as -68 dBm in adjacent channels. The Microsoft device also gave false negative indications with a DTV signal level of -28 dBm in adjacent channels and a microphone power of -80 dBm. The presence of DTV signals had little effect on the sensitivity of the devices to the Lectrosonics microphone but there was insufficient data to determine the effect of DTV signals on the sensitivity of the Microsoft device to the Shure microphone. The sensitivity of the Philips device to a modulated Shure microphone signal was not significantly affected by the presence of DTV signals. However, when the Shure microphone signal was unmodulated, the Philips device was significantly less sensitive to a microphone in the middle of the TV channel with or without DTV signals in adjacent channels.

6.2. Wireless Microphone Field Tests

6.2.1. Introduction

Field tests were conducted to evaluate the performance of the WSDs in detecting wireless microphones under real-world conditions. Arrangements were made with the National Football League (NFL) and the ESPN Network (ESPN) to perform tests before and during a pre-season football game at FedEx Field in Landover, MD and with the Majestic Theatre in New York City to perform tests before and during a performance of a Broadway play.

ESPN offered to simulate a game day broadcast at FedEx Field from 10:30 AM in the morning until 5:00 PM in the afternoon before the start of the game. During this time they would turn all their wireless microphones on or off upon request. The WSDs were operated and spectrum measurements were taken at four different locations at the venue: on the east (home field) side of the playing field from 10:30 AM to 12:00 PM, at the "tailgate" area outside the stadium from 12:40 PM to 2:00 PM, on the upper deck from 2:20 PM to 3:30 PM, and in the Press Box from 4:00 PM on. At 5:00 PM, ESPN was required to cease transmissions so the NFL could set up and test their wireless microphones for use during the game, which began at 7:00 PM. From 6:20 PM to 7:00 PM during pre-game activities and from 7:20 PM to 7:45 PM during the first quarter of the game, the WSDs were operated and spectrum measurements taken.

At the Majestic Theatre, WSD tests and spectrum measurements were taken at three different locations before the performance: on the sidewalk at the entrance to the theater, in the middle of the mezzanine and in the orchestra seating area. Measurements were also taken at one location during the performance (entrance ramp to the orchestra seating area). Before the performance, measurements were performed with the wireless microphones to be used during the show turned on and off upon request.

6.2.2. Test Procedure

At each location the Philips and I2R WSDs were set up with their receive antennas located approximately 2 meters from the wireless microphone base unit. The frequency spectrum was scanned by the WSDs from TV channels 21 to 51 with the wireless microphones turned on and off. The channels indicated as occupied by the WSD were recorded in each case. The frequency spectrum was also observed and recorded with a spectrum analyzer using a ground plane vertical antenna with a gain of approximately 0 dBi from channels 21 to 51.

6.2.3. Test Results

Tables 6-29 through 6-48 show the results of the WSD scans. "X" indicates the channels on which ESPN, the NFL or the Majestic Theatre were operating wireless microphones and the channels which the WSDs reported as occupied with the microphones off and on. The channels indicated as NFL channels are those on which it appears that there were NFL microphones based on the spectrum scans. The spectrum plots obtained with the spectrum analyzer under the same conditions are given in Appendix A for FedEx Field and Appendix B for the Majestic Theatre.

Additionally, during two of the field tests of the of the white space devices' ability to detect TV signals, brief tests were conducted to check the ability of the Philips and the I2R devices to detect wireless microphones. At the Portals location, it was found that the Philips device could detect the Shure UR1 and the Lectrosonics UM 700 microphones at distances up to 30.5 meters (100 feet) within the building with intervening walls. The I2R device reported detection of the Shure UR1 with less than 100% probability at 4.6 meters (15 feet) or less but could not detect the microphone at 30.5 meters (100 feet). The I2R device reported detection of the Lectrosonics unit with 100 % probability at distances up to 6.1 meters (20 feet) but with only 60% probability at distances of 30.5 meters (100 feet) and 61 meters (200 feet). At the #3 test site, a residential location, the Philips device could not be tested because it indicated that all channels were occupied. At this location, the I2R device could not detect the Shure UR2 even as close as 1.5 meters (5 feet) but it could detect the Lectrosonics UM 700 at distances up to 15.2 meters (50 feet) where the reported probability of detection decreased to 50%.

FedEx Field

X indicates occupied channel

Location: East (home team) side of field at the 50 yard line.

Time: 10:30 AM to 12:00 PM

Table 6-29. The Philips No. 2 WSD

Channel	**21**	**22**	**23**	**24**	**25**	**26**	**27**	**28**	**29**	**30**	**31**	**32**	**33**	**34**	**35**	**36**
ESPN	X		X				X				X		X	X		
Mic off	X	X	X	X	X	X	X	X	X		X	X	X	X	X	X
Mic on																X
Channel	**37**	**38**	**39**	**40**	**41**	**42**	**43**	**44**	**45**	**46**	**47**	**48**	**49**	**50**	**51**	
ESPN							X	X	X		X		X			
Mic off	X	X	X	X	X	X	X	X	X	X	X	X	X	X	X	
Mic on	X	X	X	X	X	X	X	X	X	X	X	X	X	X	X	

Table 6-30. The I2R WSD

Channel	**21**	**22**	**23**	**24**	**25**	**26**	**27**	**28**	**29**	**30**	**31**	**32**	**33**	**34**	**35**	**36**
ESPN	X		X				X				X		X	X		
Mic off							X	X	X	X		X	X	X		X
Mic on		X				X	X	X	X	X			X	X		X
Channel	**37**	**38**	**39**	**40**	**41**	**42**	**43**	**44**	**45**	**46**	**47**	**48**	**49**	**50**	**51**	
ESPN							X	X	X		X		X			
Mic off	X	X	X		X	X	X		X			X		X	X	
Mic on	X	X	X		X	X				X		X			X	

Location: Tailgate area

Time: 12:40 PM to 2:00 PM

Table 6-31. The Philips No. 2 WSD

Channel	**21**	**22**	**23**	**24**	**25**	**26**	**27**	**28**	**29**	**30**	**31**	**32**	**33**	**34**	**35**	**36**
ESPN	X		X				X				X		X	X		
Mic off	X	X	X	X	X	X	X	X	X	X	X	X	X	X	X	X
Mic on	X	X	X	X	X	X	X	X	X	X	X	X	X	X	X	X
Channel	**37**	**38**	**39**	**40**	**41**	**42**	**43**	**44**	**45**	**46**	**47**	**48**	**49**	**50**	**51**	
ESPN							X	X	X		X		X			
Mic off	X	X	X	X	X	X	X	X	X	X	X	X	X	X	X	
Mic on		X	X	X	X	X	X	X	X	X	X	X	X	X	X	

Table 6-32. The I2R WSD

Channel	**21**	**22**	**23**	**24**	**25**	**26**	**27**	**28**	**29**	**30**	**31**	**32**	**33**	**34**	**35**	**36**
ESPN	X		X				X				X		X	X		
Mic off		X						X	X	X		X	X			
Mic on							X	X	X	X			X			
Channel	**37**	**38**	**39**	**40**	**41**	**42**	**43**	**44**	**45**	**46**	**47**	**48**	**49**	**50**	**51**	
ESPN							X	X	X		X		X			
Mic off	X				X	X	X								X	
Mic on	X		X			X	X			X			X		X	

Location: Upper deck, Section 401
Time: 2:20 PM to 3:30 PM

Table 6-33. The Philips No. 2 WSD

Channel	**21**	**22**	**23**	**24**	**25**	**26**	**27**	**28**	**29**	**30**	**31**	**32**	**33**	**34**	**35**	**36**
ESPN	X		X				X				X		X	X		
Mic off	X	X	X	X	X	X	X	X	X	X	X	X	X	X	X	X
Mic on	X	X	X	X	X	X	X	X	X	X	X	X	X	X	X	X
Channel	**37**	**38**	**39**	**40**	**41**	**42**	**43**	**44**	**45**	**46**	**47**	**48**	**49**	**50**	**51**	
ESPN							X	X	X		X		X			
Mic off	X	X	X	X	X	X	X	X	X	X	X	X	X	X	X	
Mic on		X	X	X	X	X	X	X	X	X	X	X	X	X	X	

Table 6-34. The I2R WSD

Channel	**21**	**22**	**23**	**24**	**25**	**26**	**27**	**28**	**29**	**30**	**31**	**32**	**33**	**34**	**35**	**36**
ESPN	X		X				X				X		X	X		
Mic off			X						X	X					X	X
Mic on	X	X	X	X	X	X			X	X	X	X			X	X
Channel	**37**	**38**	**39**	**40**	**41**	**42**	**43**	**44**	**45**	**46**	**47**	**48**	**49**	**50**	**51**	
ESPN							X	X	X		X		X			
Mic off	X	X				X	X		X	X			X		X	
Mic on	X	X			X	X		X	X	X	X		X			

Location: Press box

Time: 4:00 PM to 5:00 PM

Table 6-35. The Philips No. 2 WSD

Channel	**21**	**22**	**23**	**24**	**25**	**26**	**27**	**28**	**29**	**30**	**31**	**32**	**33**	**34**	**35**	**36**
ESPN	X		X				X				X		X	X		
Mic off	X	X	X	X	X	X	X	X	X	X	X	X	X	X	X	X
Mic on	X	X	X	X	X	X	X	X	X	X	X	X	X	X	X	X
Channel	**37**	**38**	**39**	**40**	**41**	**42**	**43**	**44**	**45**	**46**	**47**	**48**	**49**	**50**	**51**	
ESPN							X	X	X		X		X			
Mic off	X	X	X	X	X	X	X	X	X	X	X	X	X	X	X	
Mic on	X	X	X	X	X	X	X	X	X	X	X	X	X	X	X	

Table 6-36. The I2R WSD

Channel	**21**	**22**	**23**	**24**	**25**	**26**	**27**	**28**	**29**	**30**	**31**	**32**	**33**	**34**	**35**	**36**
ESPN	X		X				X				X		X	X		
Mic off	X	X	X		X				X		X	X	X	X	X	X
Mic on	X	X	X		X				X		X			X		X
Channel	**37**	**38**	**39**	**40**	**41**	**42**	**43**	**44**	**45**	**46**	**47**	**48**	**49**	**50**	**51**	
ESPN							X	X	X		X		X			
Mic off		X	X		X	X	X	X	X		X	X	X	X		
Mic on		X	X		X	X		X	X		X	X	X	X	X	

Location: Press box—Pre-game
Time: 6:20 PM to 7:00 PM

Table 6-37. The Philips No. 2 WSD

Channel	**21**	**22**	**23**	**24**	**25**	**26**	**27**	**28**	**29**	**30**	**31**	**32**	**33**	**34**	**35**	**36**
NFL	X		X		X						X					
Mic on	X	X	X	X	X	X	X	X	X	X	X	X	X	X	X	X
Channel	**37**	**38**	**39**	**40**	**41**	**42**	**43**	**44**	**45**	**46**	**47**	**48**	**49**	**50**	**51**	
NFL					X			X	X		X		X			
Mic on	X	X	X	X	X	X	X	X	X	X	X	X	X	X	X	

Table 6-38. I2R

Channel	**21**	**22**	**23**	**24**	**25**	**26**	**27**	**28**	**29**	**30**	**31**	**32**	**33**	**34**	**35**	**36**
NFL	X		X		X						X					
Mic on	X	X	X		X				X	X	X		X			X
Channel	**37**	**38**	**39**	**40**	**41**	**42**	**43**	**44**	**45**	**46**	**47**	**48**	**49**	**50**	**51**	
NFL					X			X	X		X		X			
Mic on		X	X	X	X	X	X	X	X		X	X	X		X	

Location: Press box—First Quarter
Time: 7:20 PM to 7:45 PM

Table 6-39. The Philips No. 2 WSD

Channel	**21**	**22**	**23**	**24**	**25**	**26**	**27**	**28**	**29**	**30**	**31**	**32**	**33**	**34**	**35**	**36**
NFL	X		X		X						X					
Mic on	X	X	X	X	X	X	X	X	X	X	X	X	X	X	X	X
Channel	**37**	**38**	**39**	**40**	**41**	**42**	**43**	**44**	**45**	**46**	**47**	**48**	**49**	**50**	**51**	
NFL					X			X	X		X		X			
Mic on	X	X	X	X	X	X	X	X	X	X	X	X	X	X	X	

Table 6-40. The I2R WSD

Channel	**21**	**22**	**23**	**24**	**25**	**26**	**27**	**28**	**29**	**30**	**31**	**32**	**33**	**34**	**35**	**36**
NFL	X		X		X						X					
Mic on	X	X	X	X	X	X	X		X	X	X		X	X		X
Channel	**37**	**38**	**39**	**40**	**41**	**42**	**43**	**44**	**45**	**46**	**47**	**48**	**49**	**50**	**51**	
NFL					X			X	X		X		X			
Mic on		X	X	X	X		X	X	X		X	X	X			

Majestic Theatre

X indicates occupied channel

Location: Sidewalk at entrance to theater

Time: 10:00 AM to 12:00 PM

Table 6-41. The Philips No. 2 WSD

Channel	**21**	**22**	**23**	**24**	**25**	**26**	**27**	**28**	**29**	**30**	**31**	**32**	**33**	**34**	**35**	**36**
Theater			X			X									X	
Mic off	X	X	X	X	X	X	X	X	X	X	X	X	X	X	X	X
Mic on	X	X	X	X	X	X	X	X	X	X	X	X	X	X	X	X
Channel	**37**	**38**	**39**	**40**	**41**	**42**	**43**	**44**	**45**	**46**	**47**	**48**	**49**	**50**	**51**	
Theater			X							X		X				
Mic off	X	X	X	X	X	X	X	X	X	X	X	X	X	X	X	
Mic on	X	X	X	X	X	X	X	X	X	X	X	X	X	X	X	

Table 6-42. The I2R WSD

Channel	**21**	**22**	**23**	**24**	**25**	**26**	**27**	**28**	**29**	**30**	**31**	**32**	**33**	**34**	**35**	**36**
Theater			X			X									X	
Mic off		X	X			X	X	X	X	X		X	X			X
Mic on	X	X		X				X	X	X			X			X
Channel	**37**	**38**	**39**	**40**	**41**	**42**	**43**	**44**	**45**	**46**	**47**	**48**	**49**	**50**	**51**	
Theater			X							X		X				
Mic off	X	X		X				X	X			X			X	
Mic on		X		X	X			X	X		X			X	X	

Location: Middle of Mezzanine

Time: 12:00 PM to 2:00 PM

Table 6-43. The Philips No. 2 WSD

Channel	**21**	**22**	**23**	**24**	**25**	**26**	**27**	**28**	**29**	**30**	**31**	**32**	**33**	**34**	**35**	**36**
Theater			X			X									X	
Mic off	X	X	X	X	X	X	X	X	X	X	X	X	X	X	X	X
Mic on	X	X	X	X	X	X	X	X	X	X	X	X	X	X	X	X
Channel	**37**	**38**	**39**	**40**	**41**	**42**	**43**	**44**	**45**	**46**	**47**	**48**	**49**	**50**	**51**	
Theater			X							X		X				
Mic off		X	X	X	X	X	X	X	X	X	X	X	X	X	X	
Mic on	X	X	X	X	X	X	X	X	X	X	X	X	X	X	X	

Table 6-44. The I2R WSD

Channel	**21**	**22**	**23**	**24**	**25**	**26**	**27**	**28**	**29**	**30**	**31**	**32**	**33**	**34**	**35**	**36**
Theater			X			X									X	
Mic off				X				X	X	X			X			X
Mic on				X				X	X	X						X
Channel	**37**	**38**	**39**	**40**	**41**	**42**	**43**	**44**	**45**	**46**	**47**	**48**	**49**	**50**	**51**	
Theater			X							X		X				
Mic off		X		X				X	X		X				X	
Mic on		X		X				X	X						X	

Location: Orchestra, stage right, row I

Time: 2:30 PM to 4:00 PM

Table 6-45. The Philips No. 2 WSD

Channel	**21**	**22**	**23**	**24**	**25**	**26**	**27**	**28**	**29**	**30**	**31**	**32**	**33**	**34**	**35**	**36**
Theater			X			X									X	
Mic off	X	X	X	X	X	X	X	X	X	X	X		X	X	X	X
Mic on	X	X	X	X	X	X	X	X	X	X	X	X	X	X	X	X
Channel	**37**	**38**	**39**	**40**	**41**	**42**	**43**	**44**	**45**	**46**	**47**	**48**	**49**	**50**	**51**	
Theater			X							X		X				
Mic off		X	X	X	X	X	X	X	X	X	X	X	X	X	X	
Mic on	X	X	X	X	X	X	X	X	X	X	X	X	X	X	X	

Table 6-46. The I2R WSD

Channel	**21**	**22**	**23**	**24**	**25**	**26**	**27**	**28**	**29**	**30**	**31**	**32**	**33**	**34**	**35**	**36**
Theater			X			X									X	
Mic off				X				X	X	X	X		X			X
Mic on				X				X	X	X			X			X
Channel	**37**	**38**	**39**	**40**	**41**	**42**	**43**	**44**	**45**	**46**	**47**	**48**	**49**	**50**	**51**	
Theater			X							X		X				
Mic off		X		X				X	X		X				X	
Mic on		X		X				X	X		X			X	X	

Location: Ramp from lobby to orchestra, stage left

Time: 7:00 PM to 9:00 PM

Table 6-47. The Philips No. 2 WSD

Channel	**21**	**22**	**23**	**24**	**25**	**26**	**27**	**28**	**29**	**30**	**31**	**32**	**33**	**34**	**35**	**36**
Theater			X			X									X	
Mic on	X	X	X	X	X	X	X	X	X	X	X	X	X	X	X	X
Channel	**37**	**38**	**39**	**40**	**41**	**42**	**43**	**44**	**45**	**46**	**47**	**48**	**49**	**50**	**51**	
Theater			X							X		X				
Mic on	X	X	X	X	X	X	X	X	X	X	X	X	X	X	X	

6.2.4. Summary

Wireless microphone sensing tests were performed with the I2R and Philips devices at 2 locations. The tests were conducted first with microphones off, and then turned on, in pre-determined channels to determine if the devices could sense the presence of wireless microphones. At both sites and all the test locations, the Philips device reported all the channels on which the microphones were designated to transmit as occupied whether the microphone was transmitting or not. The I2R device indicated several channels as available even when the microphones were on.

Table 6-48. The I2R WSD

Channel	**21**	**22**	**23**	**24**	**25**	**26**	**27**	**28**	**29**	**30**	**31**	**32**	**33**	**34**	**35**	**36**
Theater			X			X									X	
Mic on				X				X	X	X			X			X
Channel	**37**	**38**	**39**	**40**	**41**	**42**	**43**	**44**	**45**	**46**	**47**	**48**	**49**	**50**	**51**	
Theater			X							X		X				
Mic on		X		X				X	X						X	

End Notes

[1] S. Jones and T. Philips , "Initial Evaluation of the Performance of Prototype TV-Band White Space Devices,", OET Report FCC/OET 07-TR-1 006, July 31, 2007.

[2] FCC Public Notice DA 07-4179, "Office of Engineering and Technology Announces Additional Testing of TV White Space Devices, ET Docket No. 04-186," October 5, 2007.

[3] FCC Public Notice DA 08-118, "Office of Engineering and Technology Announces Plans for Conducting Measurements of Additional Prototype TV White Space Devices, ET Docket No. 04-186," January 17, 2008.

[4] The Microsoft device also had a transmitting capability but that device was withdrawn from the test program before the transmitter tests were conducted.

[5] *First Report and Order and Further Notice of Proposed Rule Making* in the Matter of Unlicensed Operation in the TV Broadcast Bands, ET Docket No. 04-186 and 02-380, October 18, 2006 (hereinafter *FNPRM*). While the focus of this proceeding is on unlicensed operation, the Commission has also requested comment on issues relevant to whether TV band low power devices should be allowed on a licensed or hybrid licensed/unlicensed basis. It also requested comment as to whether, if unused TV spectrum were made available on a licensed basis, licensed devices should be required to incorporate the same type of interference avoidance mechanisms and be subject to the same low power limits that it proposed for unlicensed devices.

[6] *Notice of Proposed Rulemaking in ET Docket Nos. 02-380 and 04-186, 19 FCC Rcd 10018 (2004).*

[7] *Id.*

[8] OET Report FCC/OET 07-TR-1006, *Initial Evaluation of the Performance of Prototype TV-Band White Space Devices*, S. Jones and T. Philips July 31, 2007 (Phase I measurement report).

[9] FCC Public Notice, DA 07-4179, "The Office of Engineering and Technology Announces Additional Testing of TV White Space Devices, DA 07-4179, October 5, 2007.

[10] FCC Public Notice DA 08-118, "Office of Engineering and Technology Announces Plans for Conducting Measurements of Additional Prototype TV White Space Devices, ET Docket No. 04-186," January 17, 2008.

[11] Microsoft submitted two units of the same device.

[12] Adaptrum, Inc *ex-parte* filing of December 5, 2007.

[13] I2R *ex-parte* filing of June 23, 2008.

[14] Microsoft *ex-parte* filings of December 11, 2007 and January 15, 2008.

[15] Motorola *ex-parte* filings of November 16 and December 6, 2007.

[16] Philips ex-parte filing of December 10, 2007.

[17] The Adaptrum device was found to be particularly prone to false detections on channel 36, likely because of the presence of an OTA signal on the channel. Thus, channel 37 was used as the middle test channel for this device.

[18] See *ex parte* submission from Qualcomm, dated August 24, 2007.

[19] See *ex parte* filing from Adaptrum, dated June 2, 2008.

[20] For some devices, an additional measurement was performed utilizing more than 30 trials (see discussion in previous text).

[21] The Adaptrum device was found to be particularly prone to false detections on channel 36, likely because of the presence of an OTA signal on the channel. Thus, channel 37 was used as the middle test channel for this device.

[22] Filter specifications available at:
http://www.mcibroadcast.com/files/42100%20Interdigital%20Band%20Pass%20Filters.pdf.

[23] IEEE P163 1TM/D2, *Draft Recommended Practice for Measurement of 8- VSB Digital Television Mask Compliance for the USA*, RF Standards Committee G-2.2 of the IEEE Broadcast Technology Society, 2007.

[24] See *ex parte* filing of Motorola, dated May 15, 2008.

[25] Phase I measurement report.

[26] RF absorbing material was placed over the conductive floor of the chamber so that contributions from reflected waves could be ignored.

[27] Martin, Stephen R., Direct-Pickup Interference Tests of Three Consumer Digital Cable Television Receivers Available in 2005, OET Report FCC/OET 07-TR-1005, July 31, 2007.

[28] Consumer Electronics Association, "CEA Standard: Cable Television Channel Identification Plan", CEA-542-B, July, 2003.

[29] Society of Cable Telecommunications Engineers, "*Digital Cable Network Interface Standard*", ANSI/SCTE 40-2004, p.1, 17.

[30] EIA Standard 542B, Channel 72 is 510–516 MHz.

[31] –48 dBm + 48.75 = +0.75 dBmV for a 75Ω system.

[32] The set top box fed HDMI the converter box and then the monitor.

[33] -56 + 48.75 = -7.25 dBmV for a 75Ω system.

[34] See *ex parte* filing from Adaptrum, dated June 2, 2008.

[35] The modified Adaptrum WSD required approximately 2-hours to complete a full scan of channels 21-51 for both digital and analog TV signals.

[36] See *ex parte* filing from Motorola, dated May 15, 2008.

[37] It is noted that the test receiving system used herein varies from both the analog and digital TV planning factors by approximately 4.3 dB (the receive antenna used in the field tests provides 7 dBd of gain rather than the 10 dBd used in the planning factors and the line loss of the field test receive system was about 5.3 dB rather than 4 dB used in the planning factors). Therefore, the test system provided 4.3 dB less signal to the receiver than would a system adhering to the planning factors.

[38] In these cases it could not be determined with certainty that a signal of the opposite type was not present at levels below the measurement threshold but still within the detection range of the device.

[39] *See ex parte* filing by MSTV filed on July 2, 2008.

[40] *Ibid.*

[41] *Ibid.*

[42] Two FM microphones manufactured by Shure Incorporated and one digital microphone manufactured by Lectrosonics were loaned to the Commission for testing.

In: Tapping the Television White Space Spectrum: A Revolution...ISBN: 978-1-60692-956-8
Editor: Julie E. Hendriks

Chapter 2

INITIAL EVALUATION OF THE PERFORMANCE OF PROTOTYPE TV - BAND WHITE SPACE DEVICES (APPENDICES A AND B)

Steven K. Jones and Thomas W. Phillips

APPENDIX A - MEASUREMENT SYSTEMS

This appendix provides information regarding the measurement equipment used to perform this test/measurement program. Table A-1 provides a list of the equipment used, including the manufacturer and model number, a brief description and where appropriate, the most recent calibration date.

Calibration information resulting from measurements performed in the laboratory to determine the signal attenuation in various circuits used to support the test/measurements is also provided.

The following information is provided with respect to calibrations performed as a part of this program to determine the signal attenuation (or gain) associated with the various components used to support the test/measurements.

The input signal level used in the detection threshold was set at the input to the WSD scanner antenna input port and thus, no signal calibration was necessary.

A short length (2-feet) of RG-223/U coaxial cable was used to connect the WSD transmitter output port directly to the spectrum analyzer for performing the transmitter emissions measurements. The signal attenuation in this cable was measured and found to exhibit a maximum loss of 0.4 dB over the frequency range involved (5 12-698 MHz). The insertion loss associated with the channel-reject (notch) filter is 2.9 dB, including the additional coaxial cable used to integrate into the measurements system.

Table A-1. List of Measurement Equipment

Equipment	Quantity	Manufacturer and Model	Specifications	Last Calibrated
Spectrum Analyzer	1	Agilent E4448 PSA	3 Hz-50 GHz	06/09/06
Spectrum Analyzer	1	Agilent E4440 PSA	3 Hz – 26.5 GHz (w/ internal PA)	07/18/06
Spectrum Analyzer	1	Agilent E7405A EMC	9 kHz – 26.5 GHz	09/01/06
Modulation Analyzer	1	Agilent 8901A		02/27/06
Audio Analyzer	1	Agilent 8903B		09/25/03
Signal Generator	1	Agilent E4437B ESG	250 kHz-4 GHz	05/25/06
Signal Generator	1	Agilent 4438C ESG	250 kHz – 6 GHz	02/22/05
Signal Generator	1	Agilent 8640B	500 kHz – 1024 MHz	09/06
Broadcast Test System	2	Rhode and Schwarz SFU	w/ ATSC Signal Generator	-
Step Attenuator	1	Hewlett-Packard 355D	10-dB steps	-
Step Attenuator	1	Hewlett-Packard 355 C	1-dB steps	-
Step Attenuator	1	Agilent 8494B	0 to 11 dB DC – 18 GHz	11/09/04
Step Attenuator	1	Agilent 8495B	0 to 70 dB DC – 18 GHz	08/05/05
TV Channel Rejection Filter	1	TLE Model CE7569-N30	566-572 MHz	-
Signal Combiner	2	MiniCircuits ZFSC-2-1W	50 Ω	-
Log Periodic Antenna	1	A.R.A. LPB-2520/A	25 MHz – 2.0 GHz	-
Impedance Matching Transformer	1	Trilithic ZMT-57	75-50 Ω	-

The OTA interference test utilized a log periodic antenna as the test antenna and it was connected to the DTV receiver (or spectrum analyzer) with a 20-ft length of RG-55 coaxial cable. The signal attenuation in the cable run was measured and determined to be a maximum of 2.3 dB. In addition, there was a measured signal loss associated with the impedance-matching transformer of 0.9 dB and the antenna gain over the 5 12-698 MHz frequency range is 7.1-7.5 dBi.

APPENDIX B - SERVICE CONTOURS OF TV STATIONS AT FIELD TEST SITES

This appendix contains plots depicting the service contours of each full service and low power TV station on channels within the prototype WSD tuning range that have been authorized for use by a licensed television broadcaster relative to each site where field measurements were performed. The test location is represented on these plots by a red star. All broadcast licensees and construction permits within a 1 50-km (94-mile) of the test site were included in the contour plots. Red circles indicate DTV stations; blue circles indicate analog TV stations. There are a total of thirty plots provided for each test site, one for each channel in the WSD's tuning range.

Test Site 1

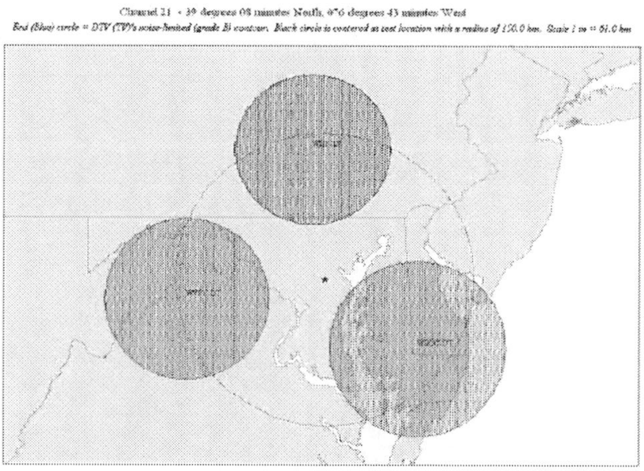

Figure 1. Channel 21 Coverage Contours Relative to Test Site

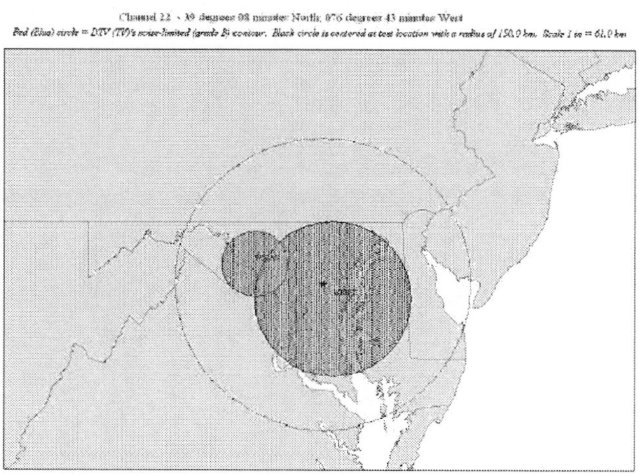

Figure 2. Channel 22 (WMPT-TV) Grade-B Contour Relative to Test Site 1

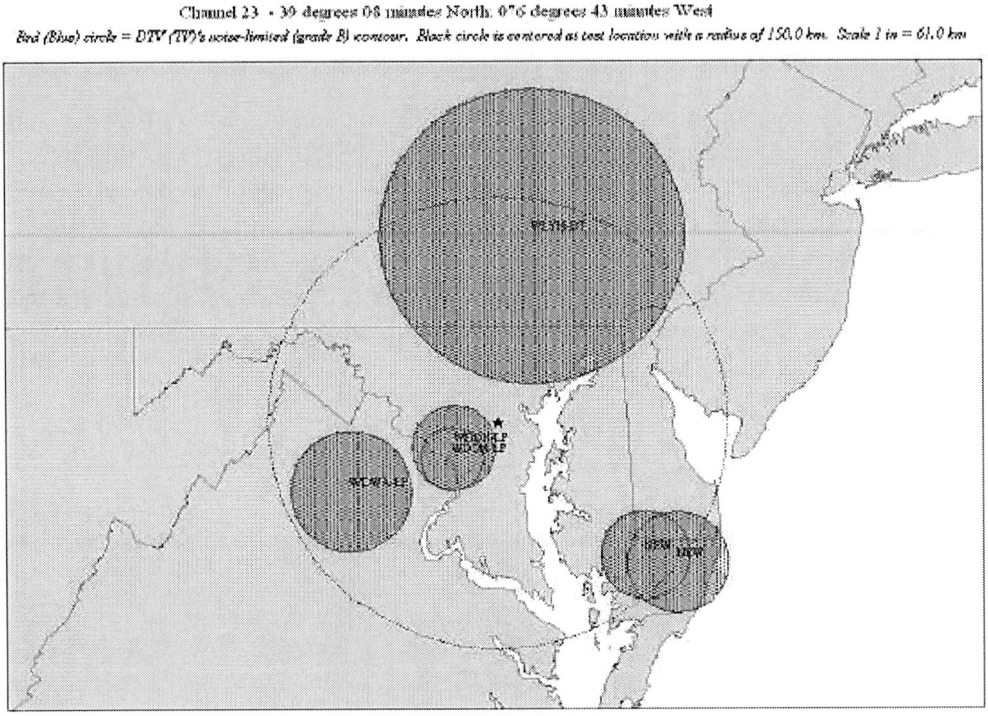

Figure 3. Channel 23 Coverage Contours Relative to Test Site 1

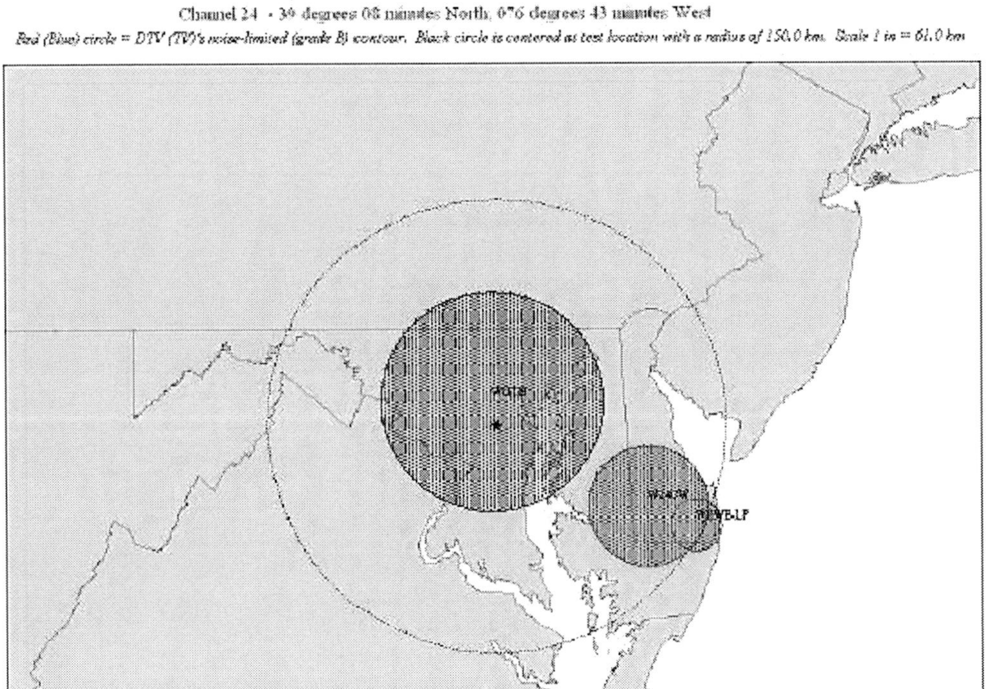

Figure 4. Channel 24 (WUTB-TV) Grade-B Contour Relative to Test Site 1

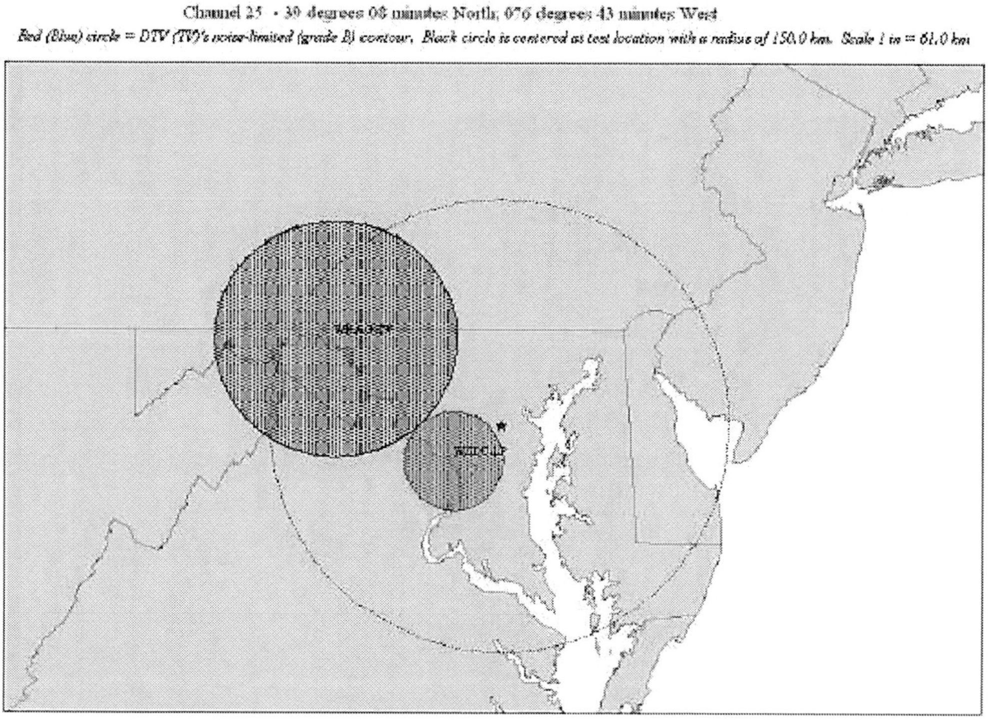

Figure 5. Channel 25 Coverage Contours Relative to Test Site 1

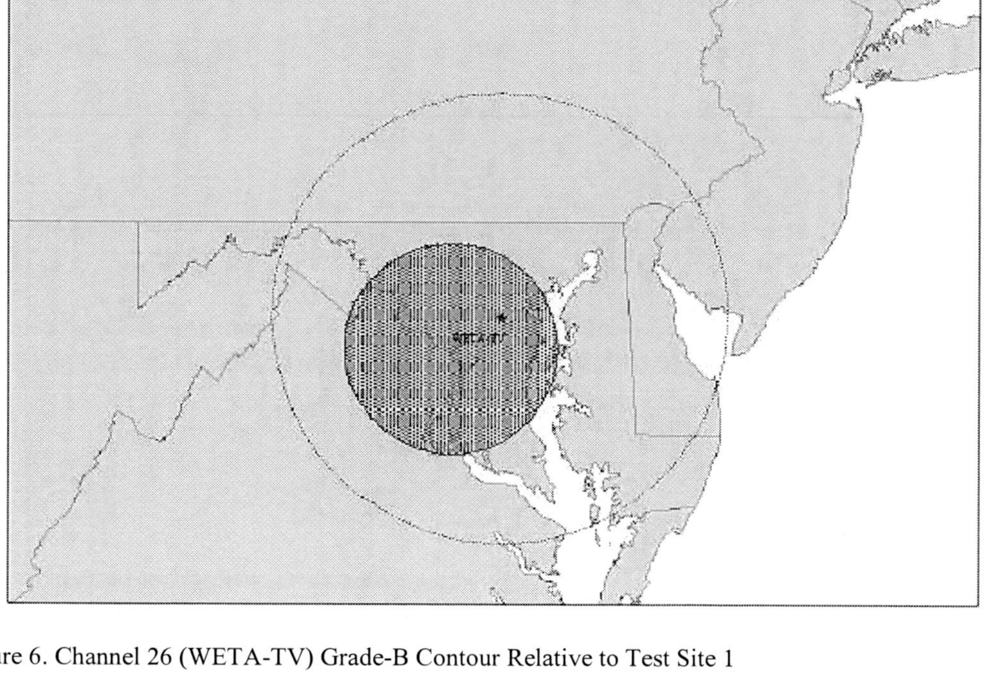

Figure 6. Channel 26 (WETA-TV) Grade-B Contour Relative to Test Site 1

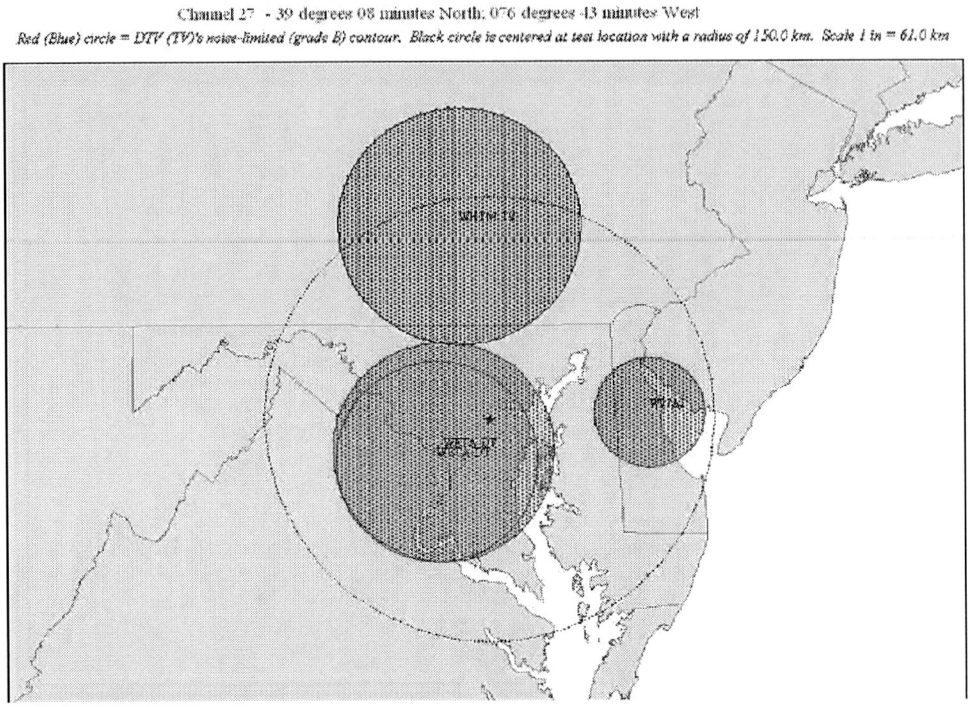

Figure 7. Channel 27 (WETA-DT) Coverage Contour Relative to Test Site 1

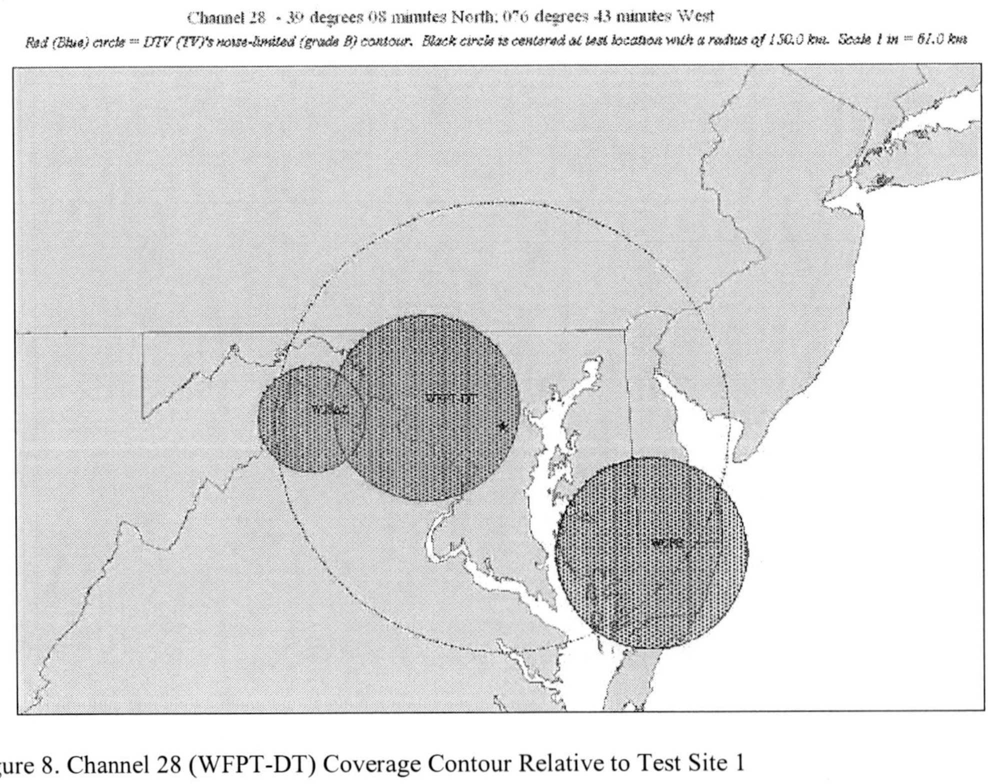

Figure 8. Channel 28 (WFPT-DT) Coverage Contour Relative to Test Site 1

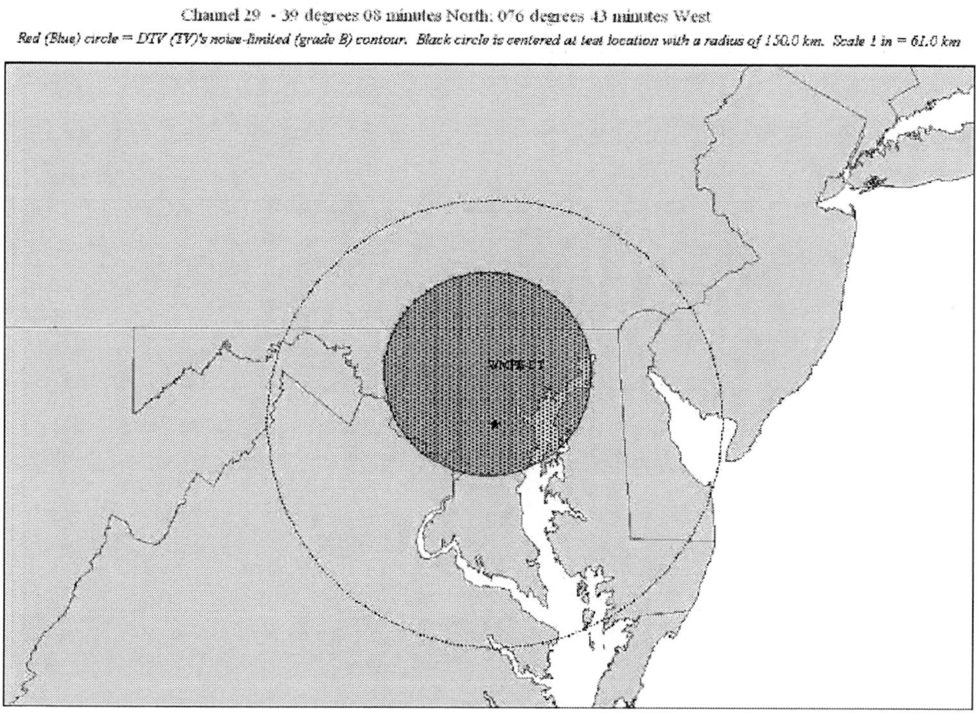

Figure 9. Channel 29 (WMPB-DT) Coverage Contour Relative to Test Site 1

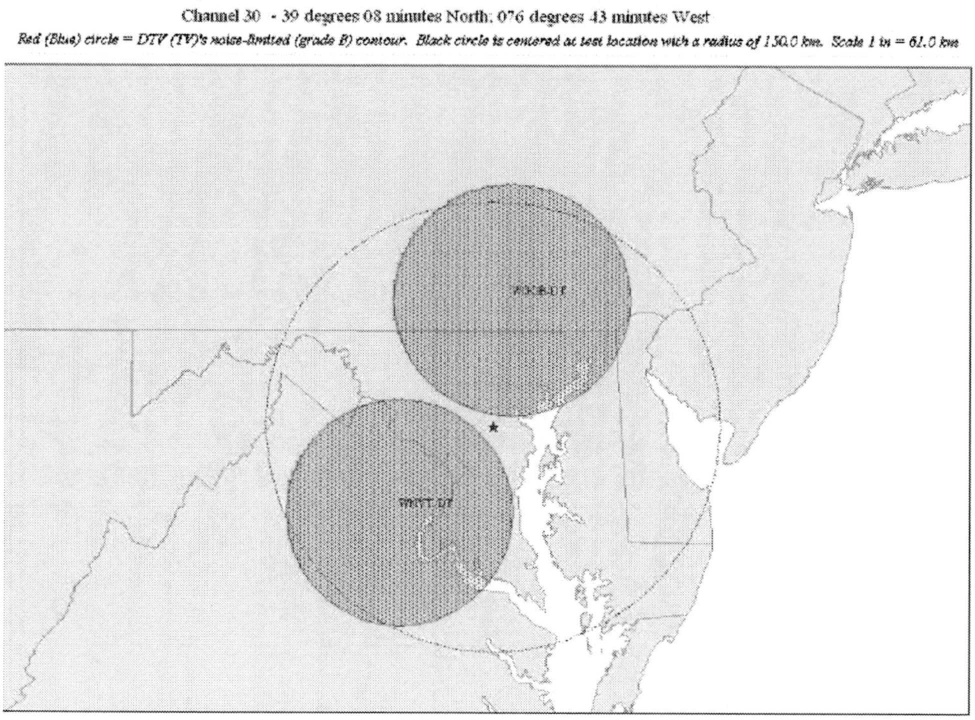

Figure 10. Channel 30 Coverage Contours Relative to Test Site 1

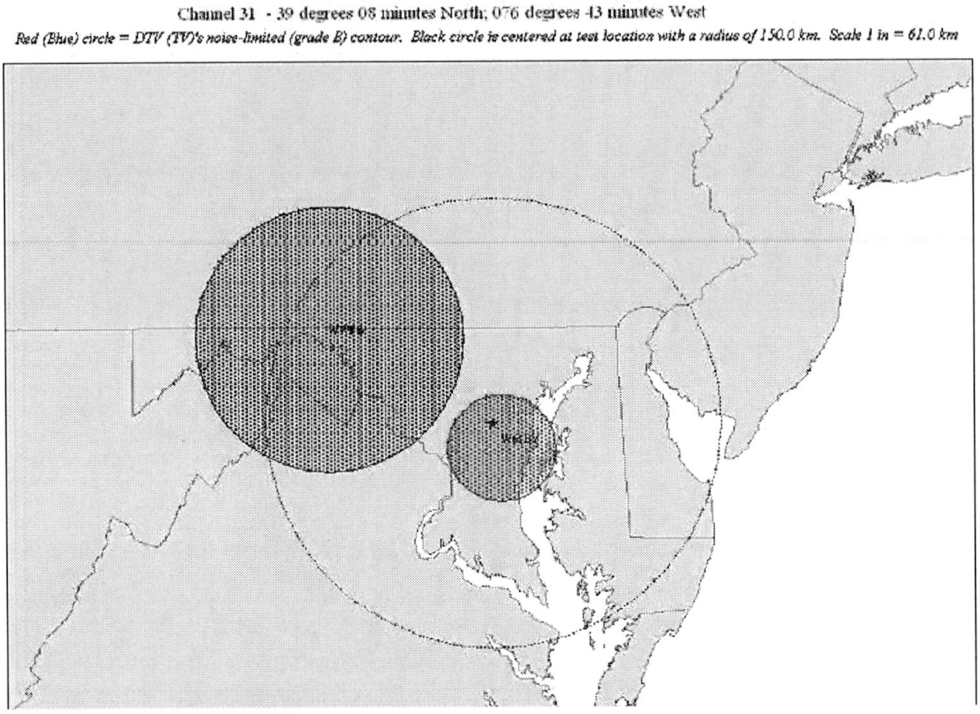

Figure 11. Channel 31 Coverage Contours Relative to Test Site 1

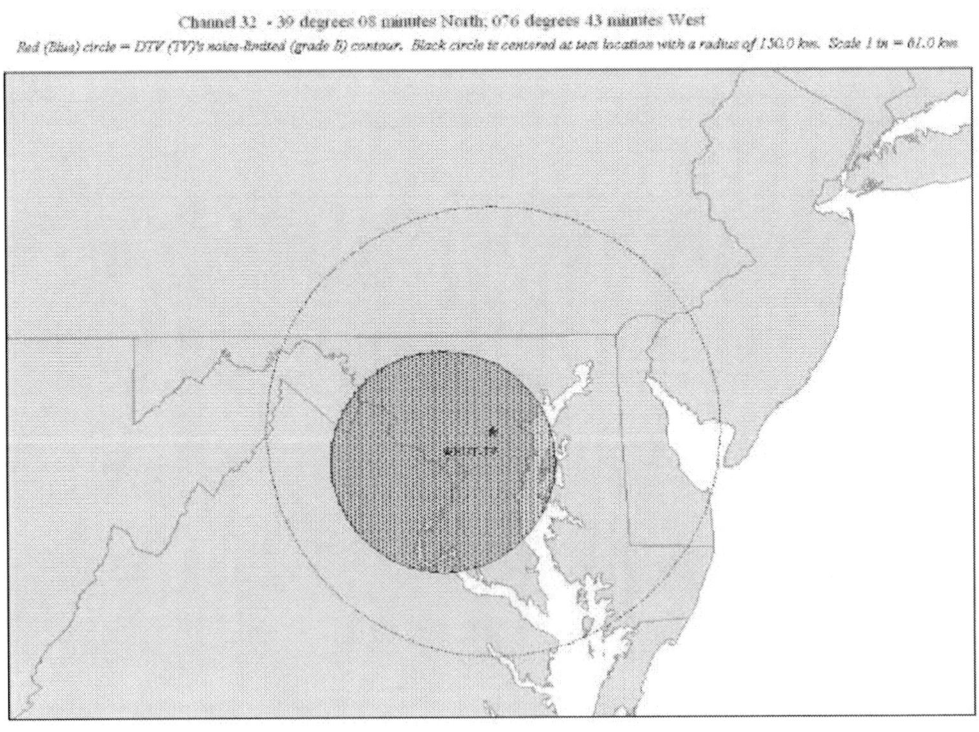

Figure 12. Channel 32 (WHUT-TV) Grade-B Contour Relative to Test Site 1

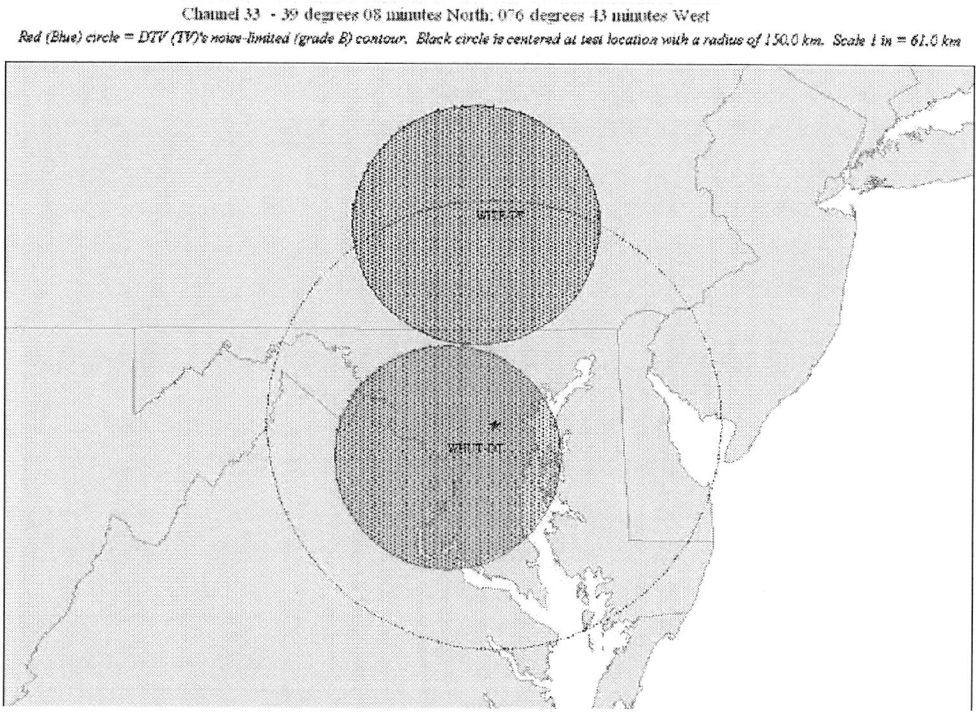

Figure 13. Channel 33 (WHUT-DT) Coverage Contour Relative to Test Site 1

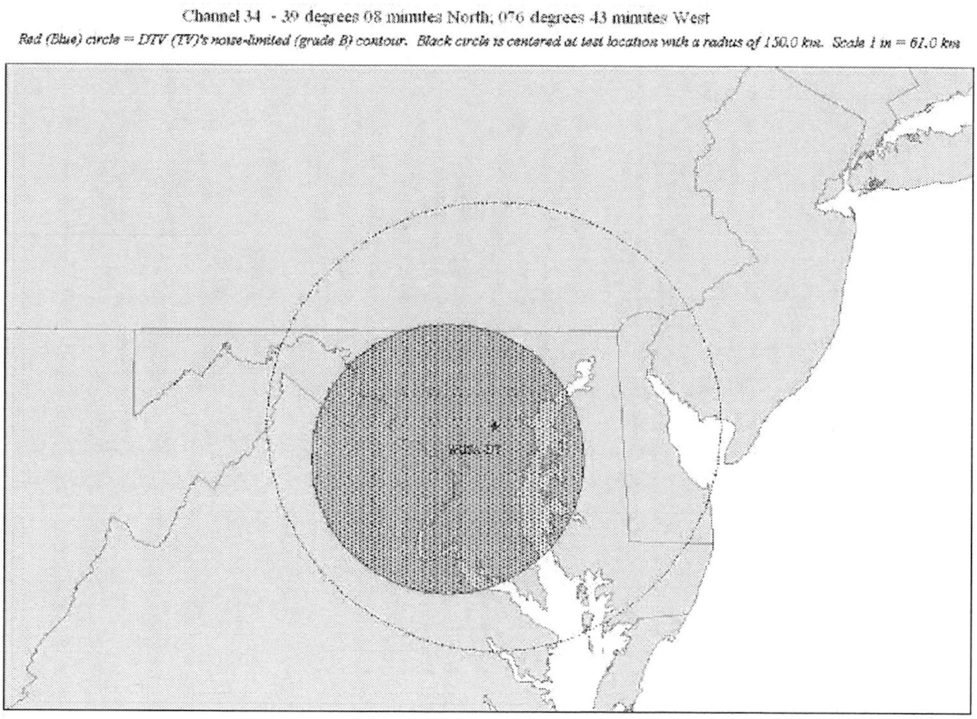

Figure 14. Channel 34 (WUSA-DT) Coverage Contour Relative to Test Site 1

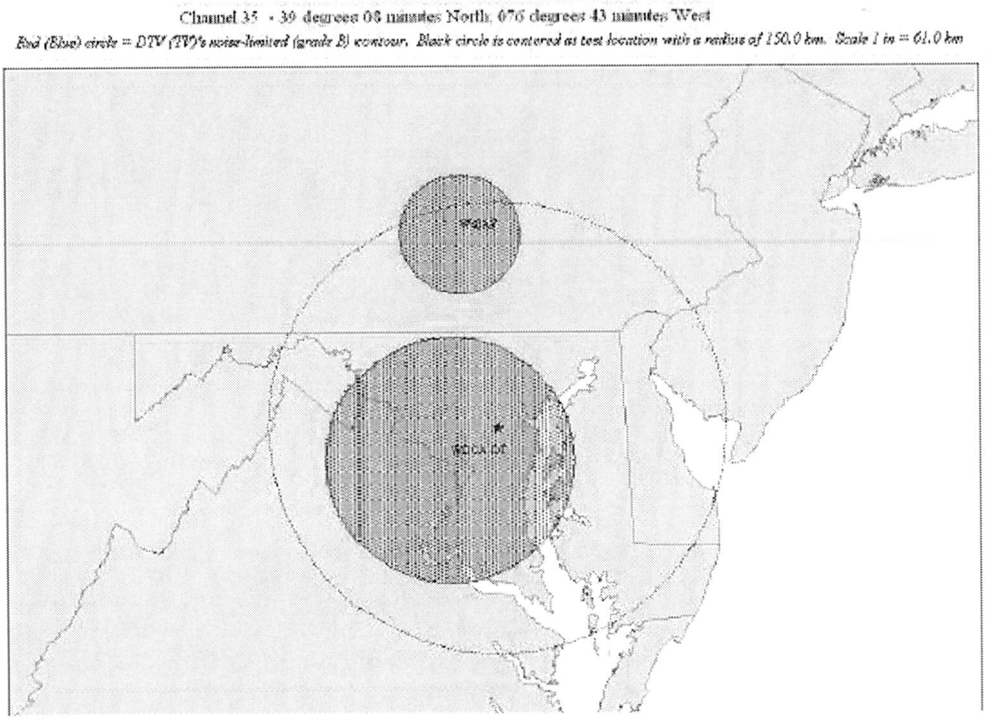

Figure 15. Channel 35 (WDCA-DT) Coverage Contour Relative to Test Site 1

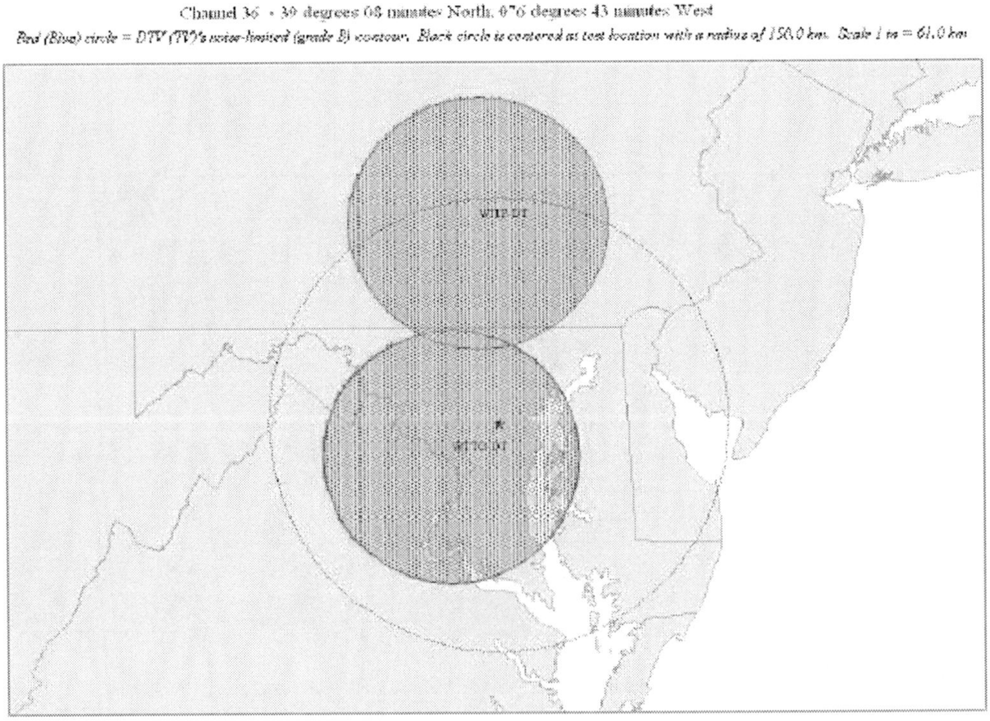

Figure 16. Channel 36 (WTTG-DT) Coverage Contour Relative to Test Site 1

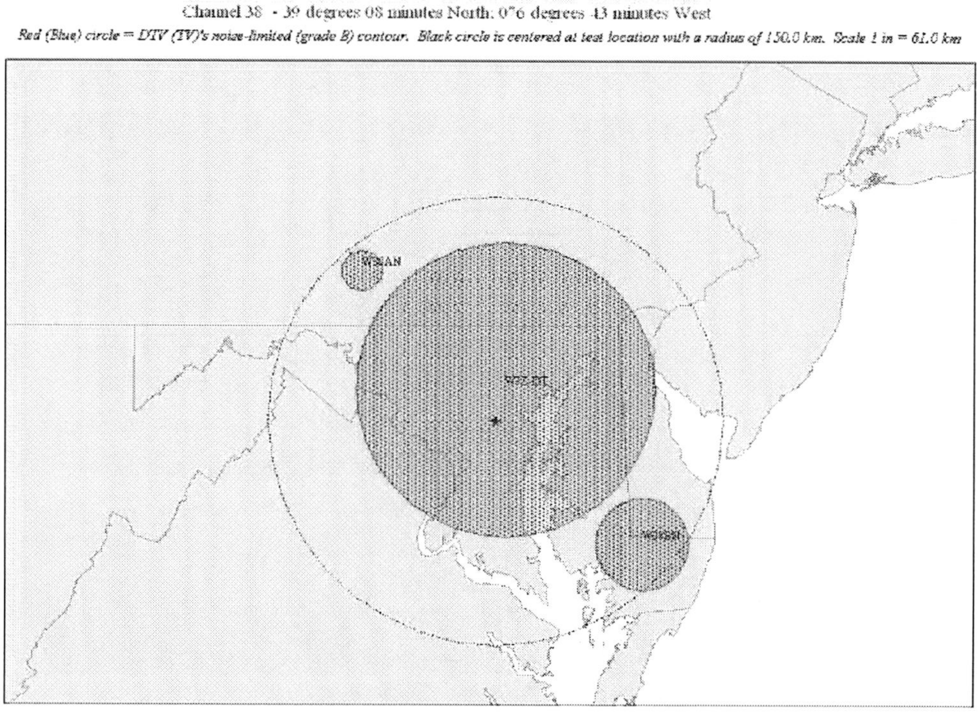

Figure 17. Channel 38 (WJZ-DT) Coverage Contour Relative to Test Site 1

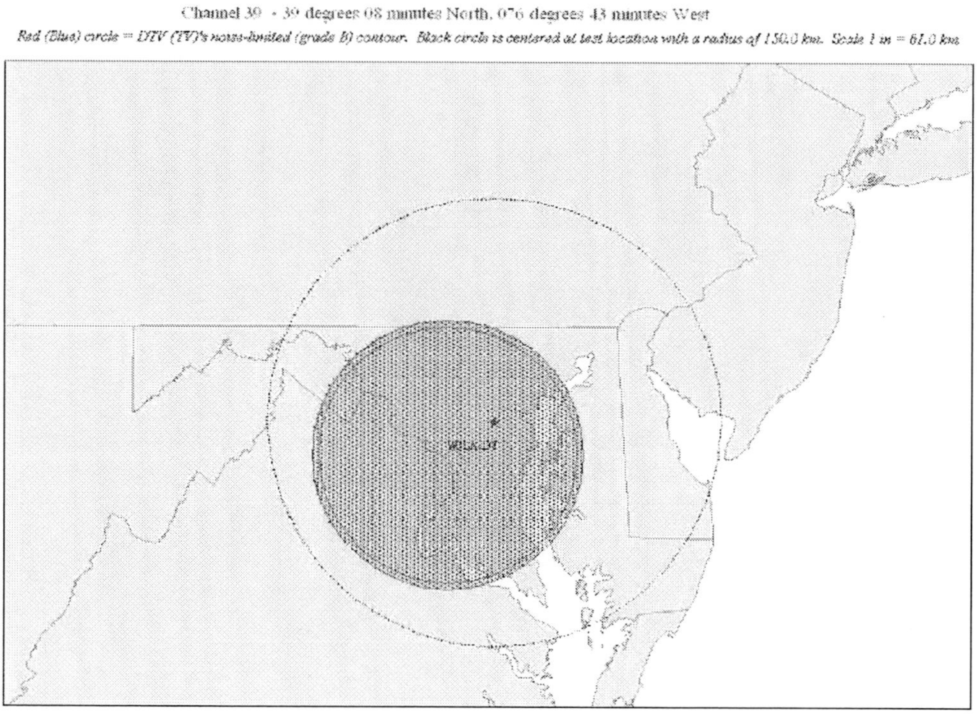

Figure 18. Channel 39 (WJLA-DT) Coverage Contour Relative to Test Site 1

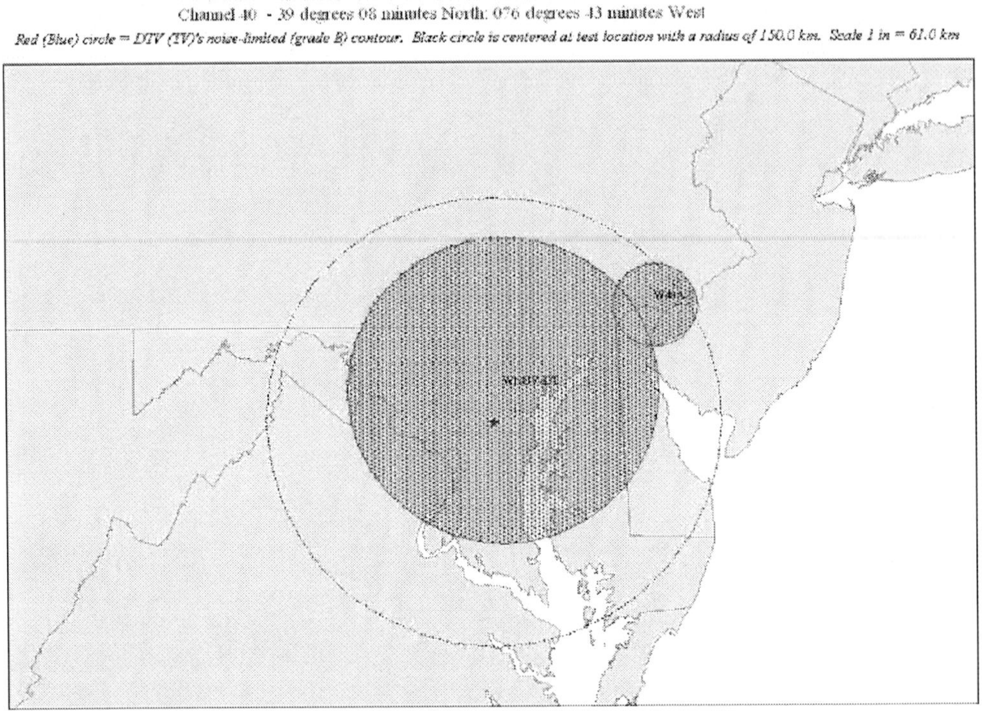

Figure 19. Channel 40 (WNUV-DT) Coverage Contour Relative to Test Site 1

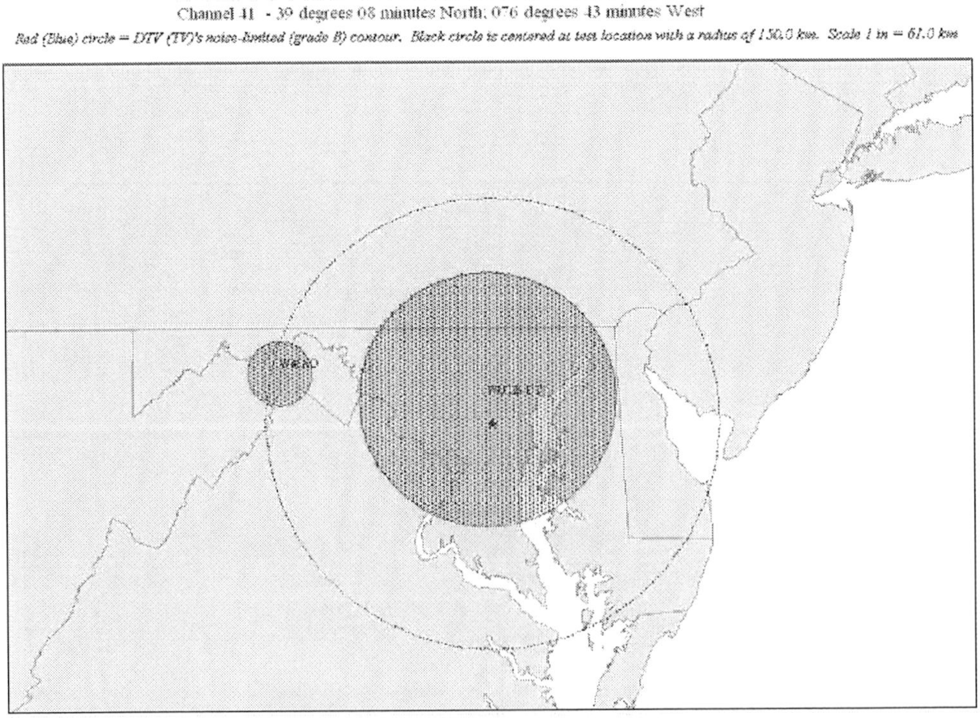

Figure 20. Channel 41 (WUTB-DT) Coverage Contour Relative to Test Site 1

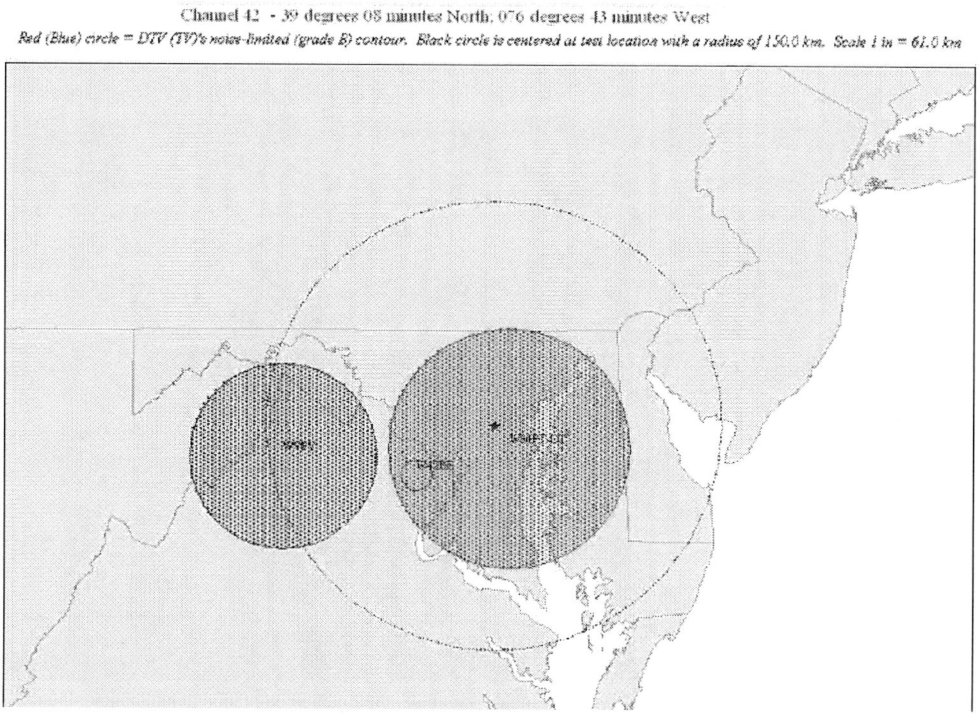

Figure 21. Channel 42 (WMPT-DT) Coverage Contour Relative to Test Site 1

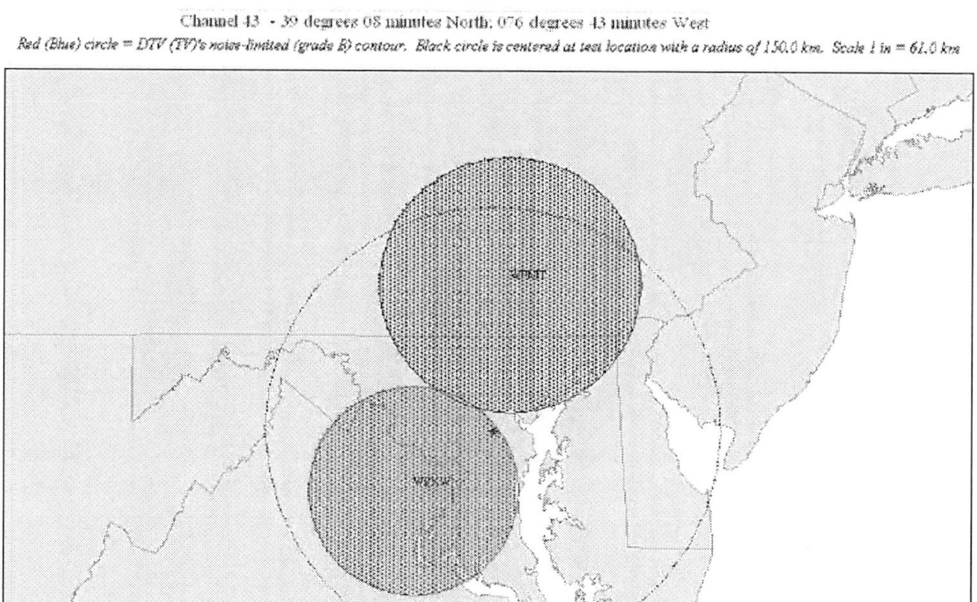

Figure 22. Channel 43 Coverage Contour Relative to Test Site 1

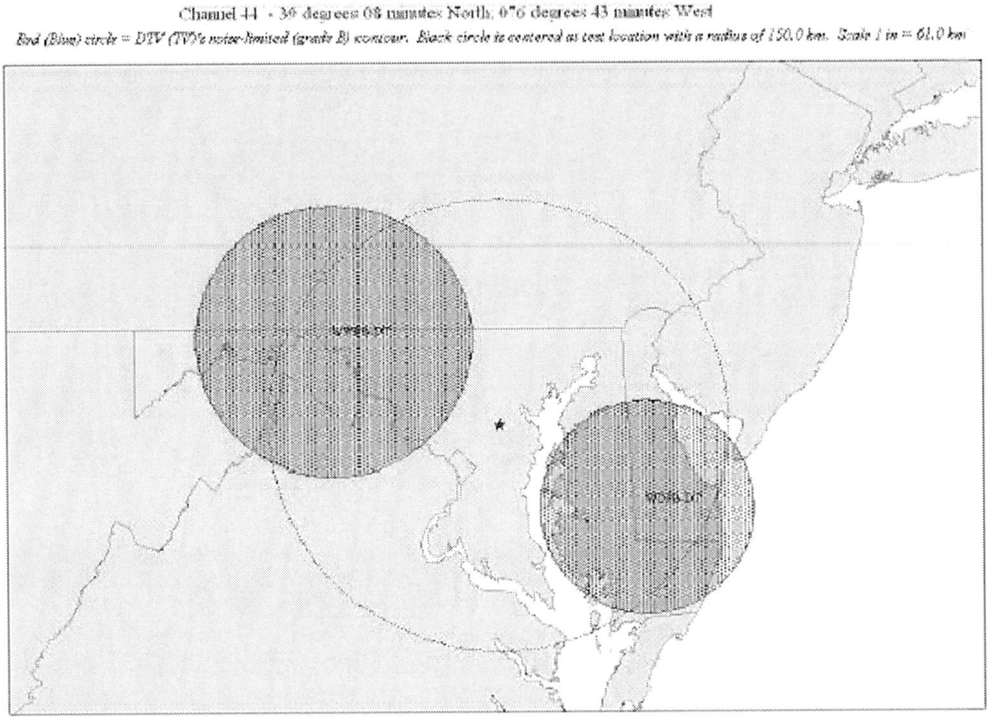

Figure 23. Channel 44 Coverage Contours Relative to Test Site 1

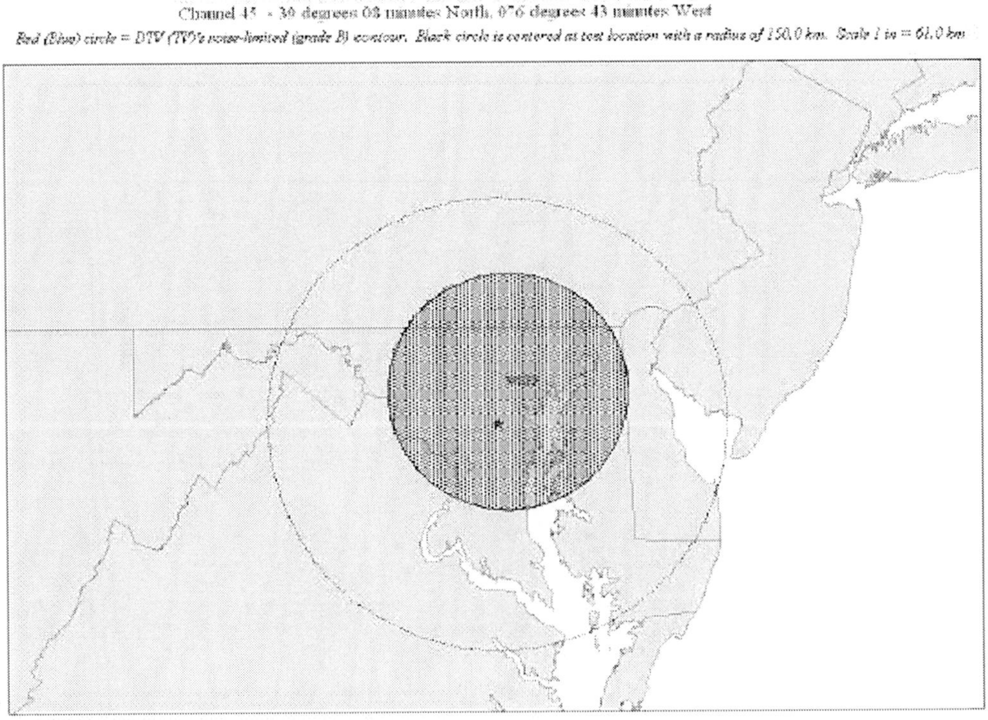

Figure 24. Channel 45 (WBFF-TV) Grade-B Contour Relative to Test Site 1

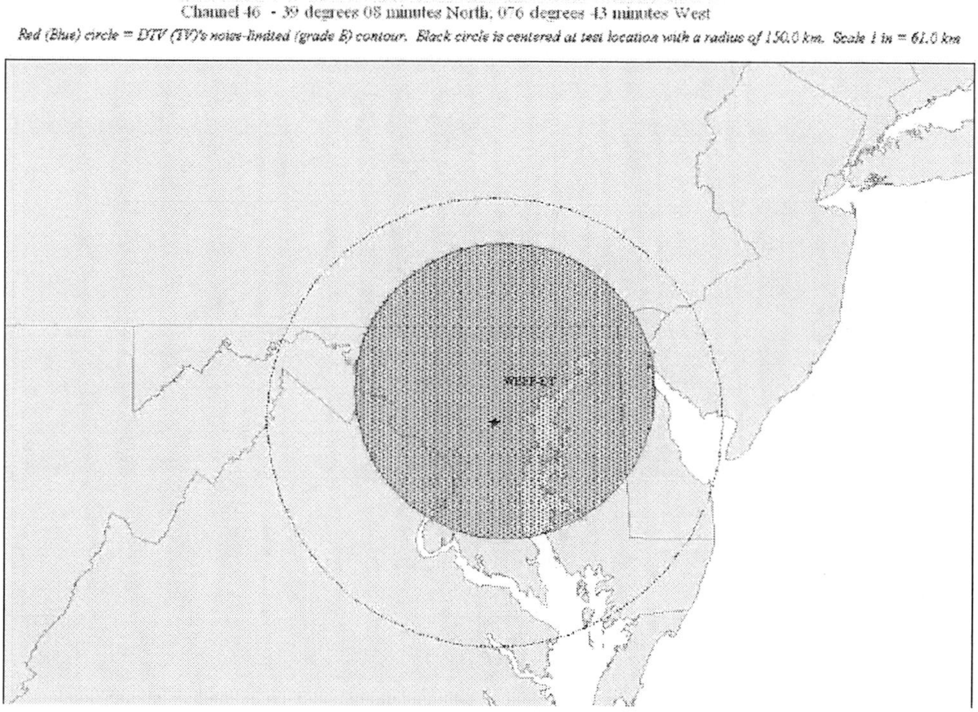

Figure 25. Channel 46 (WBFF-DT) Coverage Contour Relative to Test Site 1

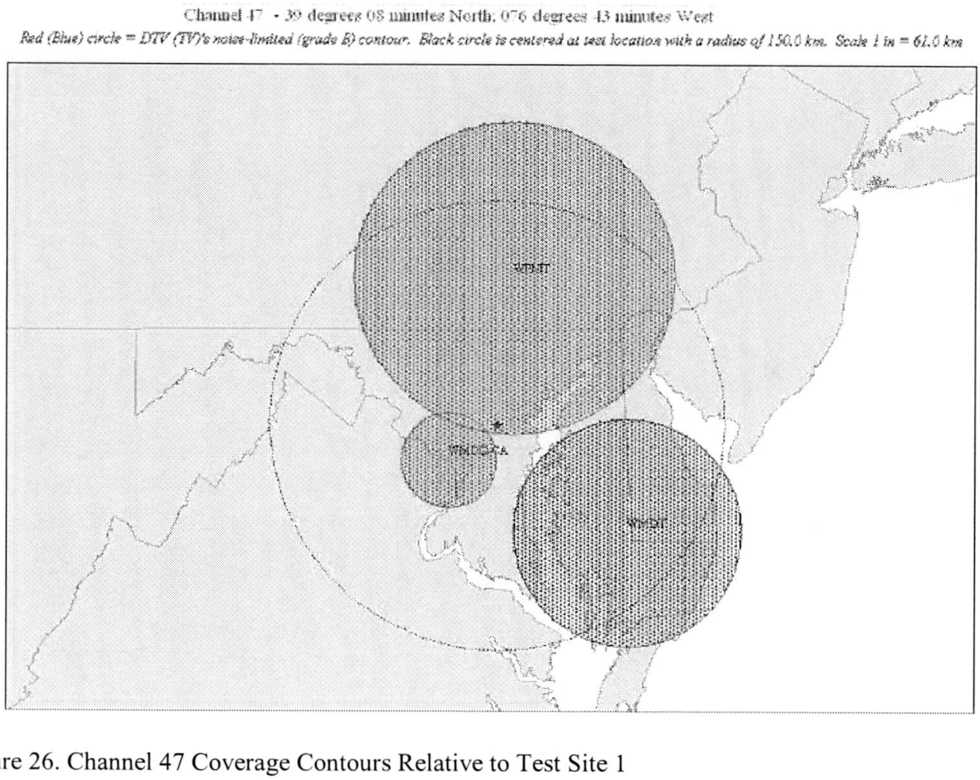

Figure 26. Channel 47 Coverage Contours Relative to Test Site 1

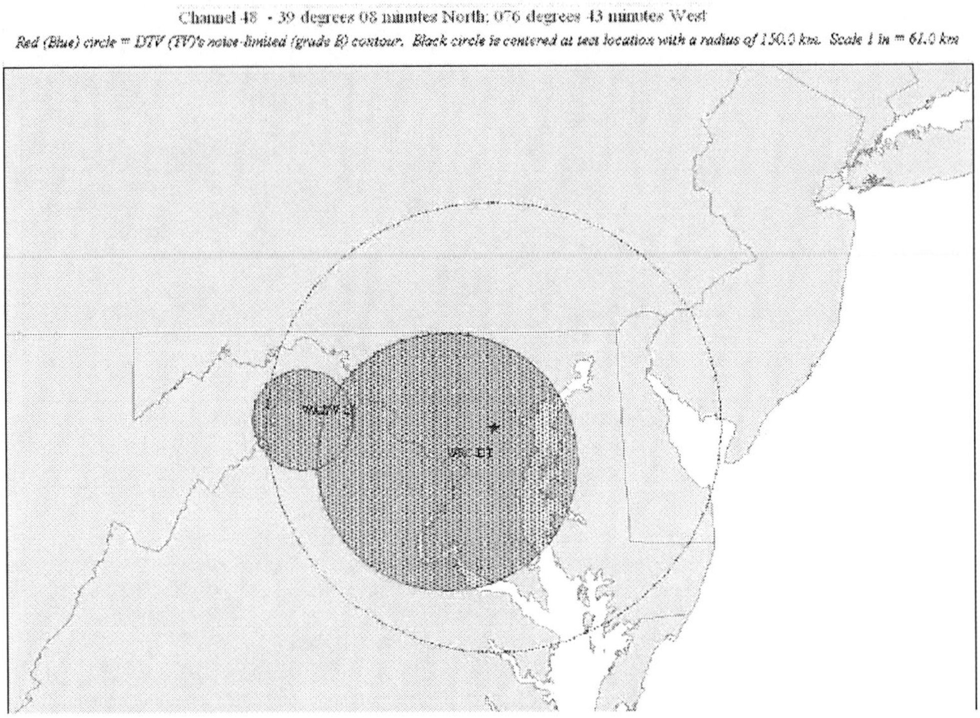

Figure 27. Channel 48 (WRC-DT) Coverage Contour Relative to Test Site 1

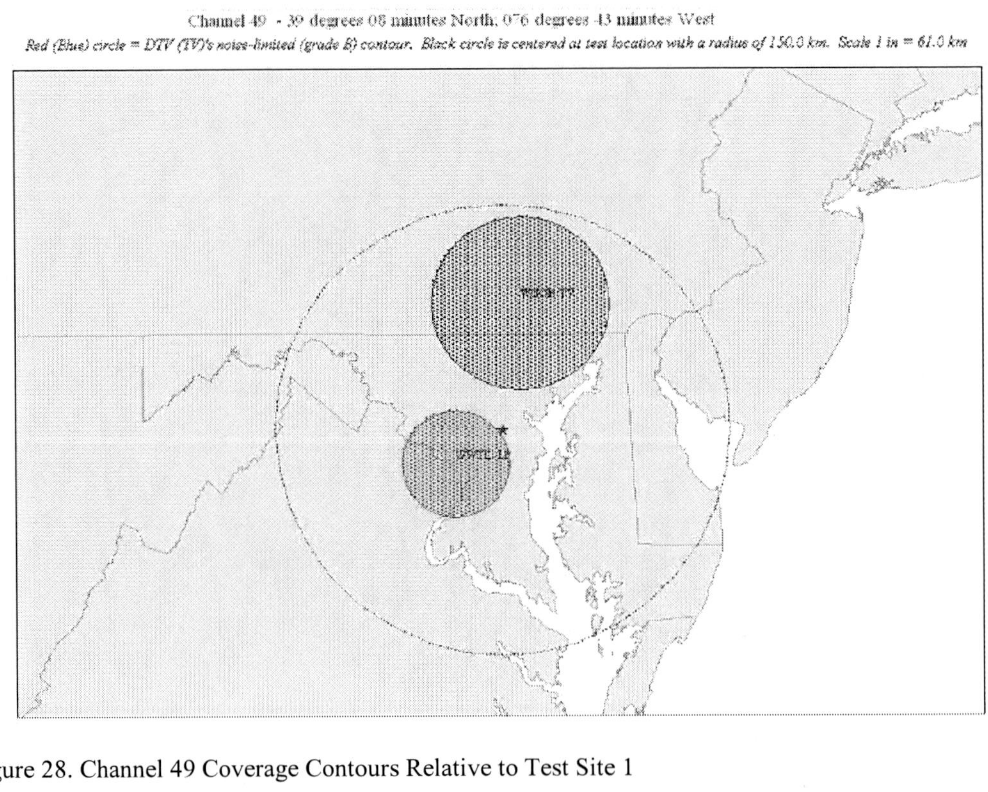

Figure 28. Channel 49 Coverage Contours Relative to Test Site 1

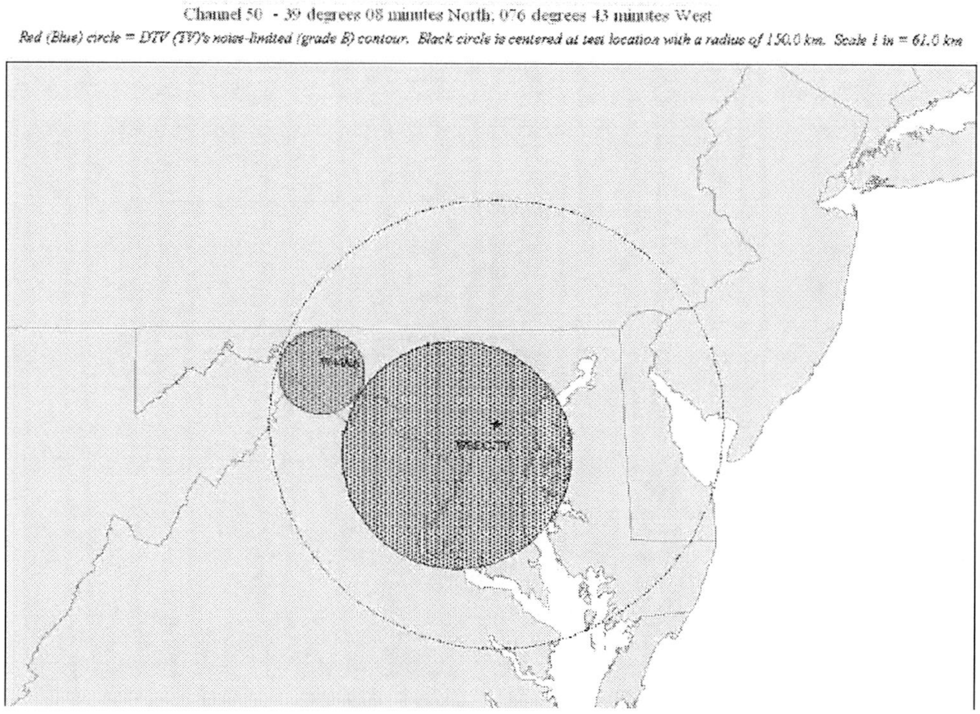

Figure 29. Channel 50 (WDCW-TV) Grade-B Contour Relative to Test Site 1

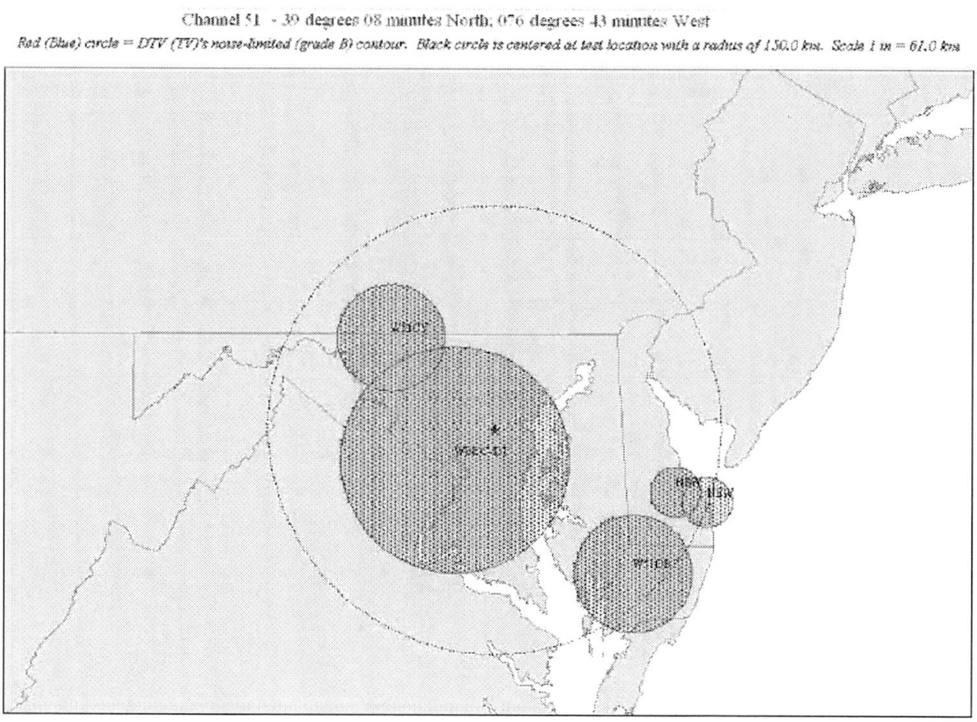

Figure 30. Channel 51 (WDCW-DT) Coverage Contour Relative to Test Site 1

Test Site 2

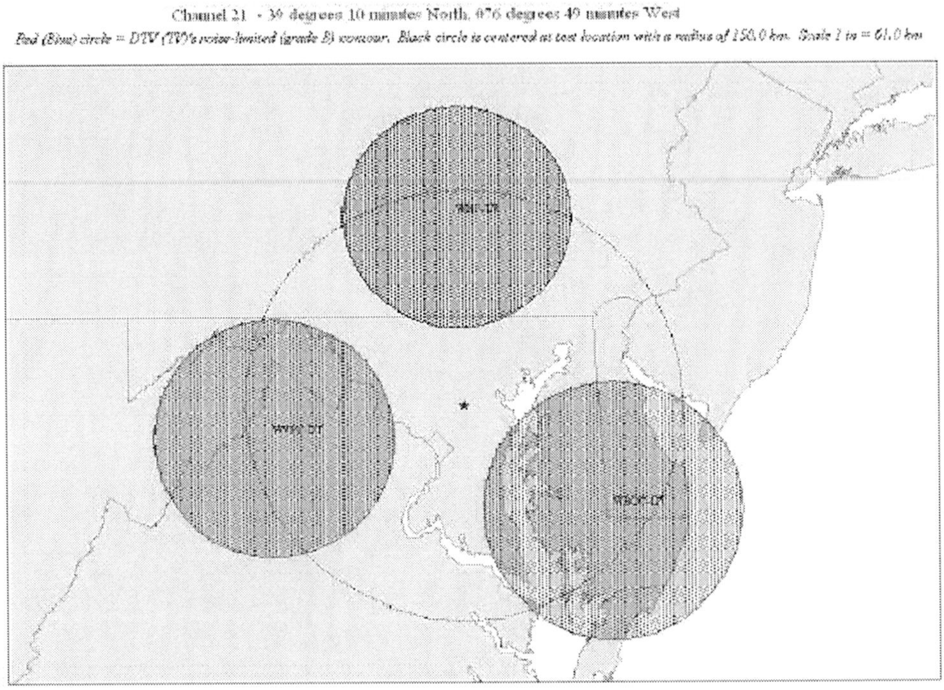

Figure 31. Channel 21 Service Contours Relative to Test Site

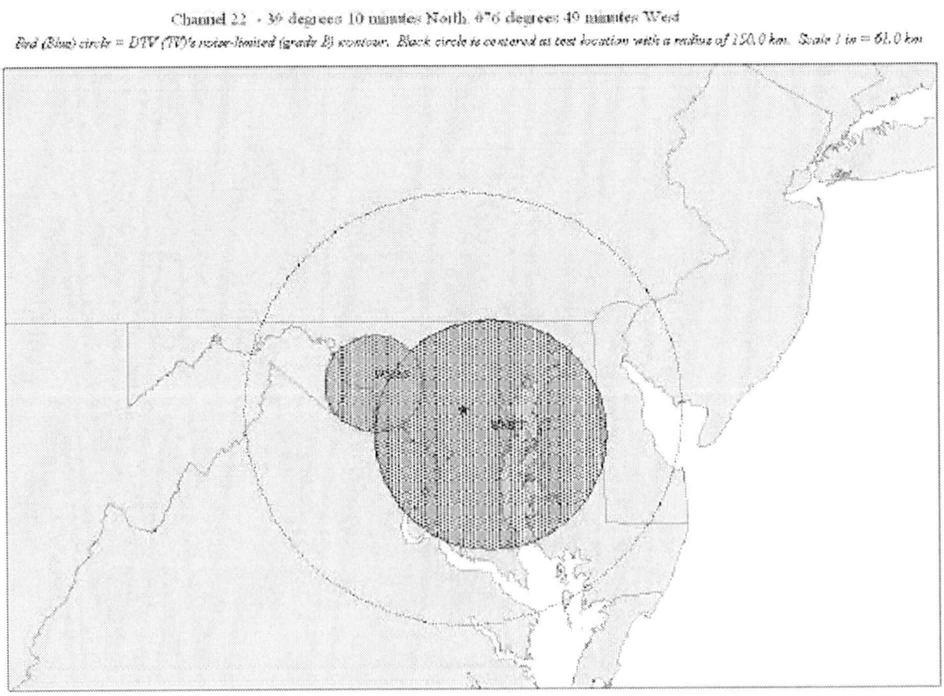

Figure 32. Channel 22 (WMPT-TV) Service Contour Relative to Test Site

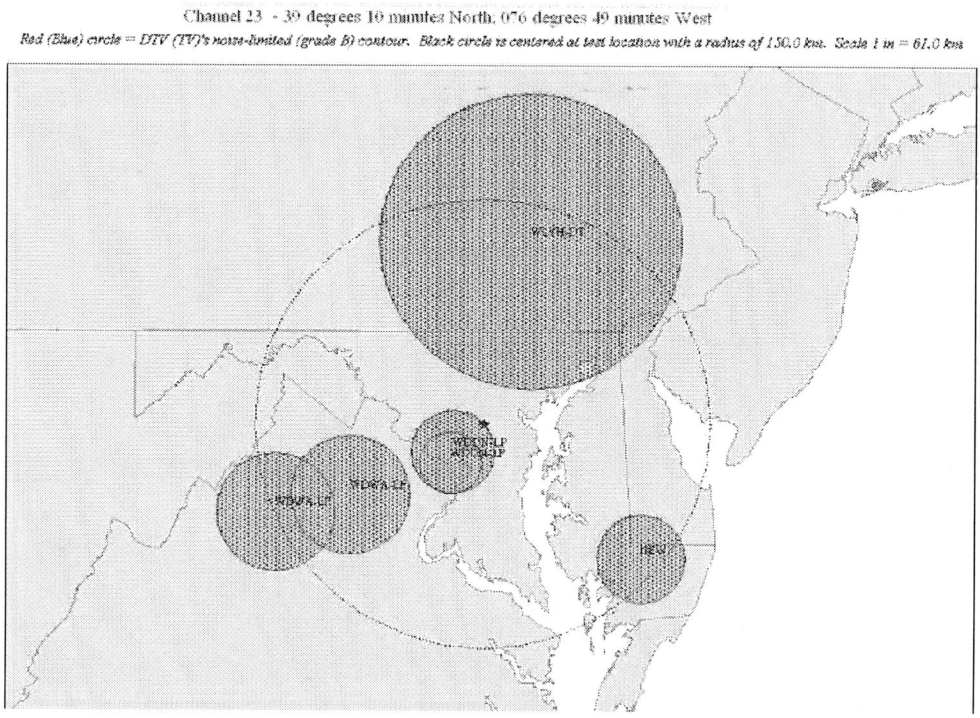

Figure 33. Channel 23 Service Contours Relative to Test Site 2

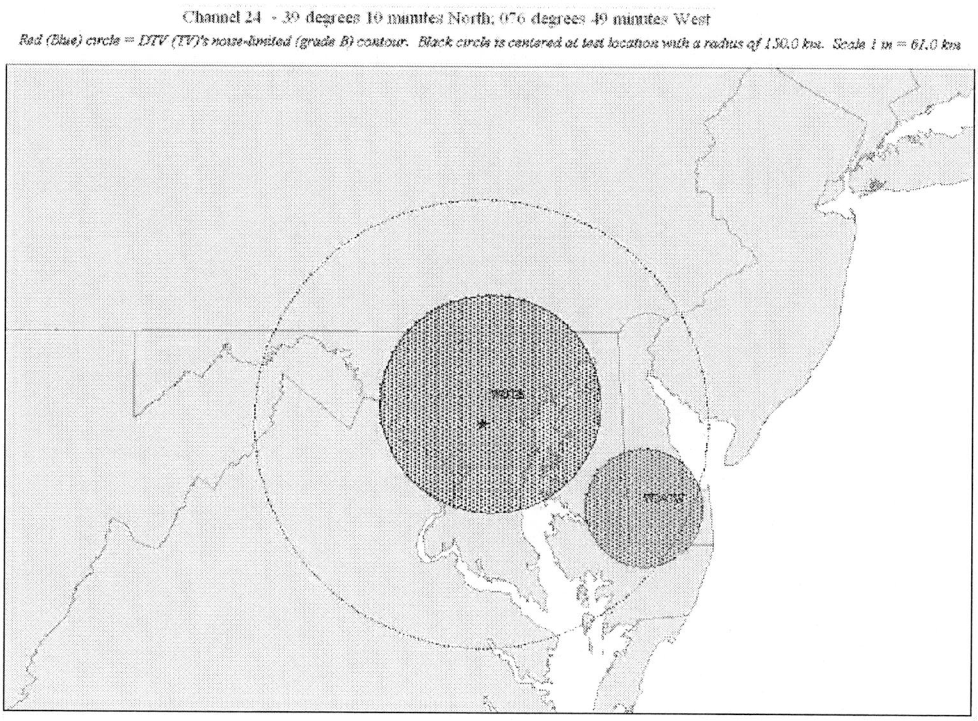

Figure 34. Channel 24 (WUTB-TV) Service Contour Relative to Test Site 2

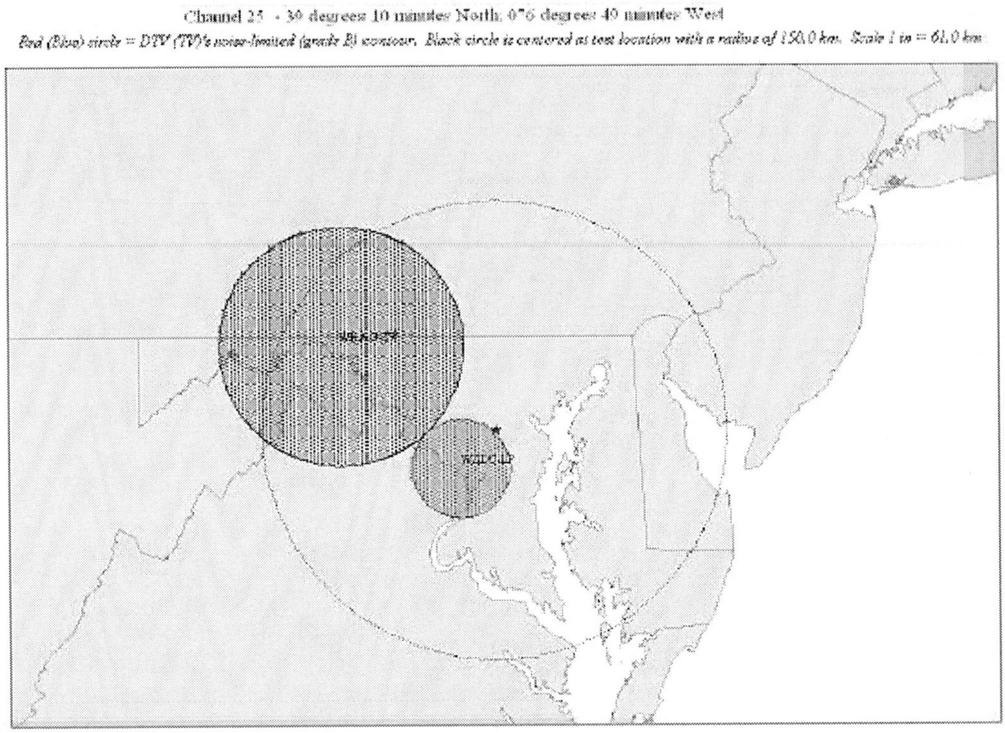

Figure 35. Channel 25 Service Contours Relative to Test Site 2

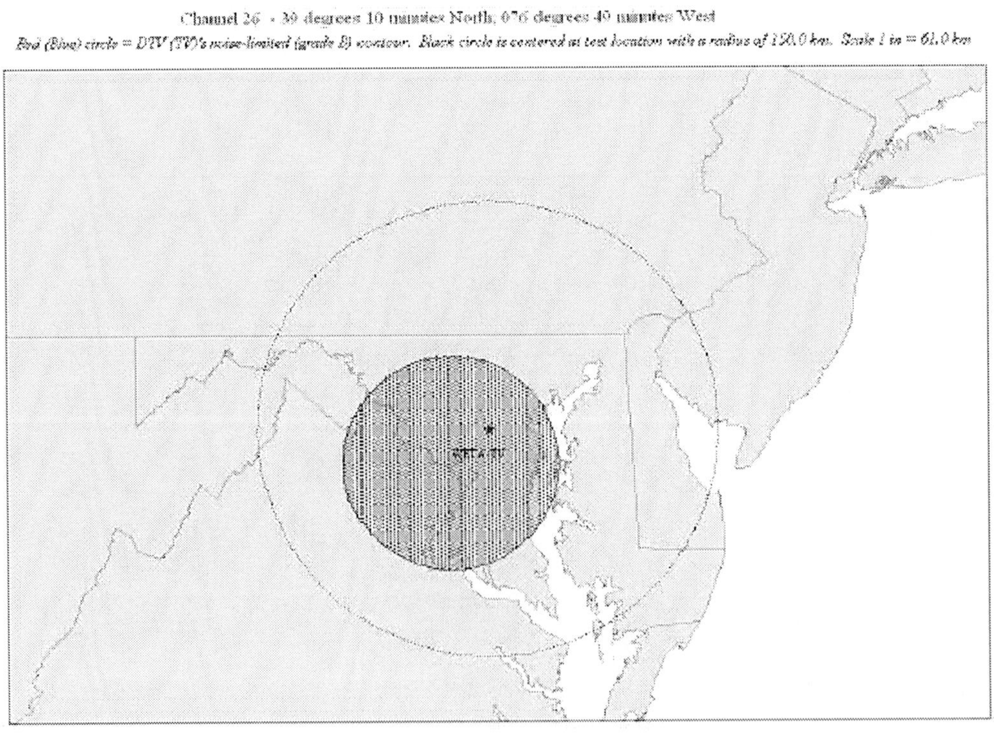

Figure 36. Channel 26 (WETA-TV) Service Contour Relative to Test Site 2

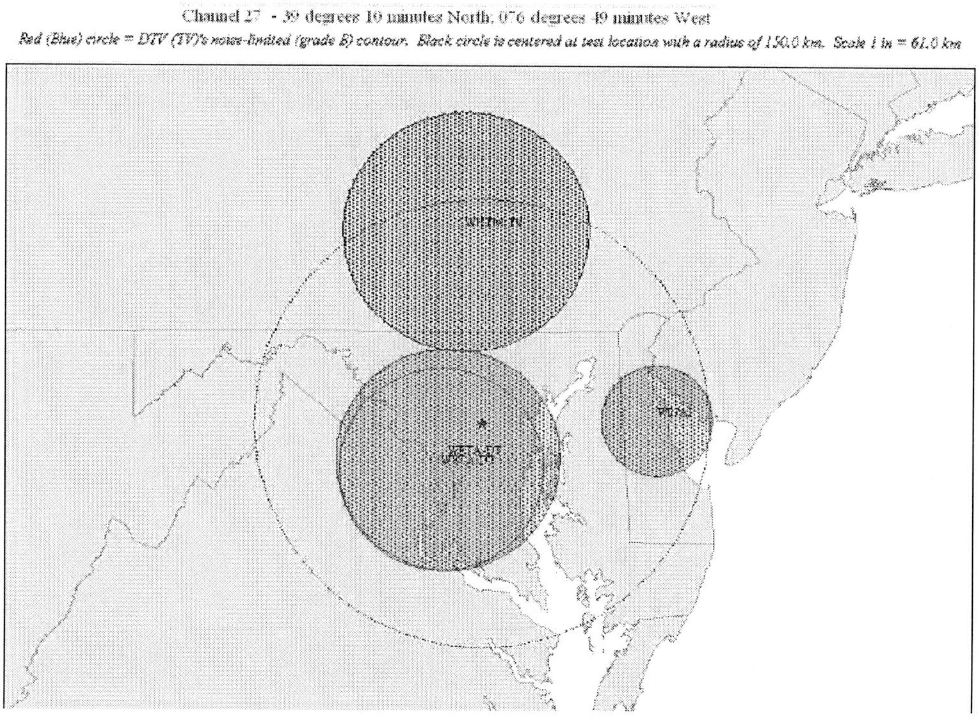

Figure 37. Channel 27 (WETA-DT) Service Contour Relative to Test Site 2

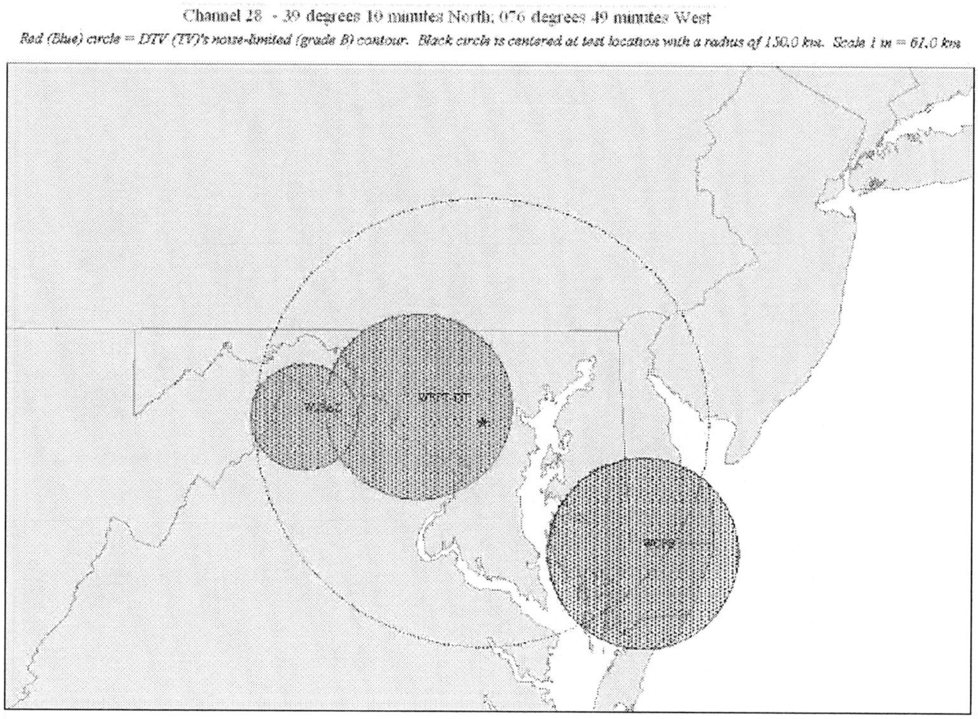

Figure 38. Channel 28 (WFPT-DT) Service Contour Relative to Test Site 2

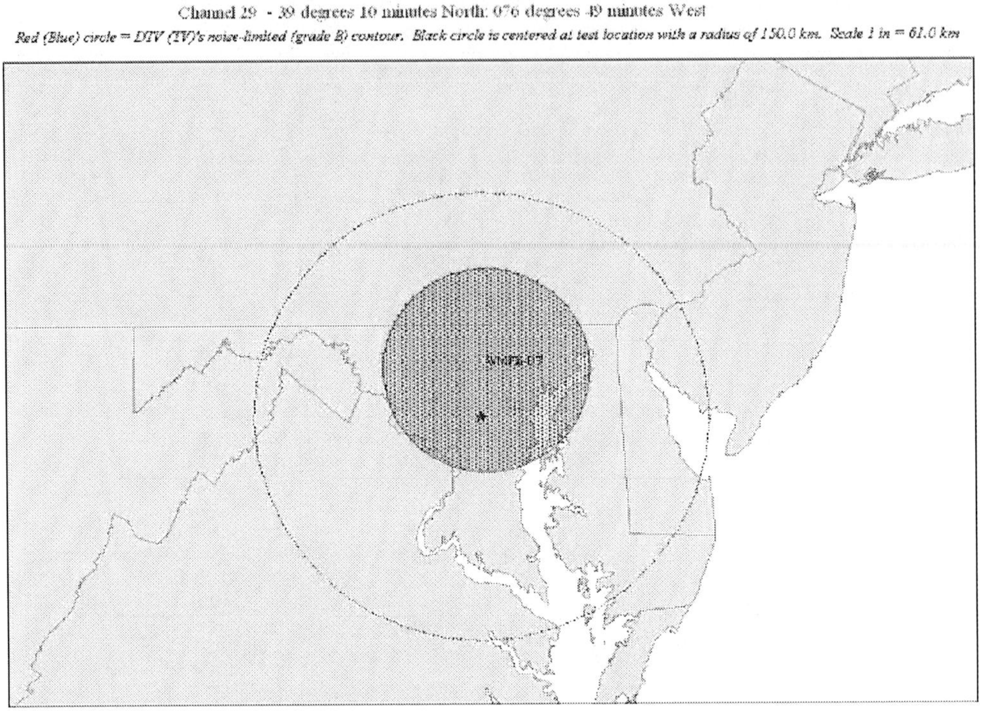

Figure 39. Channel 29 (WMPB-DT) Service Contour Relative to Test Site 2

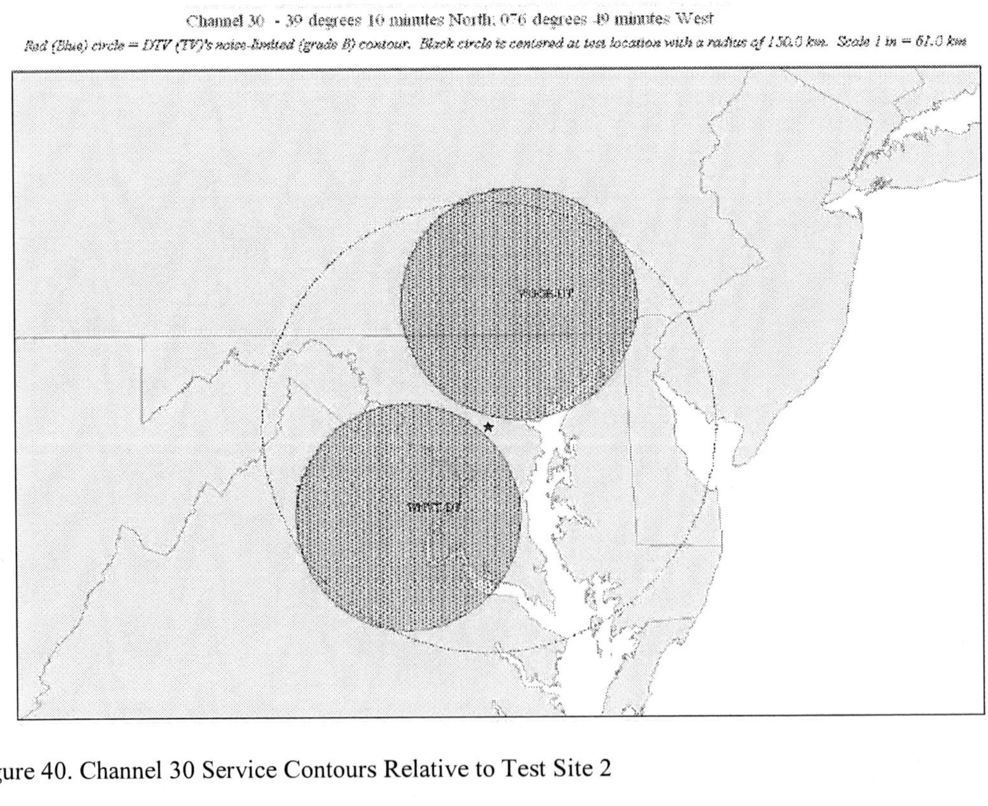

Figure 40. Channel 30 Service Contours Relative to Test Site 2

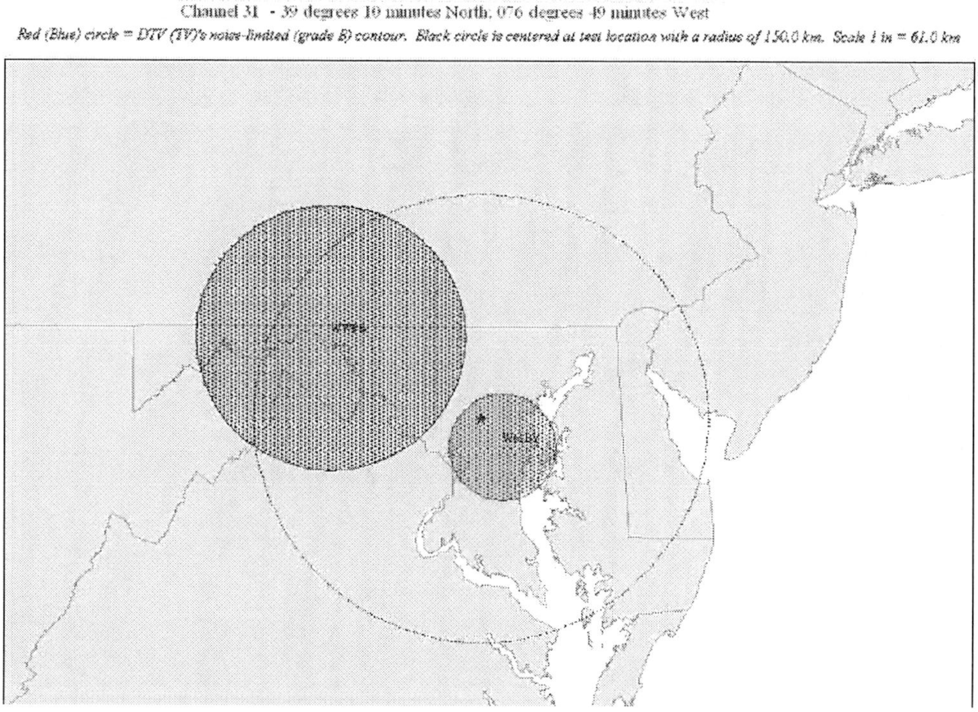

Figure 41. Channel 31 Service Contours Relative to Test Site 2

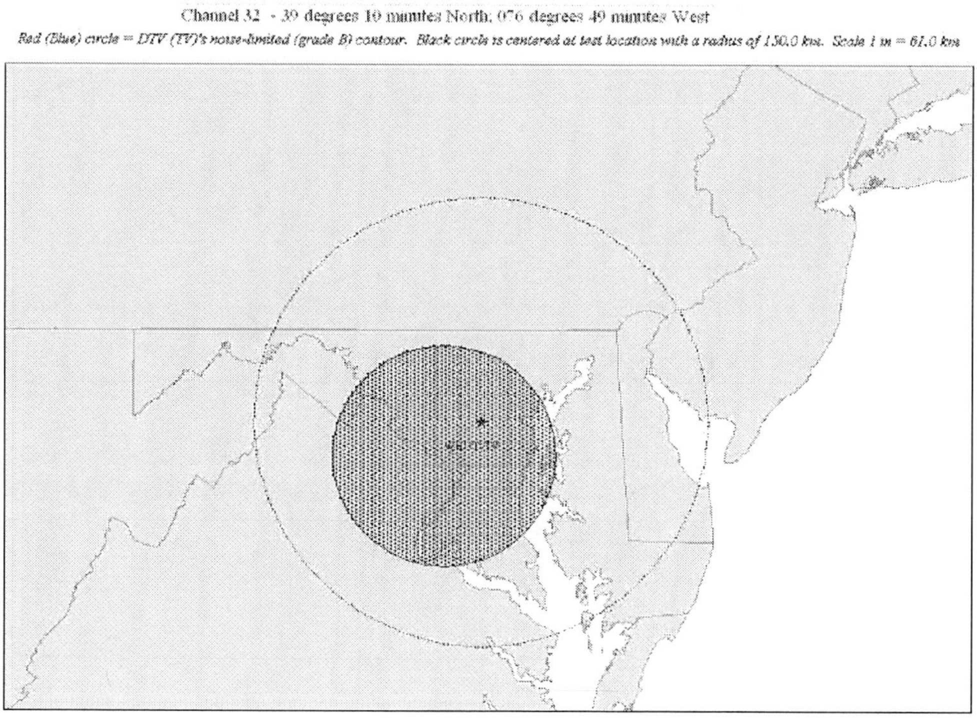

Figure 42. Channel 32 (WHUT-TV) Service Contour Relative to Test Site 2

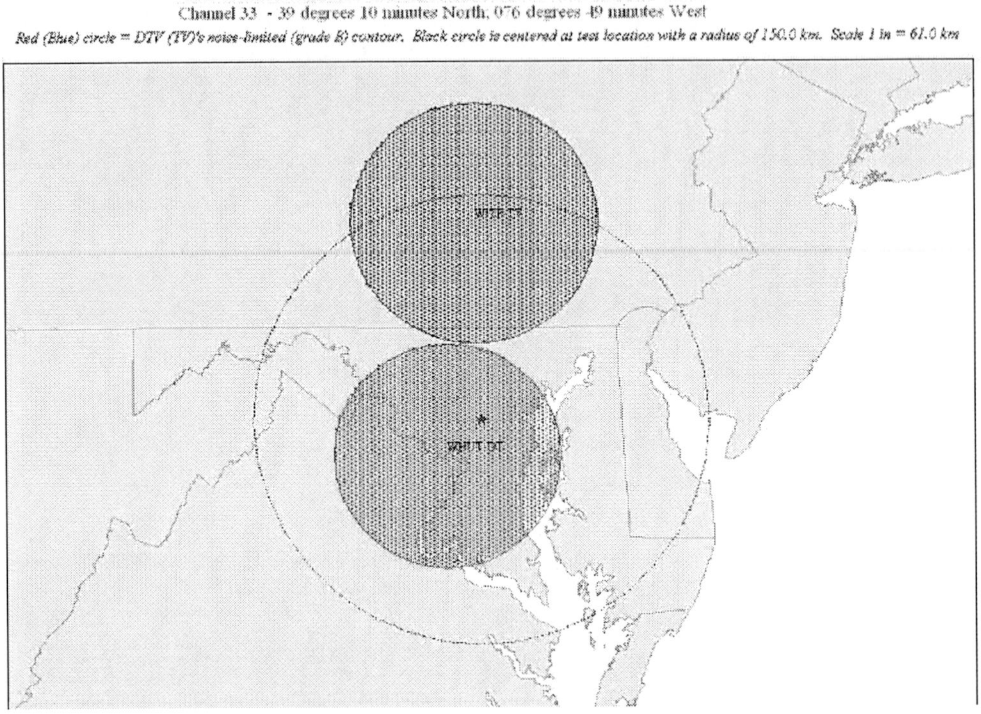

Figure 43. Channel 33 (WHUT-DT) Service Contour Relative to Test Site 2

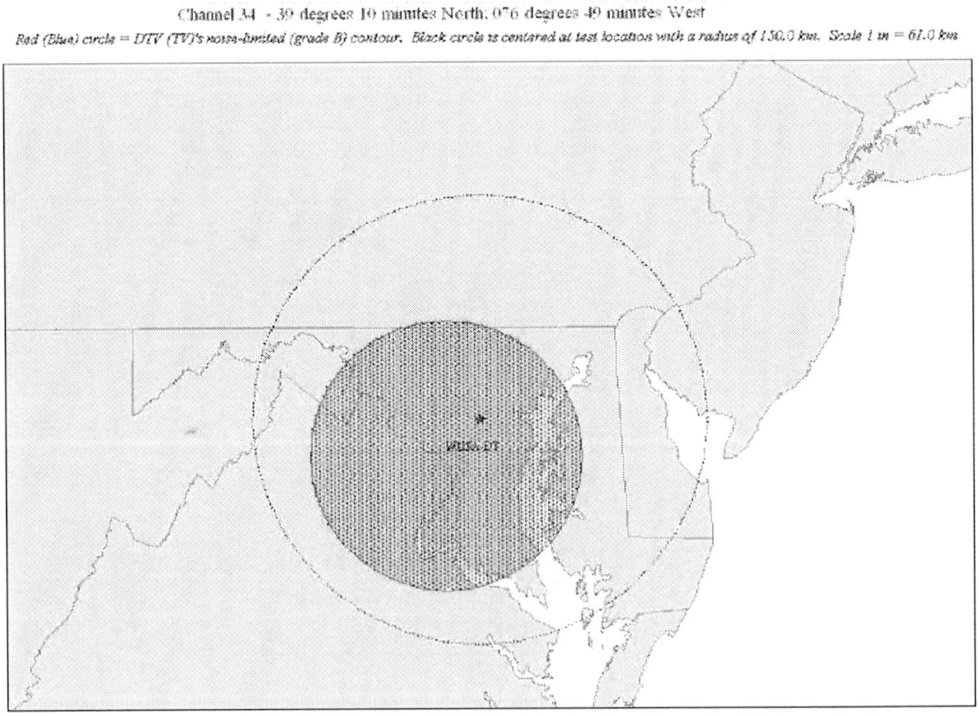

Figure 44. Channel 34 (WUSA-DT) Service Contour Relative to Test Site 2

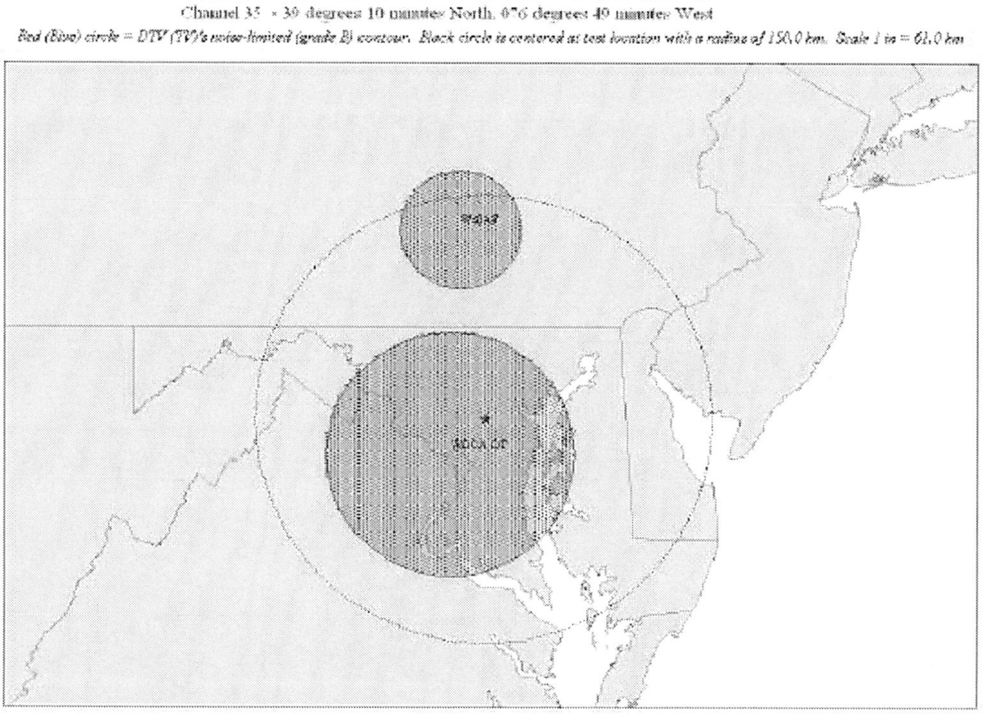

Figure 45. Channel 35 (WDCA-DT) Service Contour Relative to Test Site 2

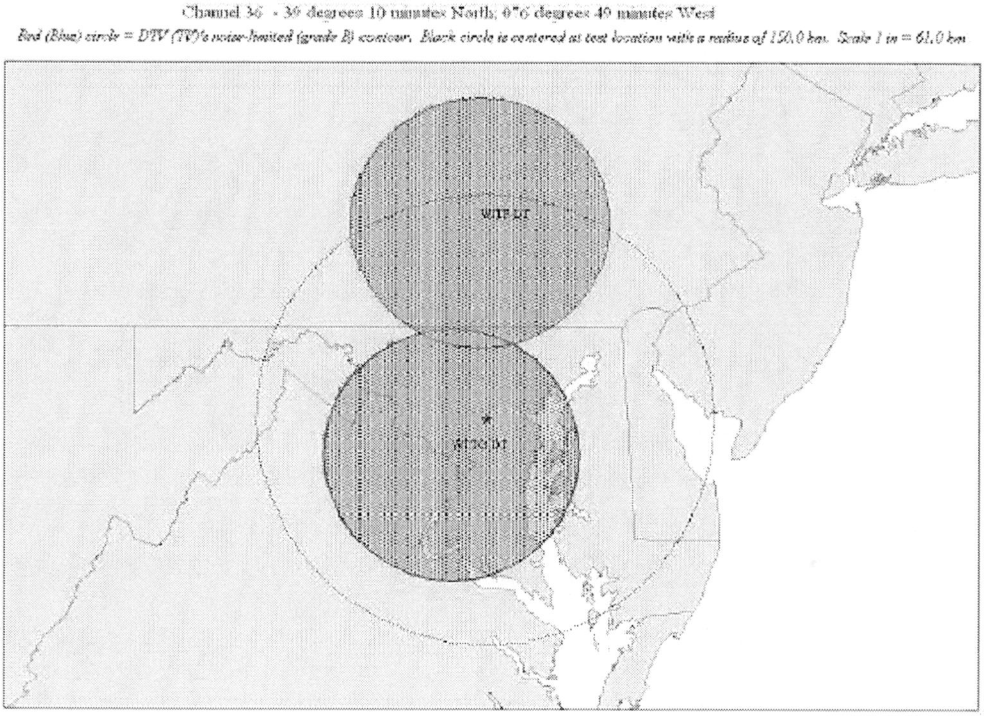

Figure 46. Channel 36 (WTTG-DT) Service Contour Relative to Test Site 2

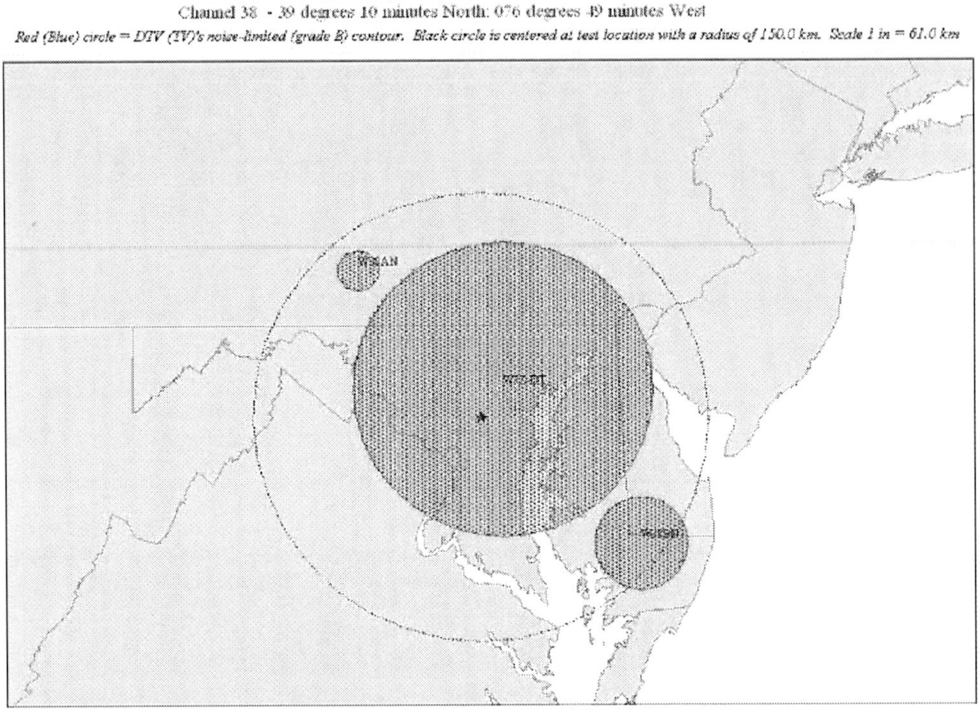

Figure 47. Channel 38 (WJZ-DT) Service Contour Relative to Test Site 2

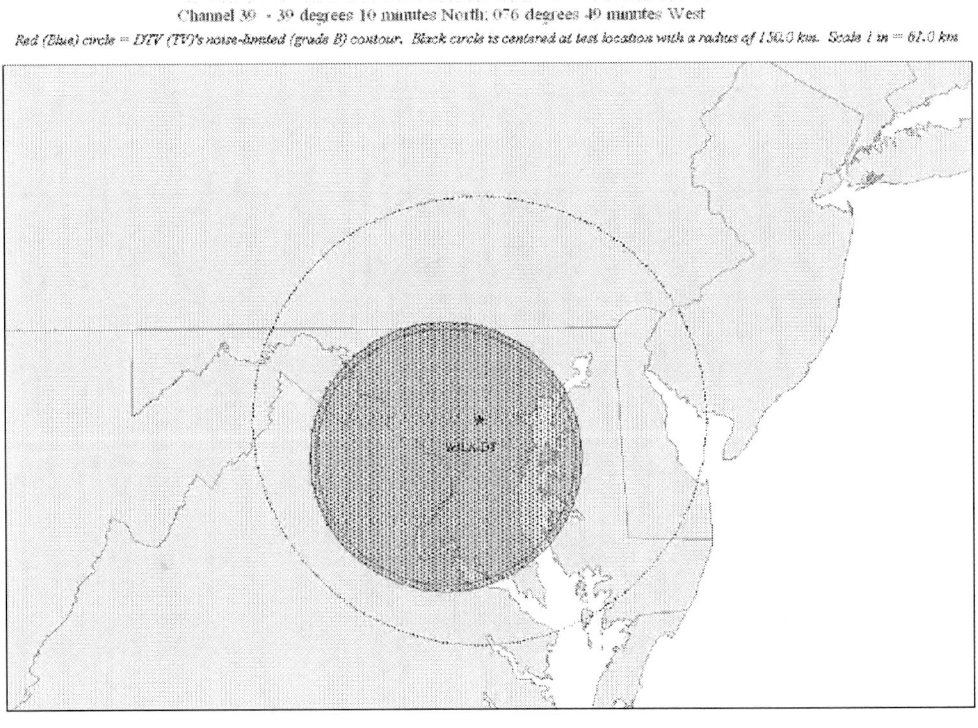

Figure 48. Channel 39 (WJLA-DT) Service Contour Relative to Test Site 2

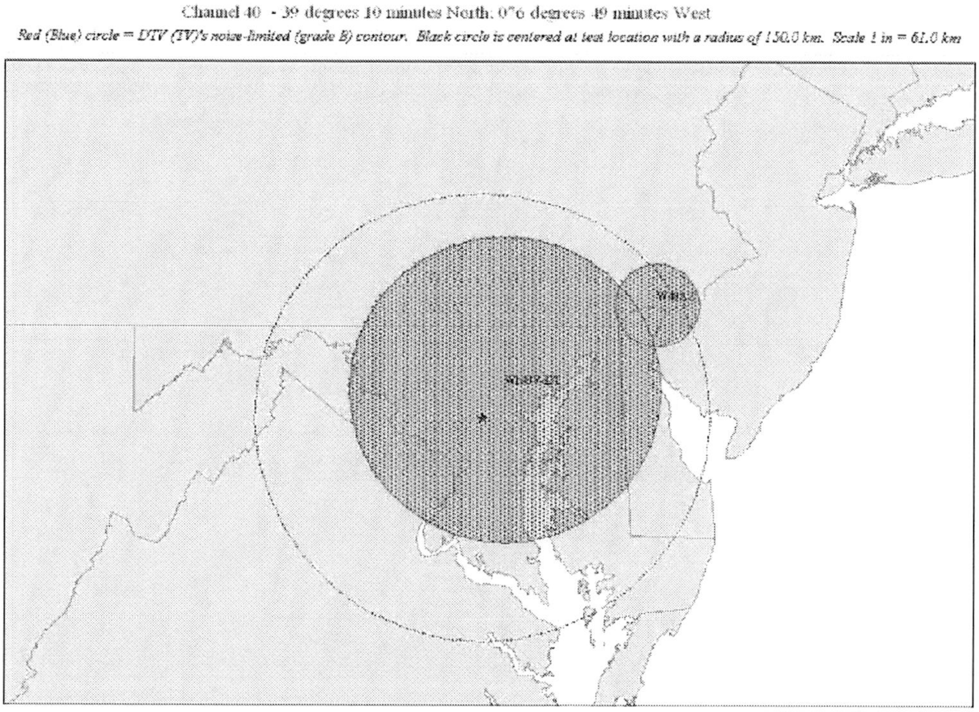

Figure 49. Channel 40 (WNUV-DT) Service Contour Relative to Test Site 2

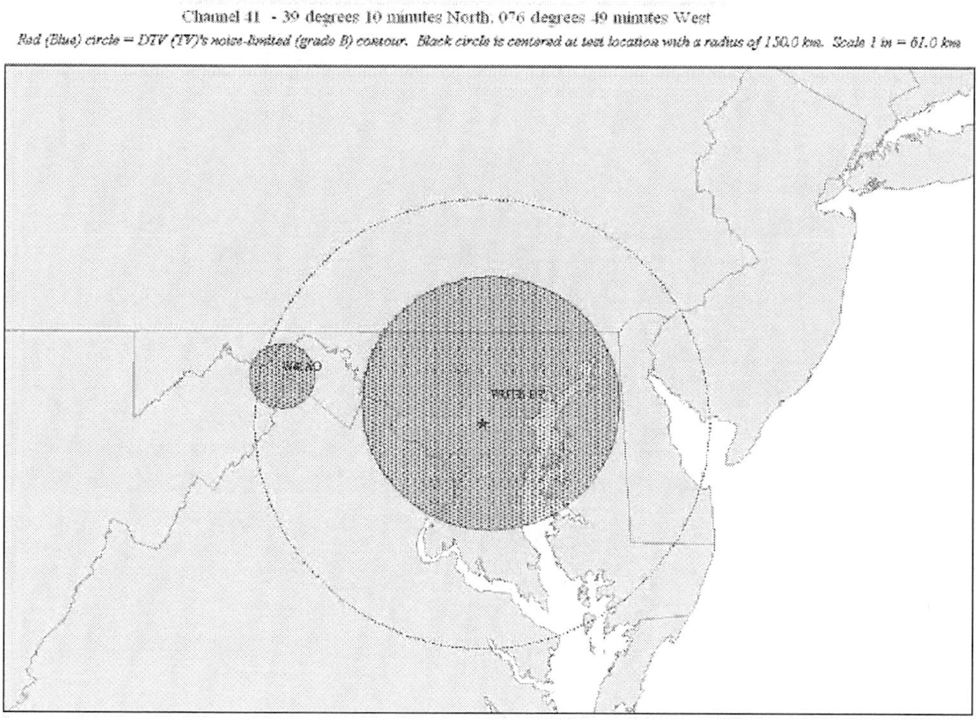

Figure 50. Channel 41 (WUTB-DT) Service Contour Relative to Test Site 2

Channel 42 - 39 degrees 10 minutes North, 076 degrees 49 minutes West

Red (Blue) circle = DTV (TV)'s noise-limited (grade B) contour. Black circle is centered at test location with a radius of 150.0 km. Scale 1 in = 61.0 km

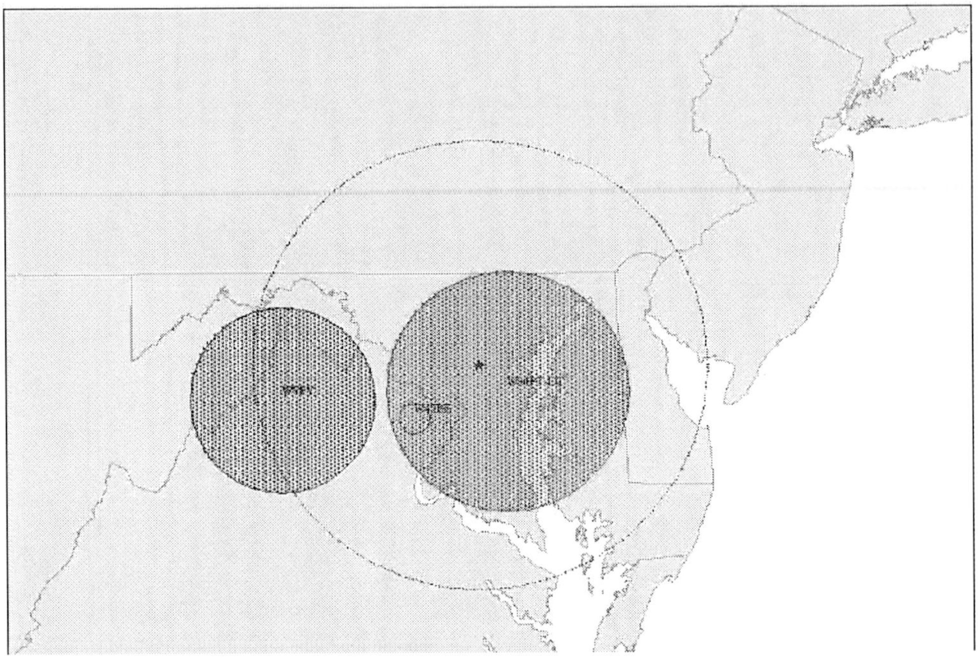

Figure 51. Channel 42 (WMPT-DT) Service Contour Relative to Test Site 2

Channel 43 - 39 degrees 10 minutes North, 076 degrees 49 minutes West

Red (Blue) circle = DTV (TV)'s noise-limited (grade B) contour. Black circle is centered at test location with a radius of 150.0 km. Scale 1 in = 61.0 km

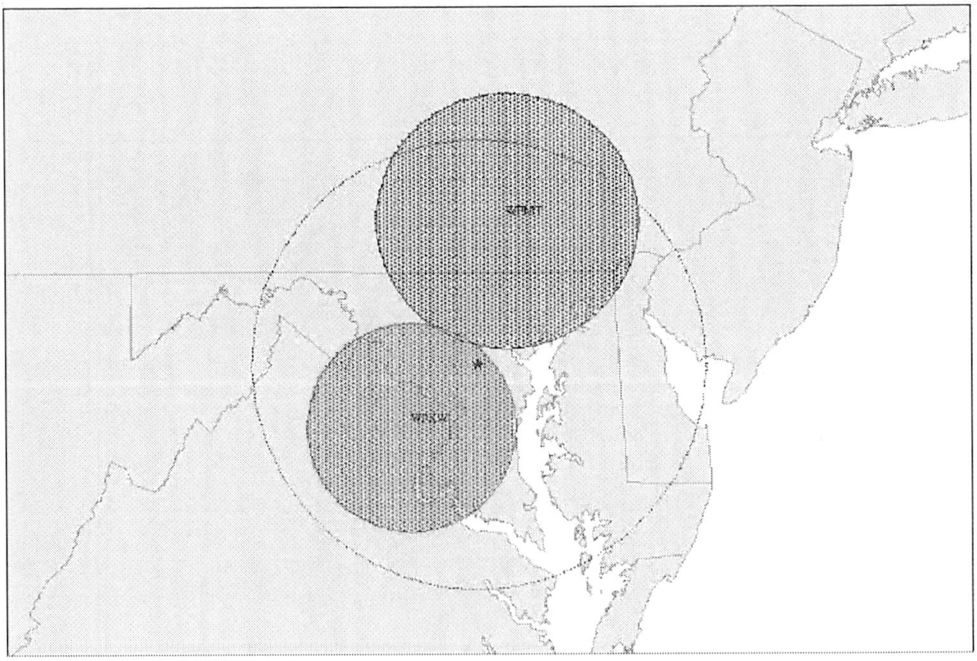

Figure 52. Channel 43 (WPXW-DT) Service Contour Relative to Test Site 2

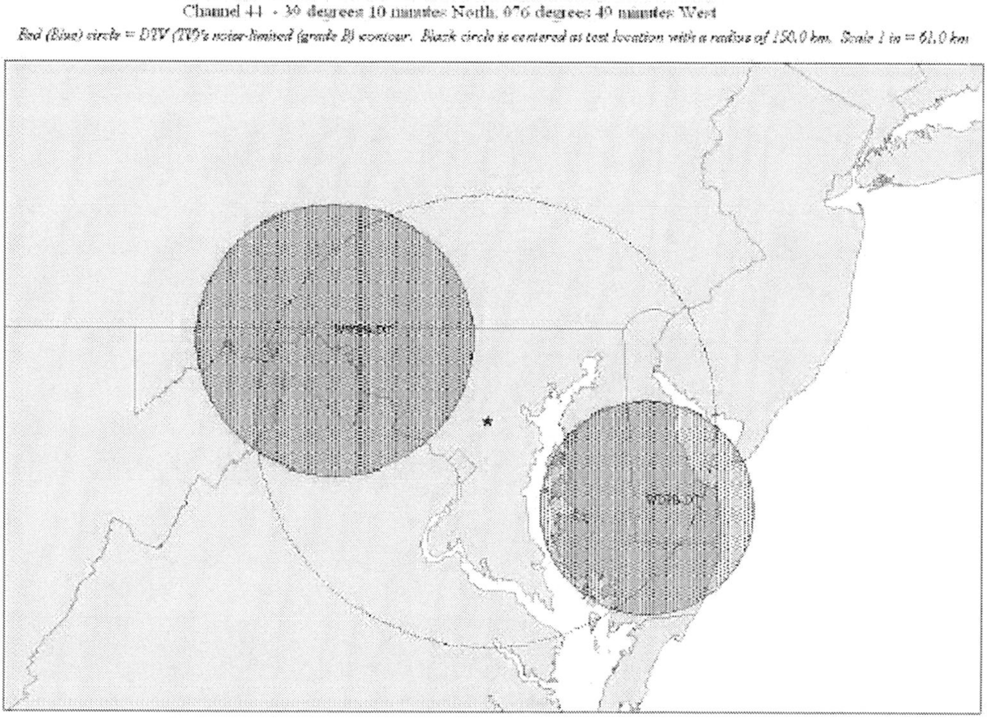

Figure 53. Channel 44 Service Contours Relative to Test Site 2

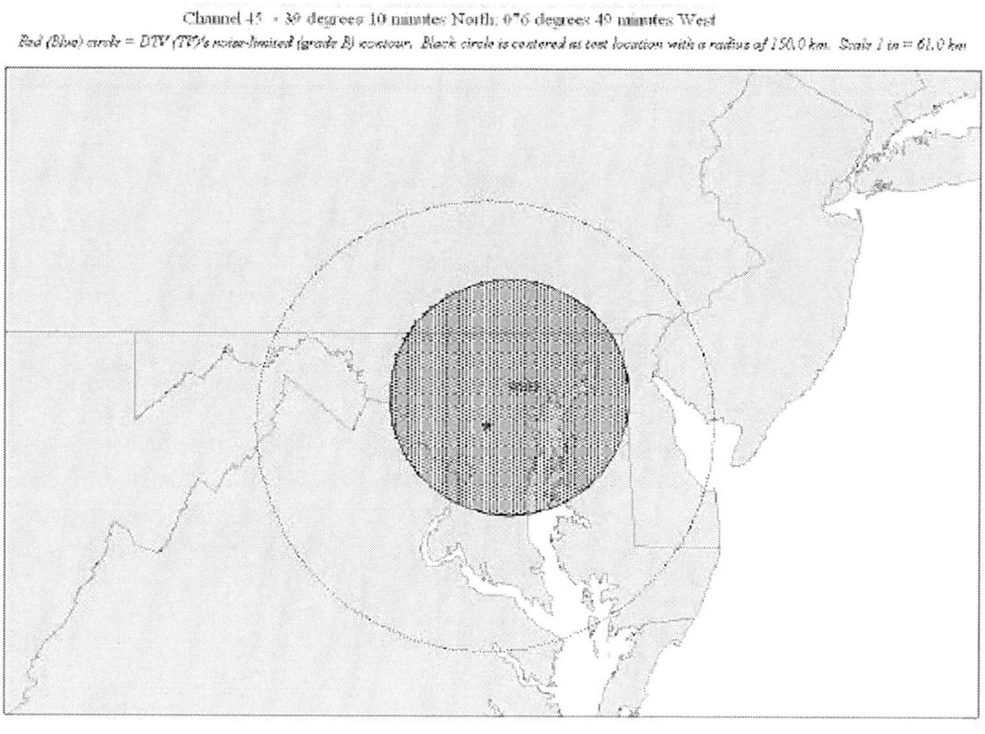

Figure 54. Channel 45 (WBFF-TV) Service Contour Relative to Test Site 2

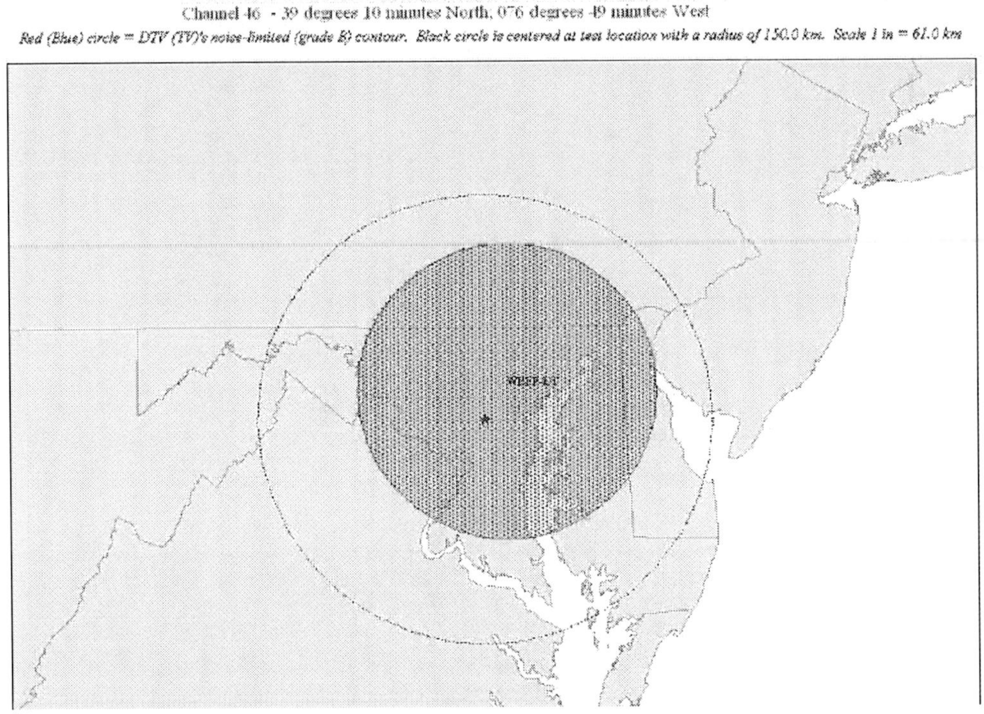

Figure 55. Channel 46 (WBFF-DT) Service Contour Relative to Test Site 2

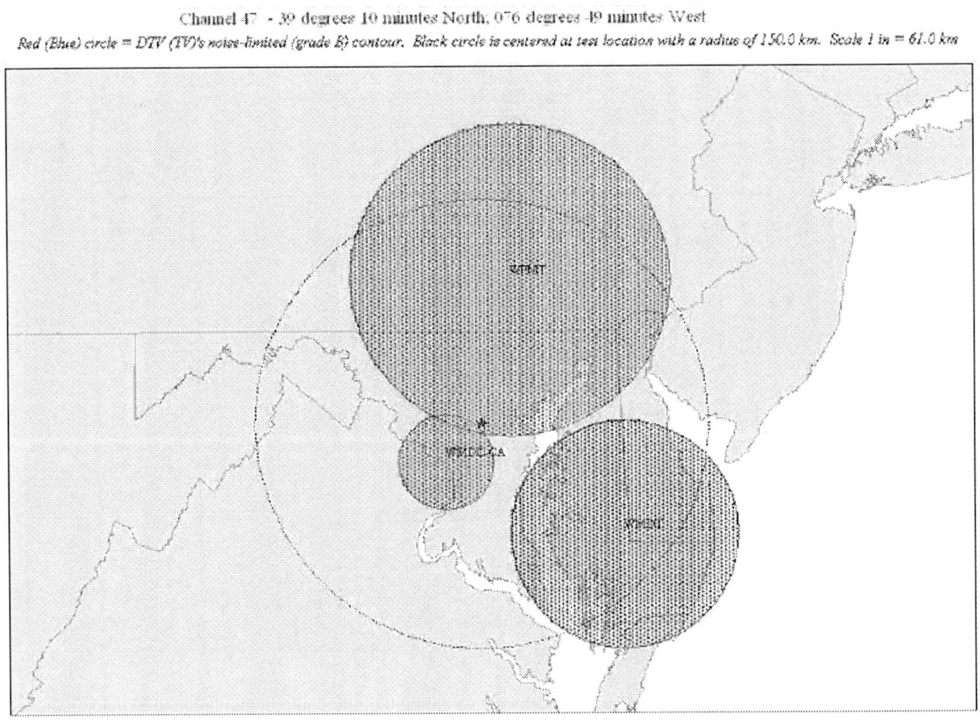

Figure 56. Channel 47 (WPMT-DT) Service Contour Relative to Test Site 2

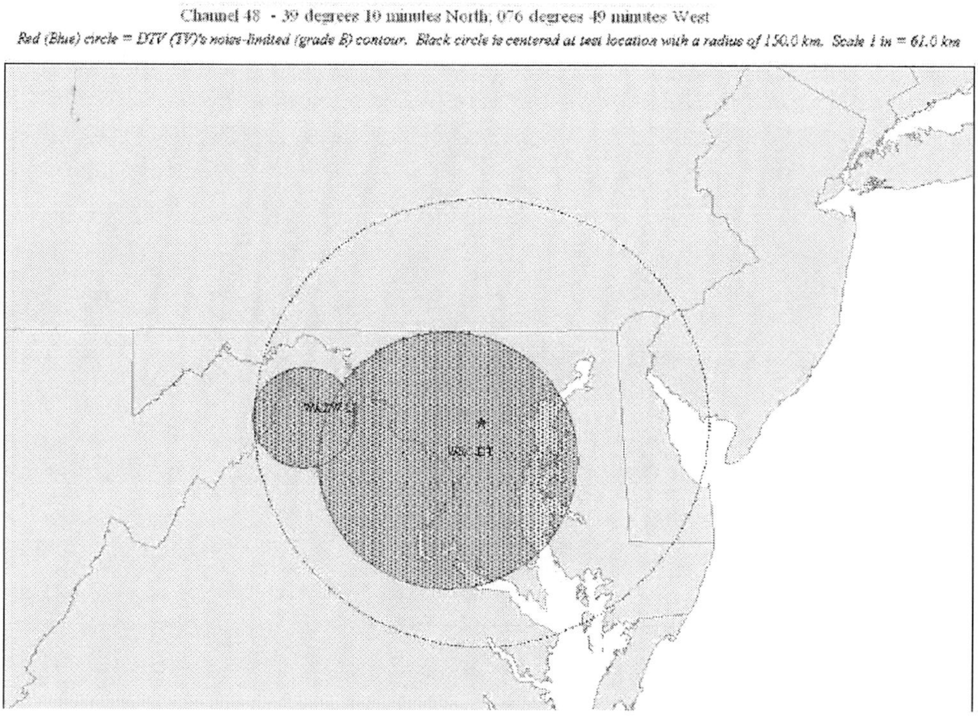

Figure 57. Channel 48 (WRC-DT) Service Contour Relative to Test Site 2

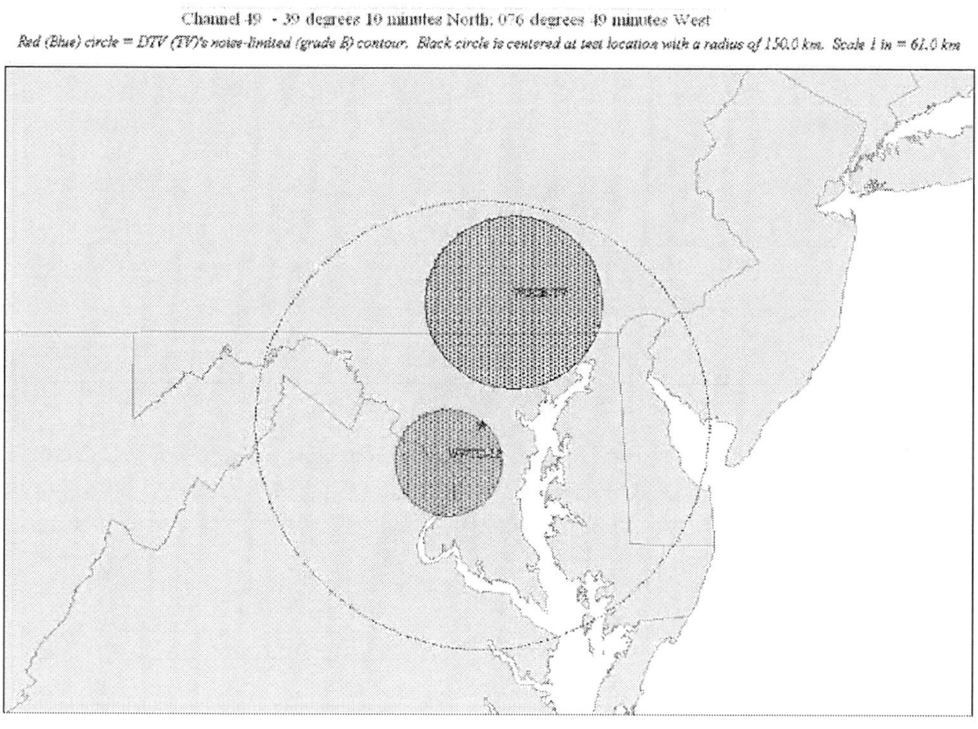

Figure 58. Channel 49 Service Contours Relative to Test Site 2

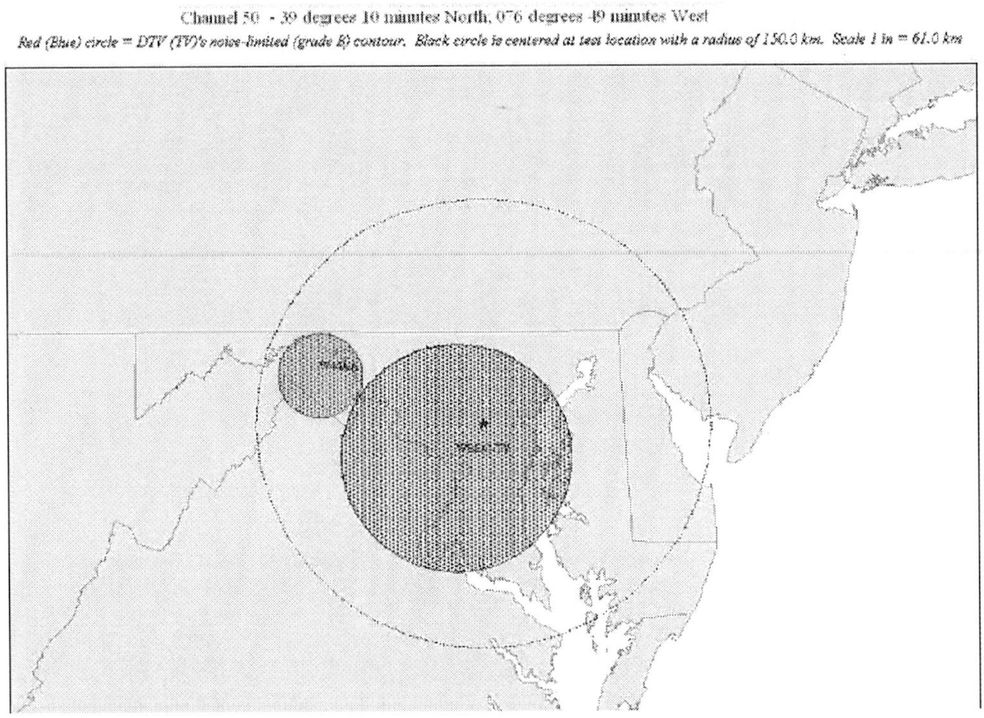

Figure 59. Channel 50 (WDCW-TV) Service Contour Relative to Test Site 2

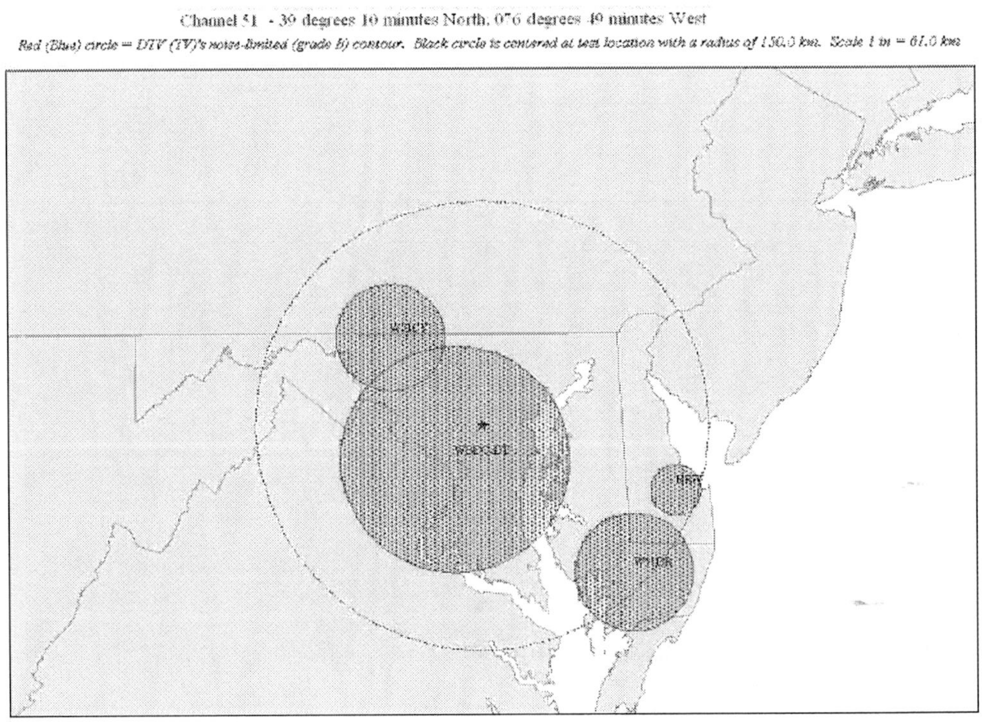

Figure 60. Channel 51 (WDCW-DT) Service Contour Relative to Test Site 2

Test Site 3

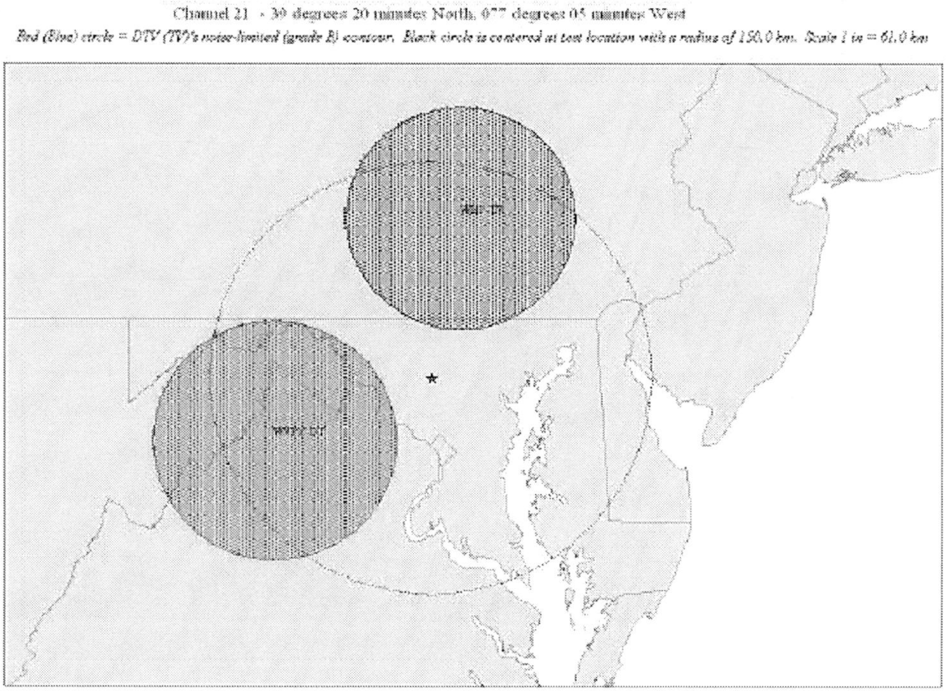

Figure 61. Channel 21 Service Contours Relative to Test Site 3

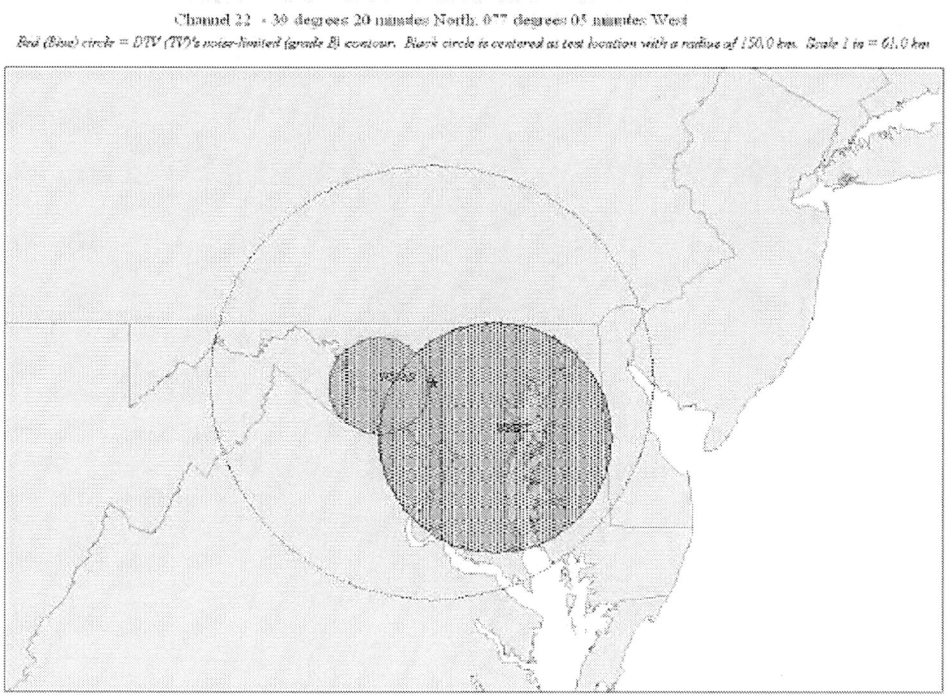

Figure 62. Channel 22 (WMPT-TV) Service Contour Relative to Test Site 3

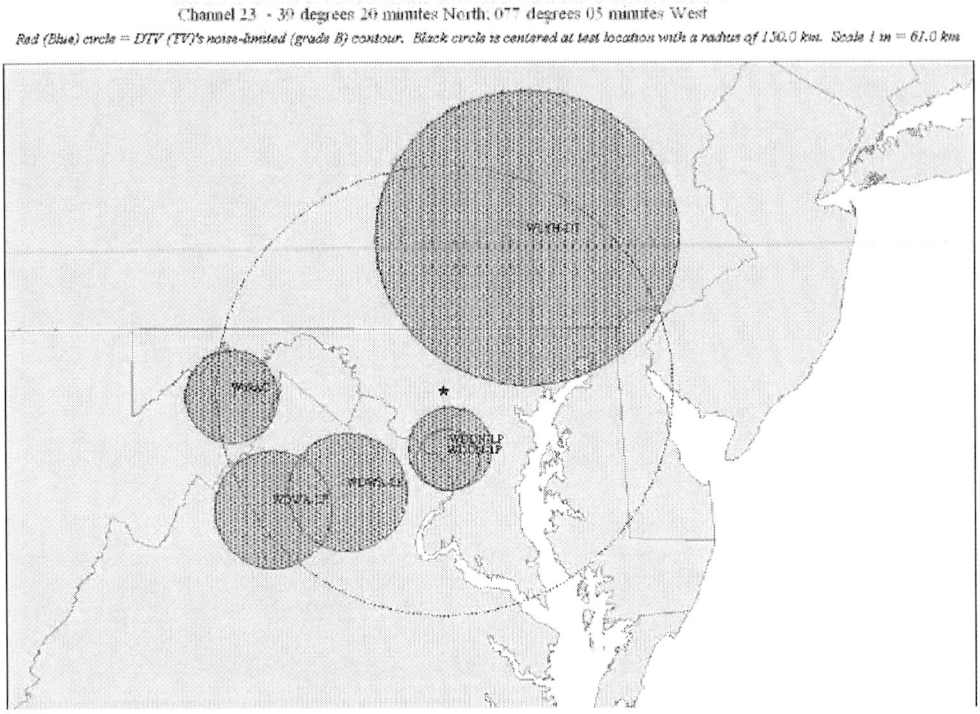

Figure 63. Channel 23 Service Contours Relative to Test Site 3

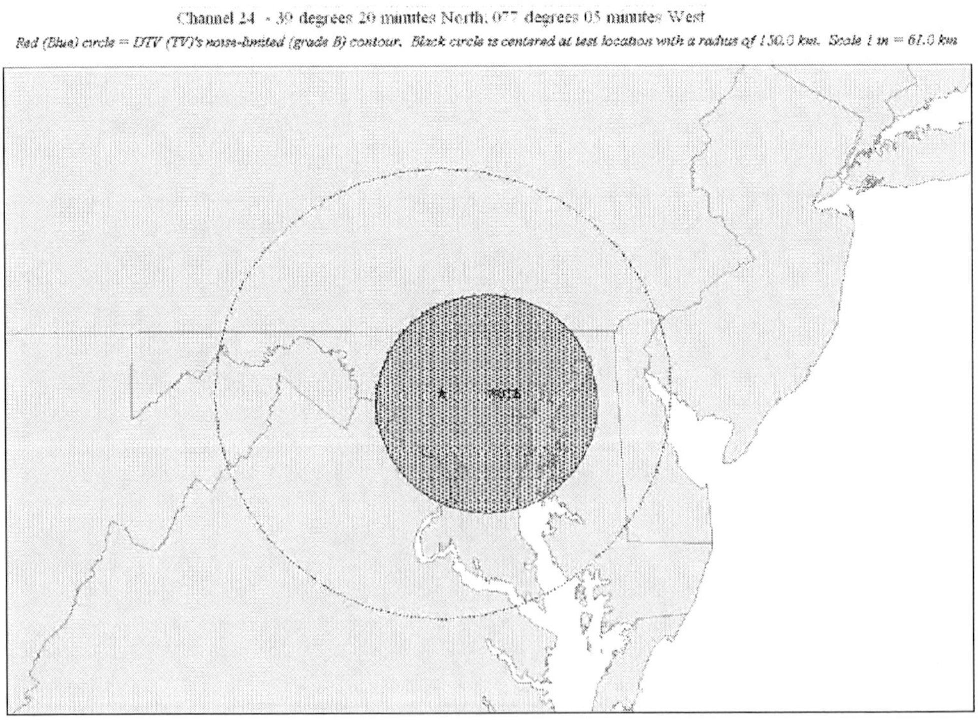

Figure 64. Channel 24 (WUTB-TV) Service Contour Relative to Test Site 3

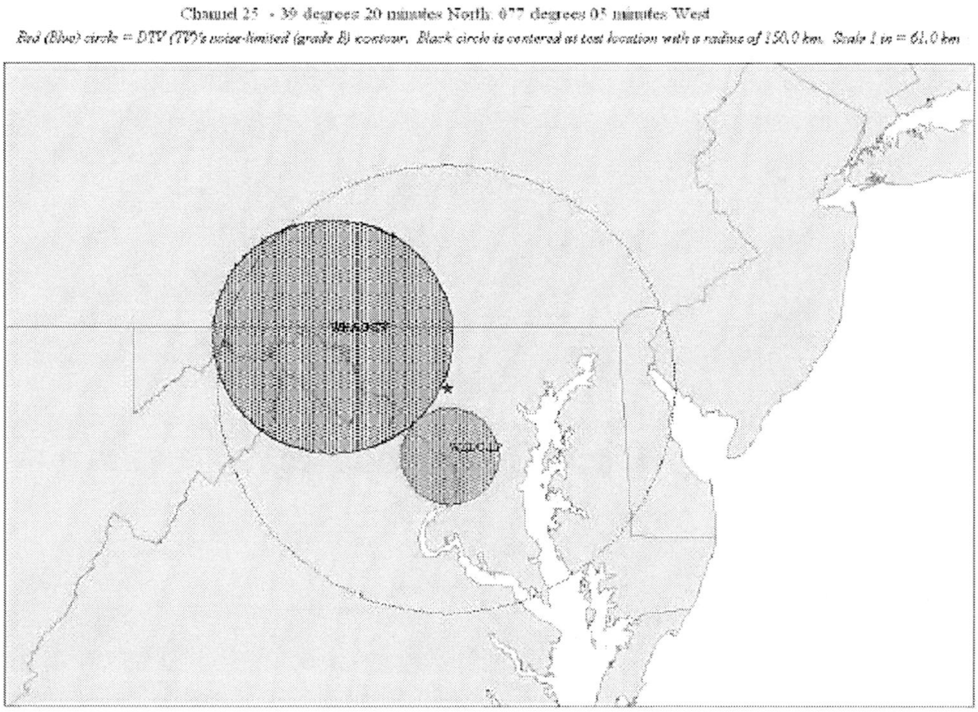

Figure 65. Channel 25 Service Contours Relative to Test Site 3

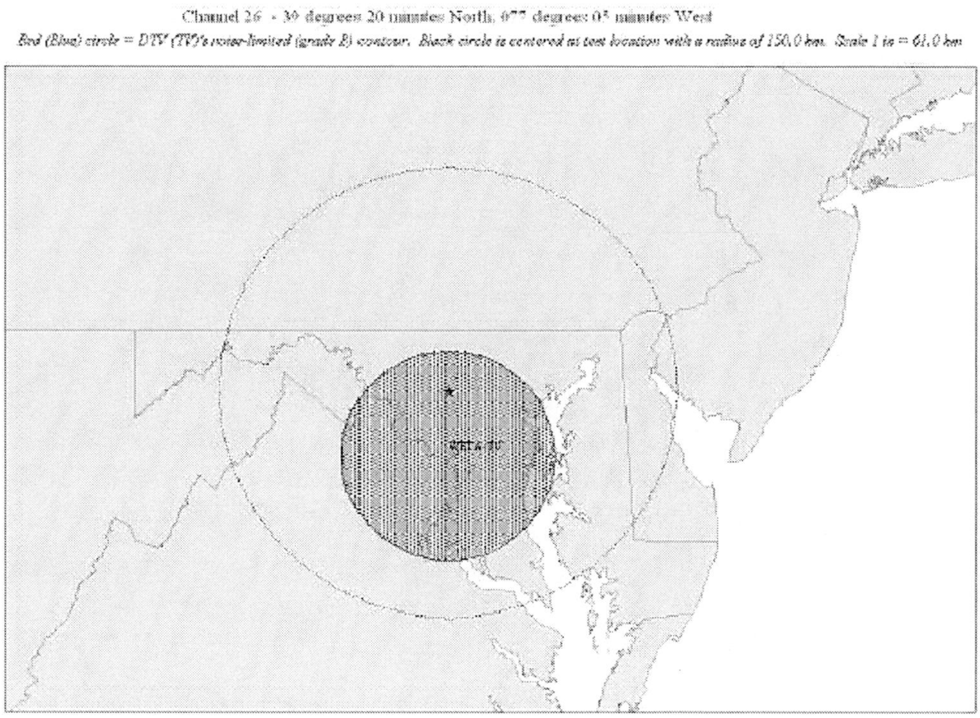

Figure 66. Channel 26 (WETA-TV) Service Contour Relative to Test Site 3

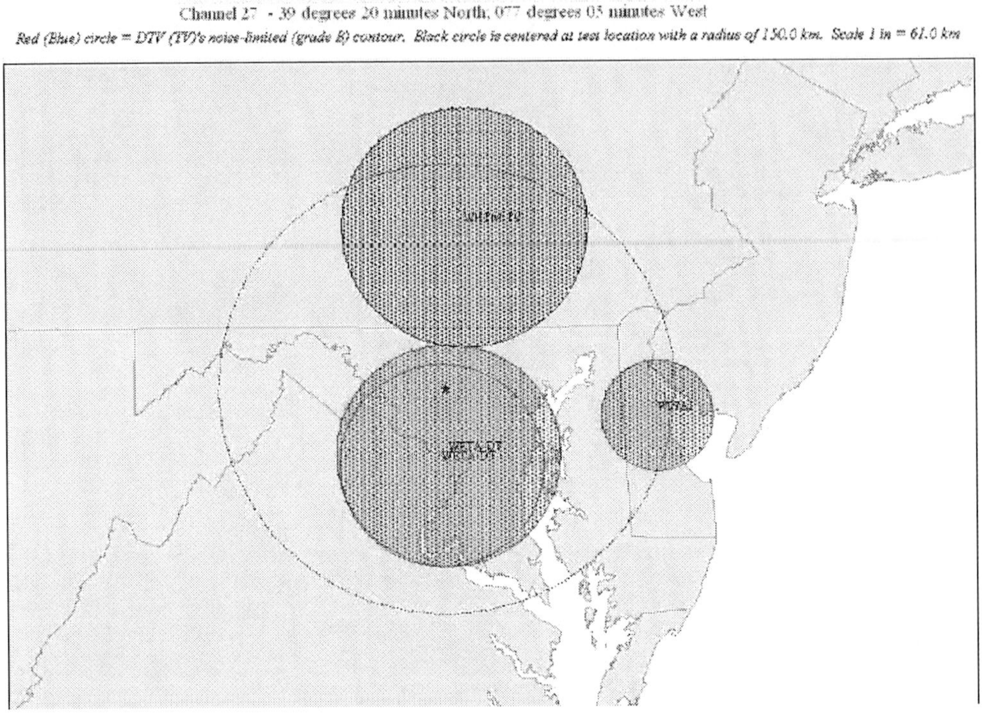

Figure 67. Channel 27 (WETA-DT) Service Contour Relative to Test Site 3

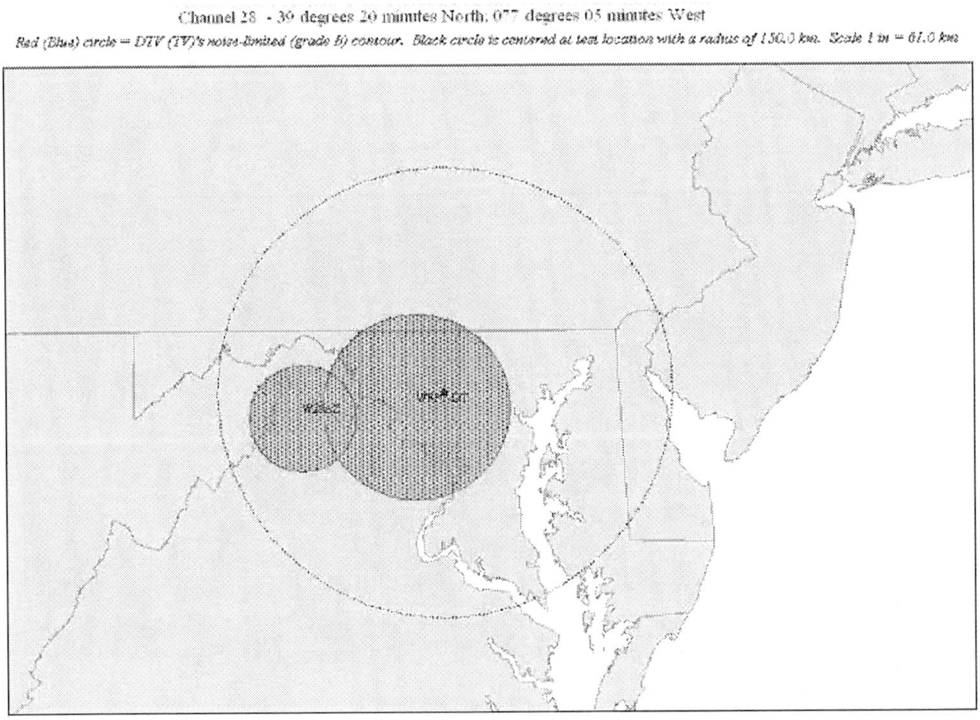

Figure 68. Channel 28 (WFPT-DT) Service Contour Relative to Test Site 3

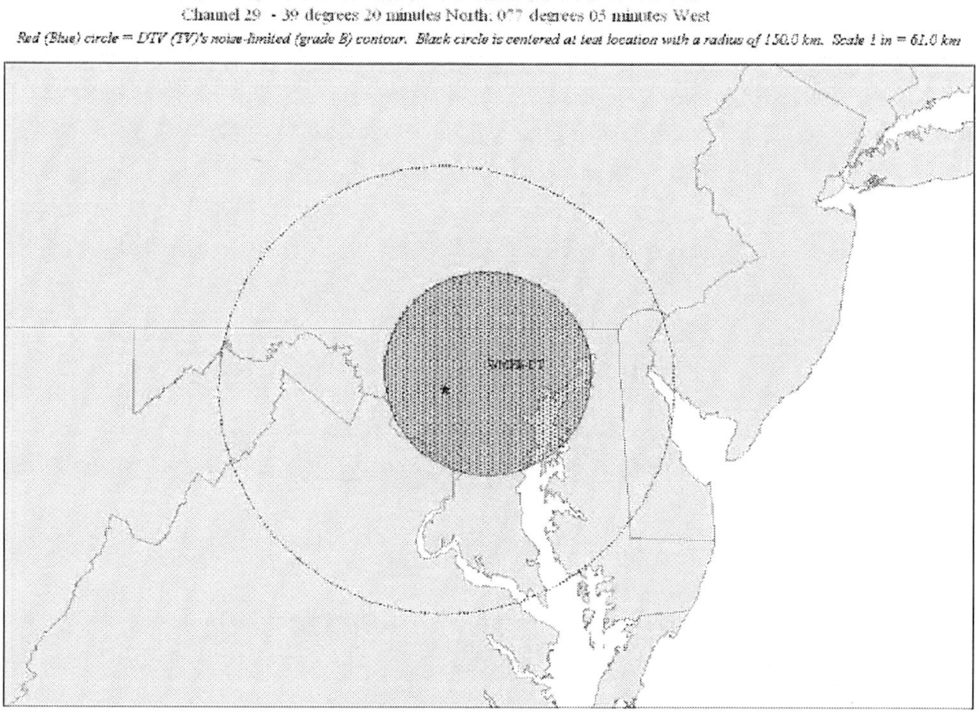

Figure 69. Channel 29 (WMPB-DT) Service Contour Relative to Test Site 3

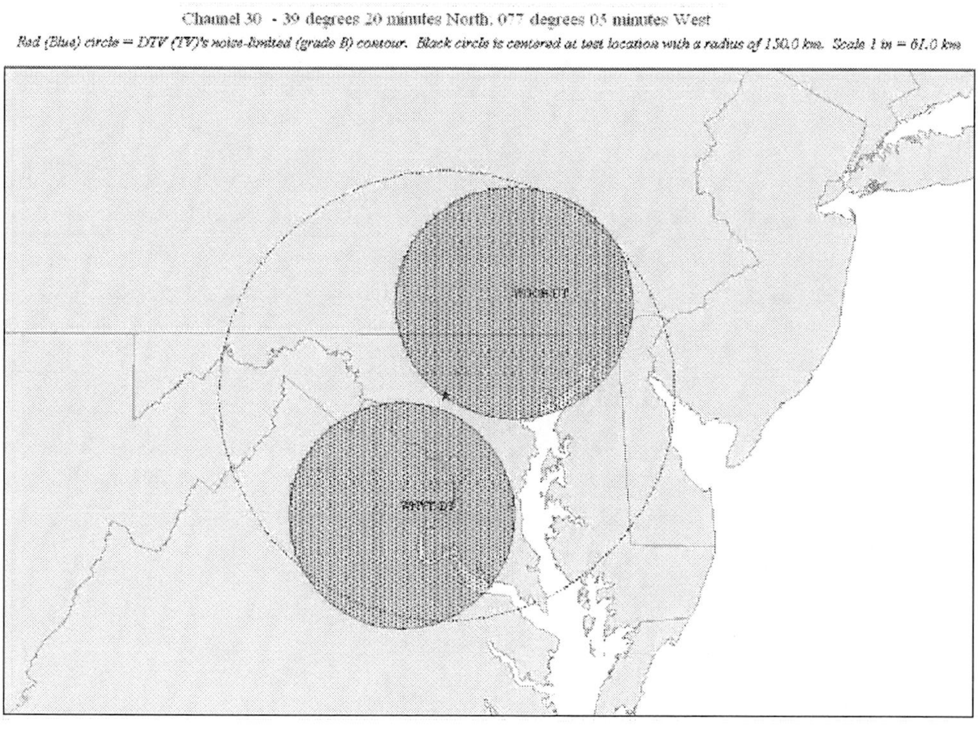

Figure 70. Channel 30 Service Contours Relative to Test Site 3

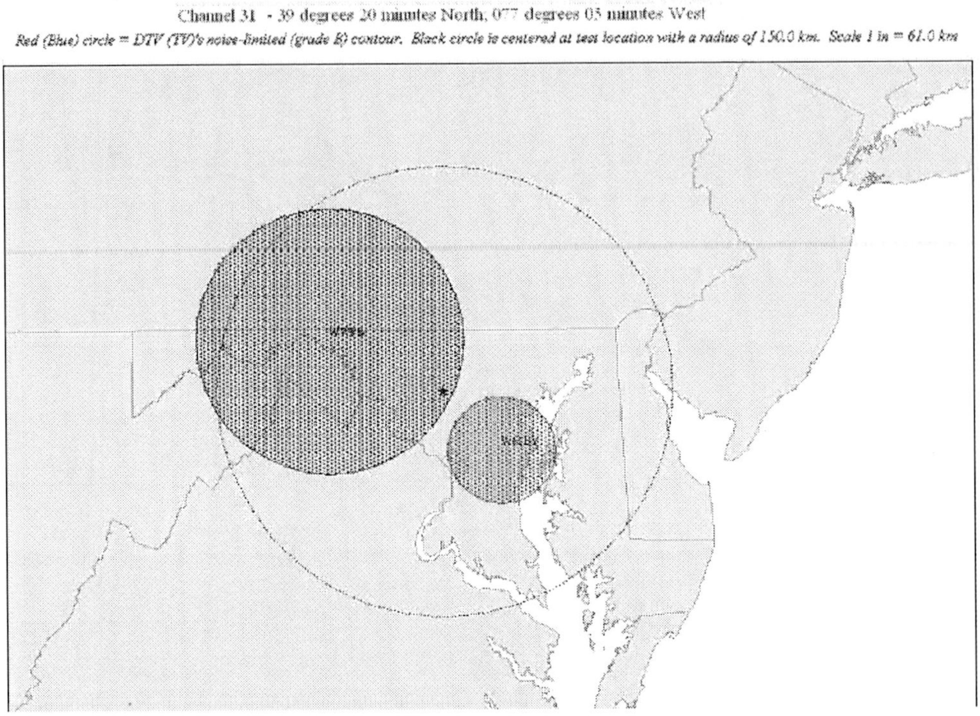

Figure 71. Channel 31 Service Contours Relative to Test Site 3

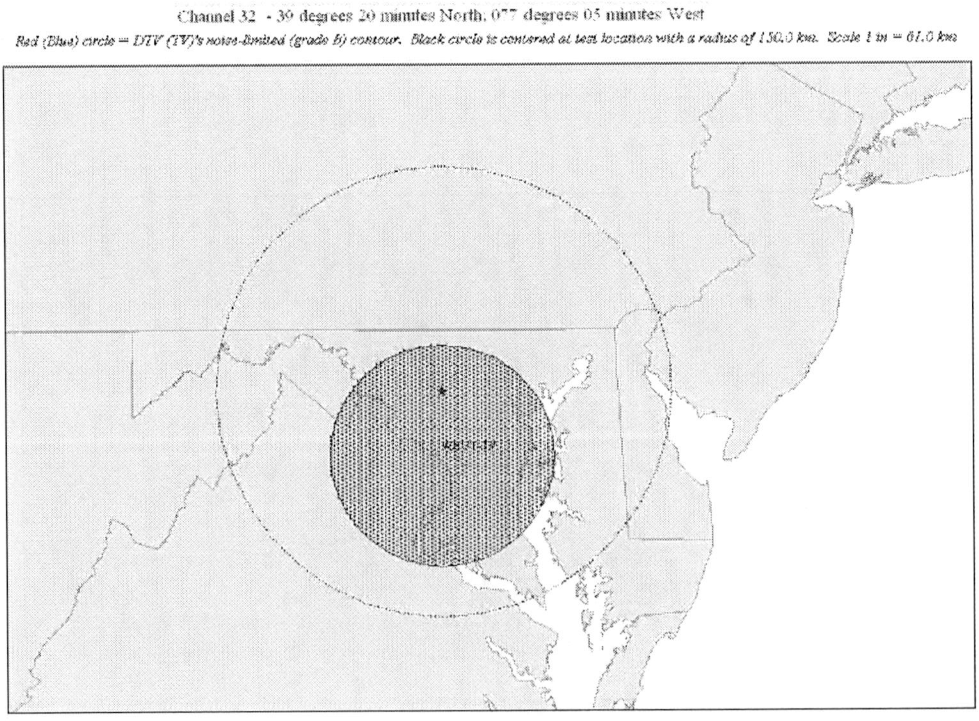

Figure 72. Channel 32 (WHUT-TV) Service Contour Relative to Test Site 3

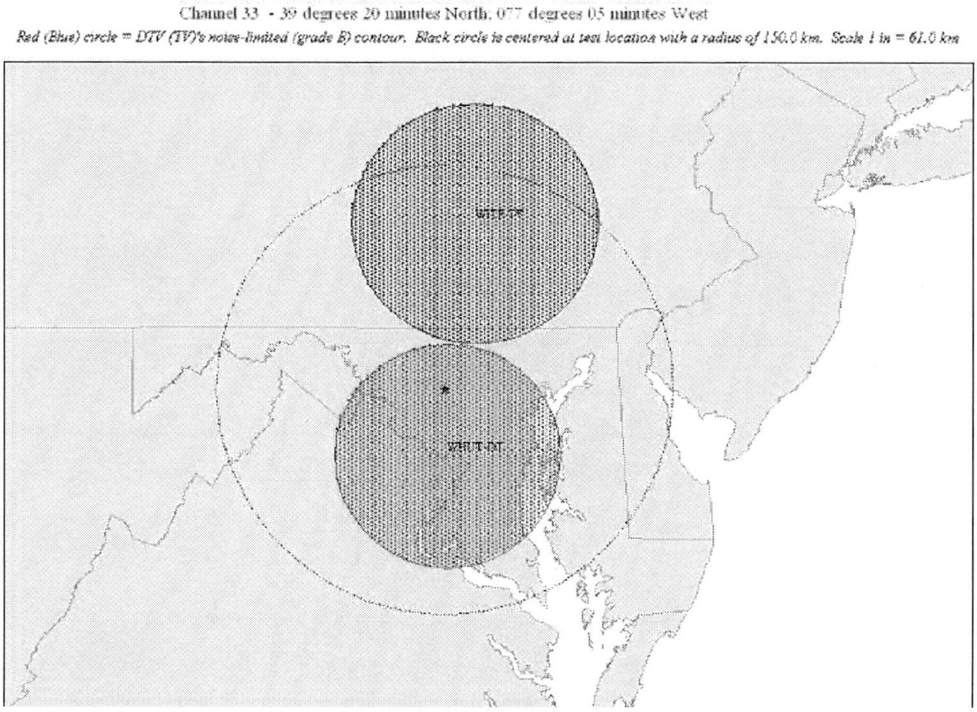

Figure 73. Channel 33 (WHUT-DT) Service Contour Relative to Test Site 3

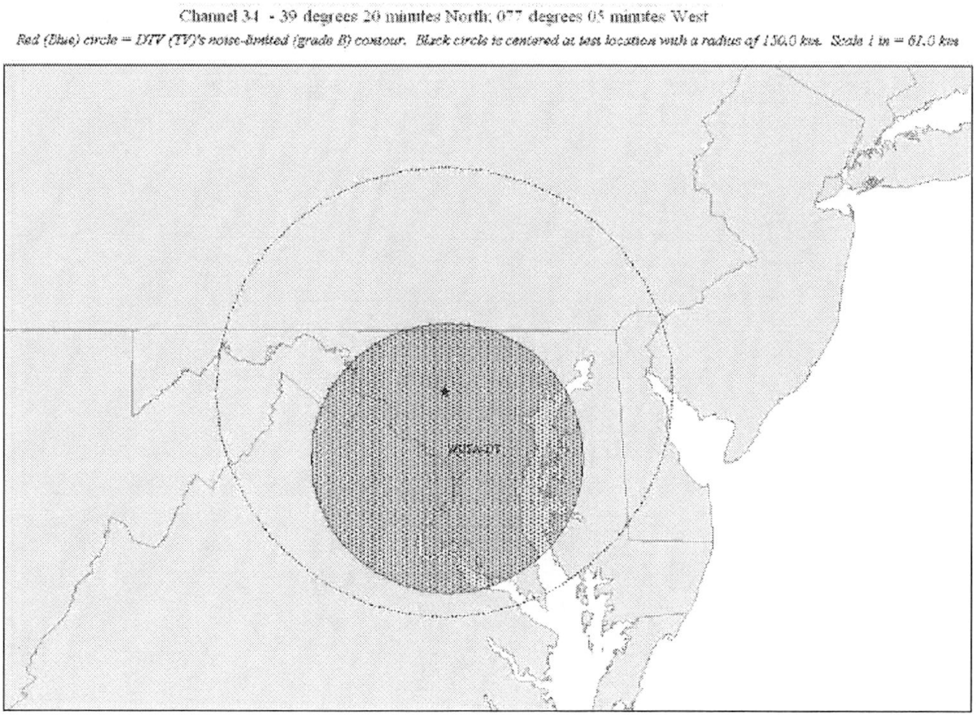

Figure 74. Channel 34 (WUSA-DT) Service Contour Relative to Test Site 3

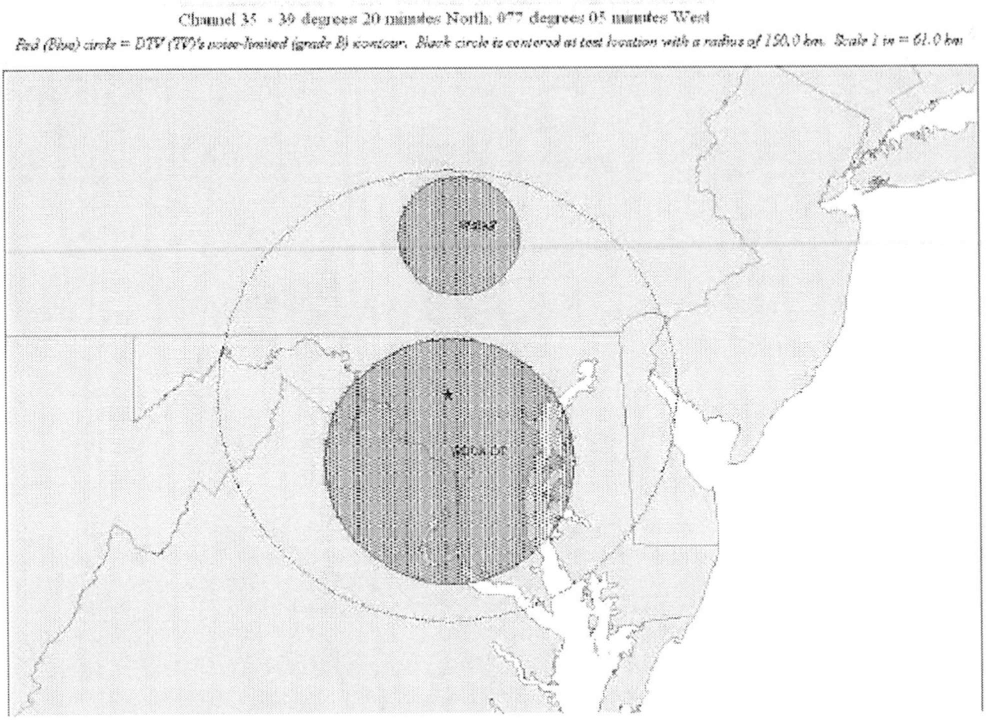

Figure 75. Channel 35 (WDCA-DT) Service Contour Relative to Test Site 3

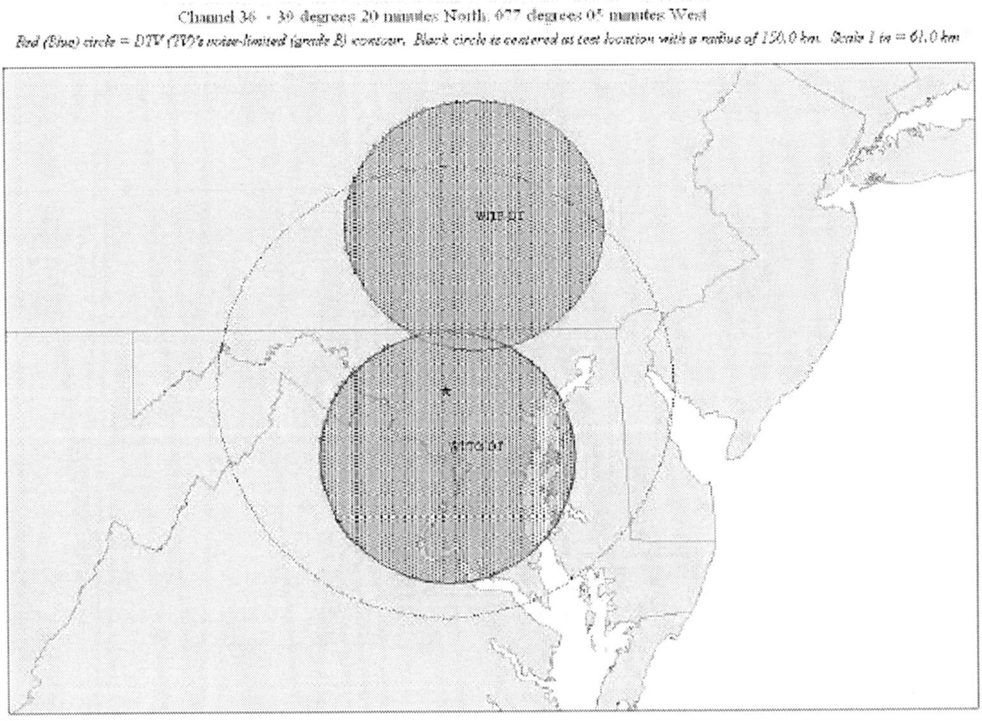

Figure 76. Channel 36 (WTTG-DT) Service Contour Relative to Test Site 3

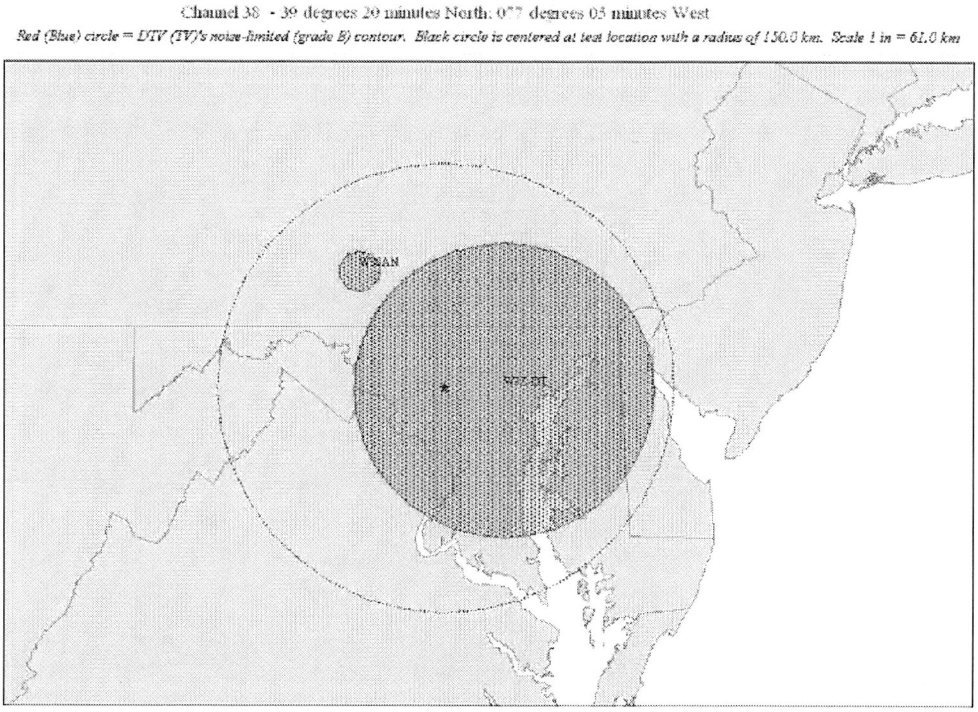

Figure 77. Channel 38 (WJZ-DT) Service Contour Relative to Test Site 3

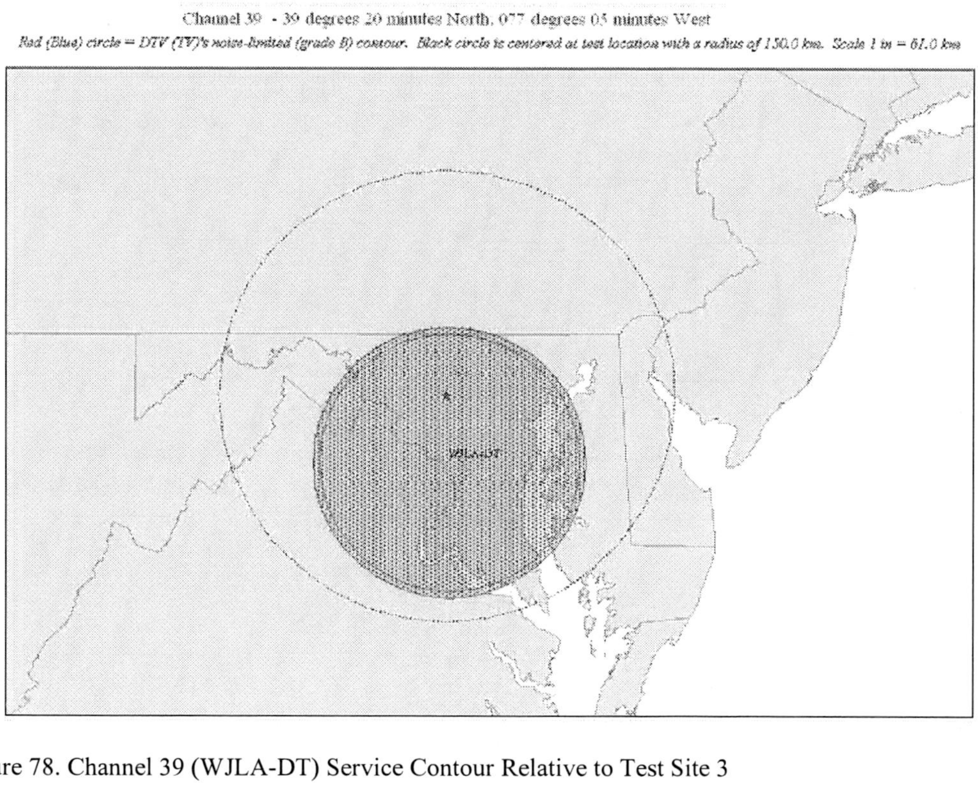

Figure 78. Channel 39 (WJLA-DT) Service Contour Relative to Test Site 3

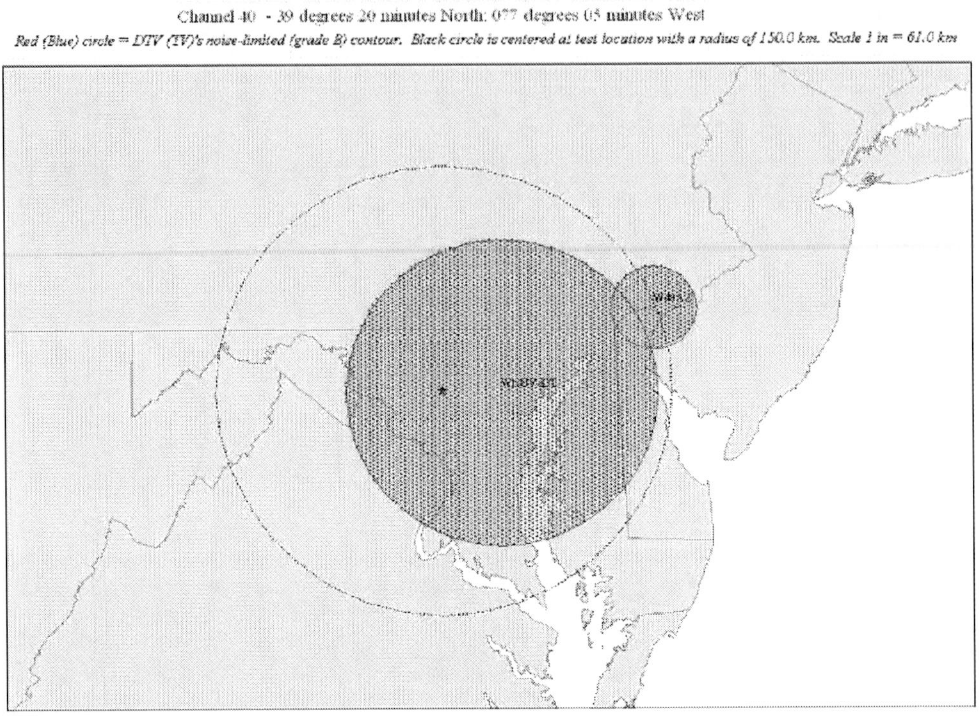

Figure 79. Channel 40 (WNUV-DT) Service Contour Relative to Test Site 3

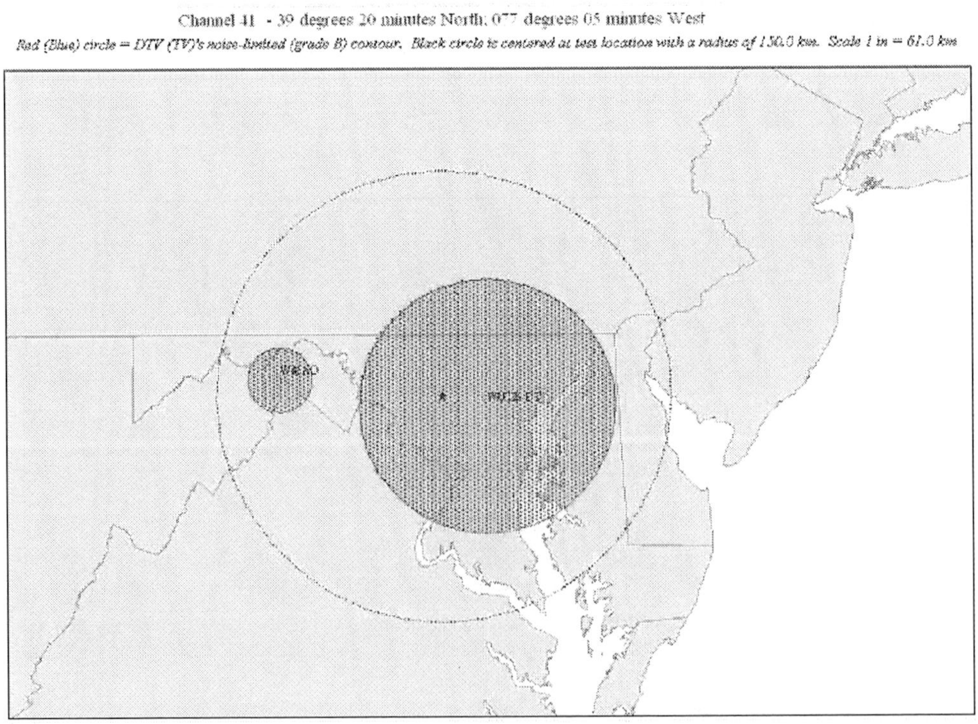

Figure 80. Channel 41 (WUTB-DT) Service Contour Relative to Test Site 3

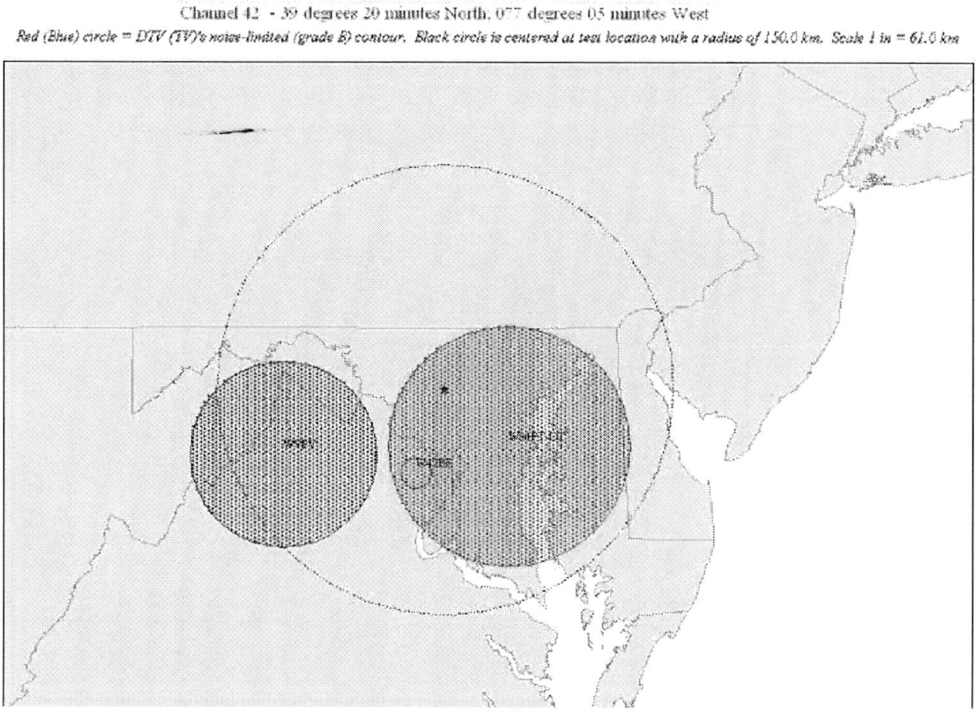

Figure 81. Channel 42 (WMPT-DT) Service Contour Relative to Test Site 3

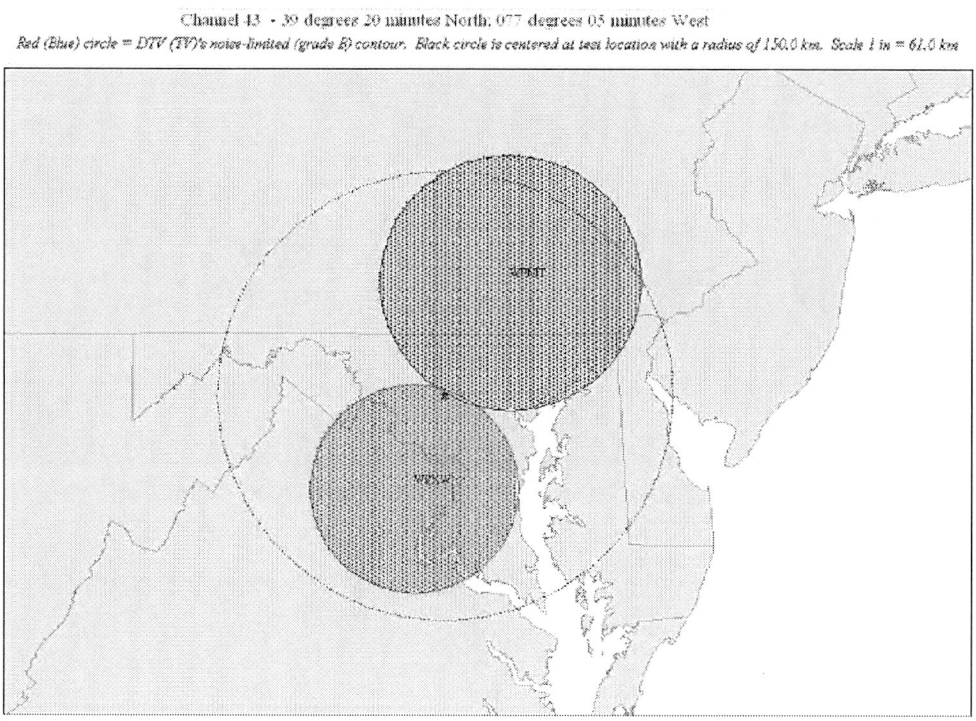

Figure 82. Channel 43 (WPXW-DT) Service Contour Relative to Test Site 3

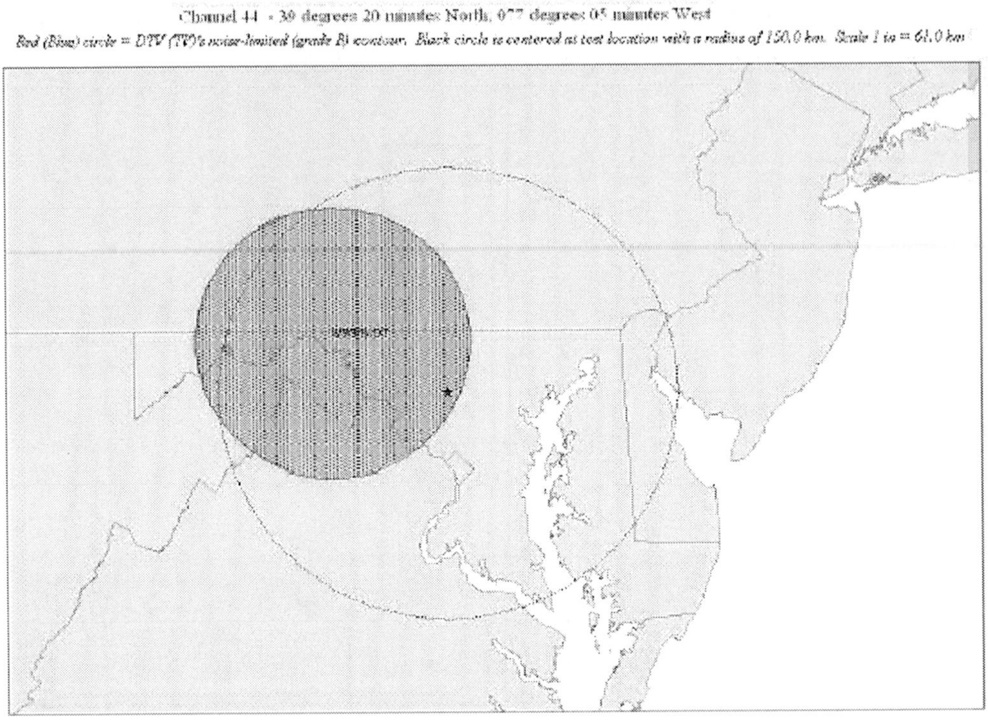

Figure 83. Channel 44 Service Contours Relative to Test Site 3

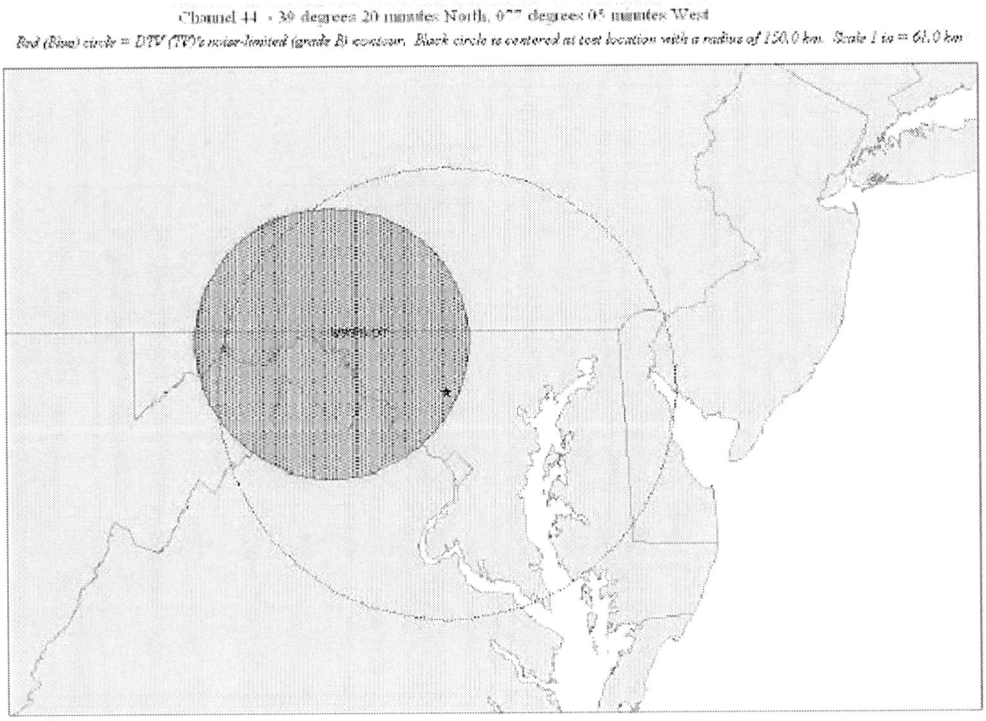

Figure 84. Channel 45 (WBFF-TV) Service Contour Relative to Test Site 3

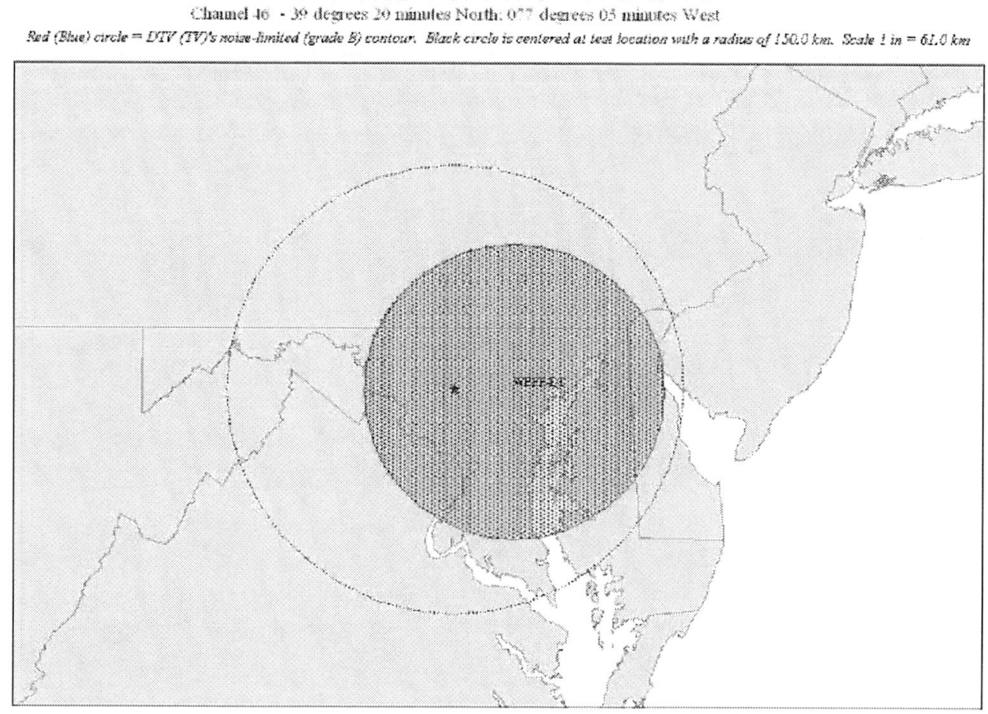

Figure 85. Channel 46 (WBFF-DT) Service Contour Relative to Test Site 3

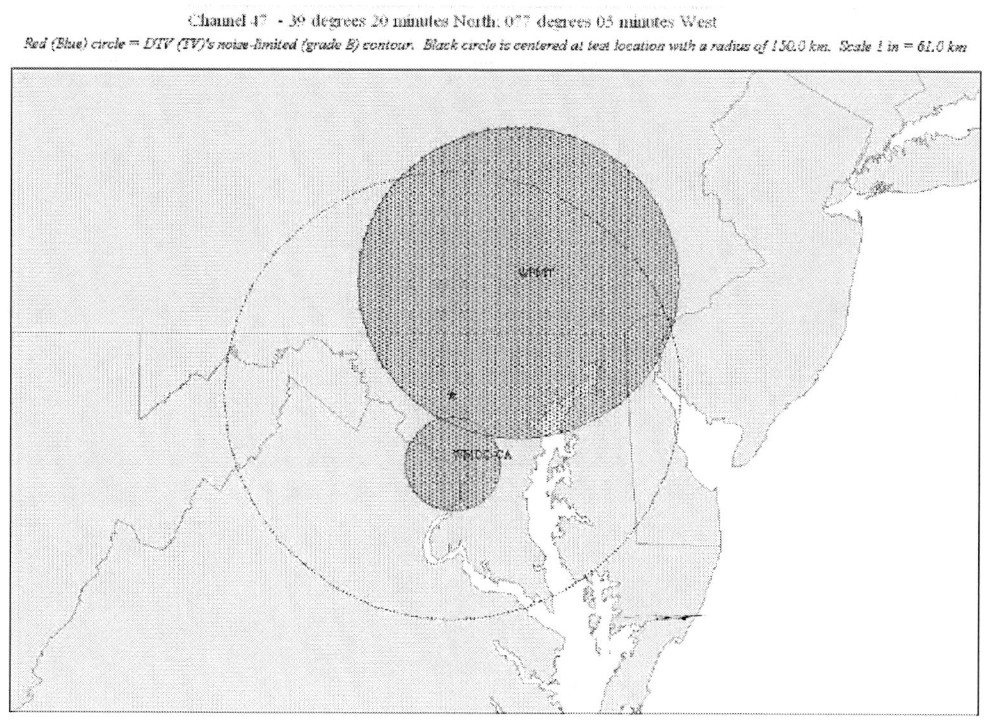

Figure 86. Channel 47 (WPMT-DT) Service Contour Relative to Test Site 3

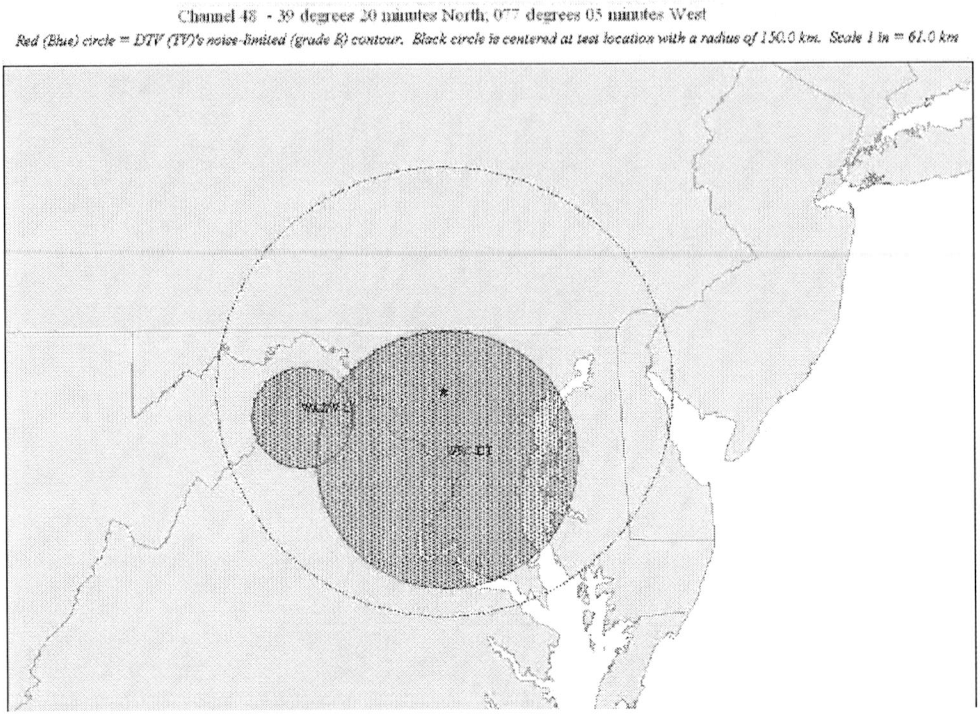

Figure 87. Channel 48 (WRC-DT) Service Contour Relative to Test Site 3

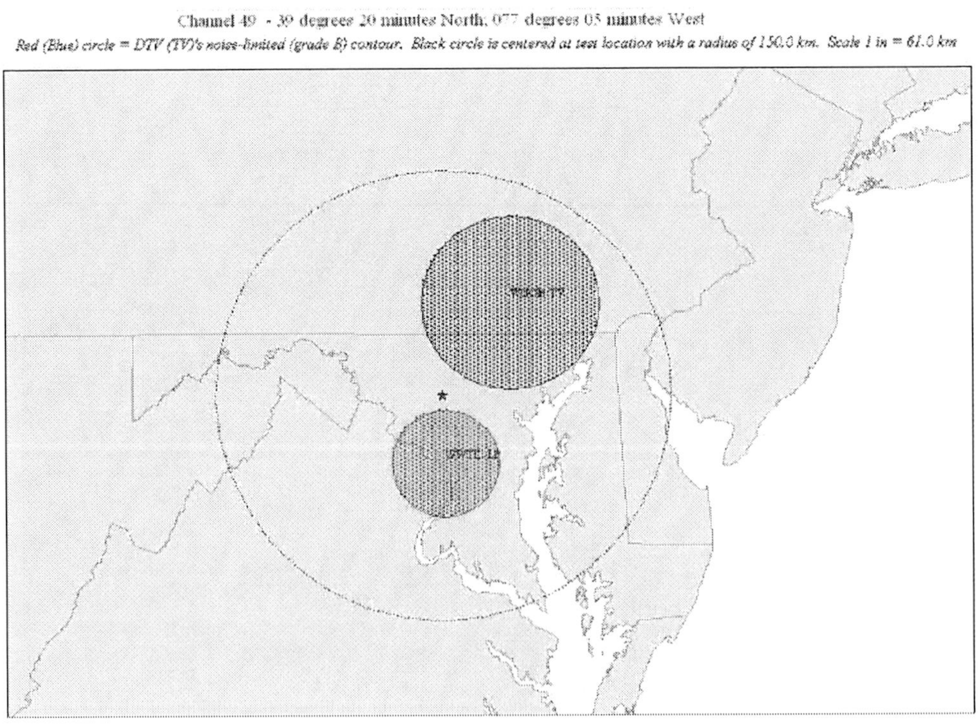

Figure 88. Channel 49 Service Contours Relative to Test Site 3

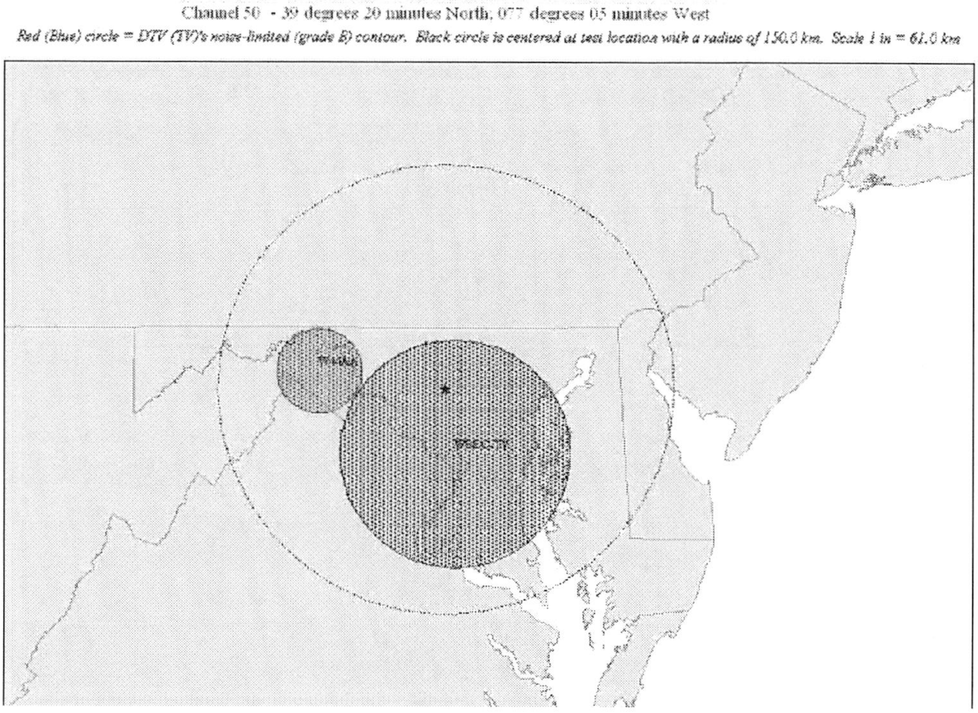

Figure 89. Channel 50 (WDCW-TV) Service Contour Relative to Test Site 3

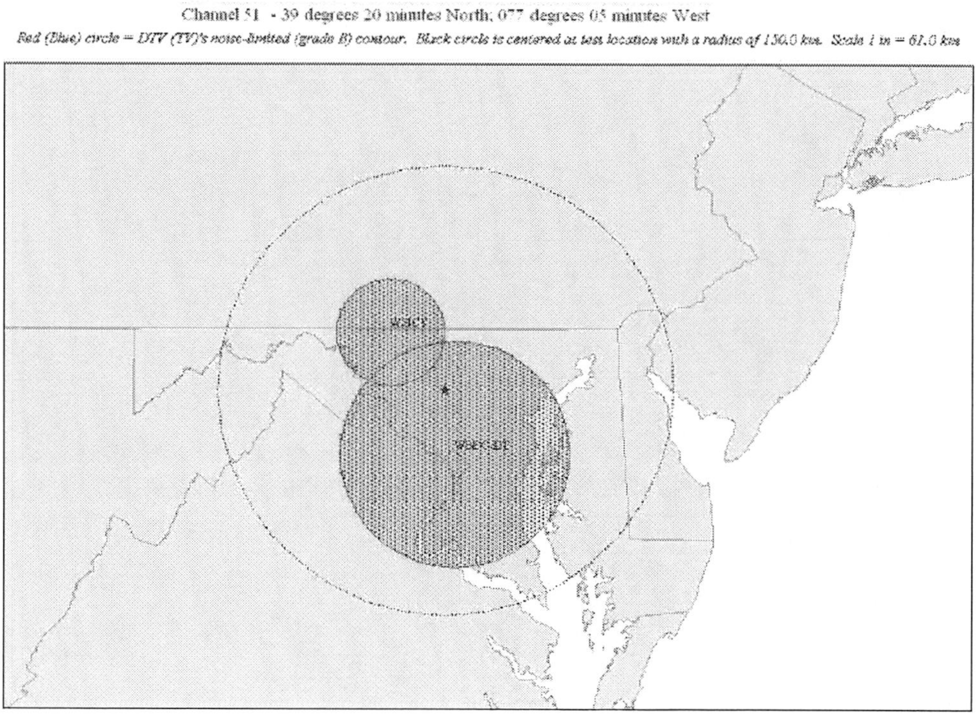

Figure 90. Channel 51 (WDCW-DT) Service Contour Relative to Test Site 3

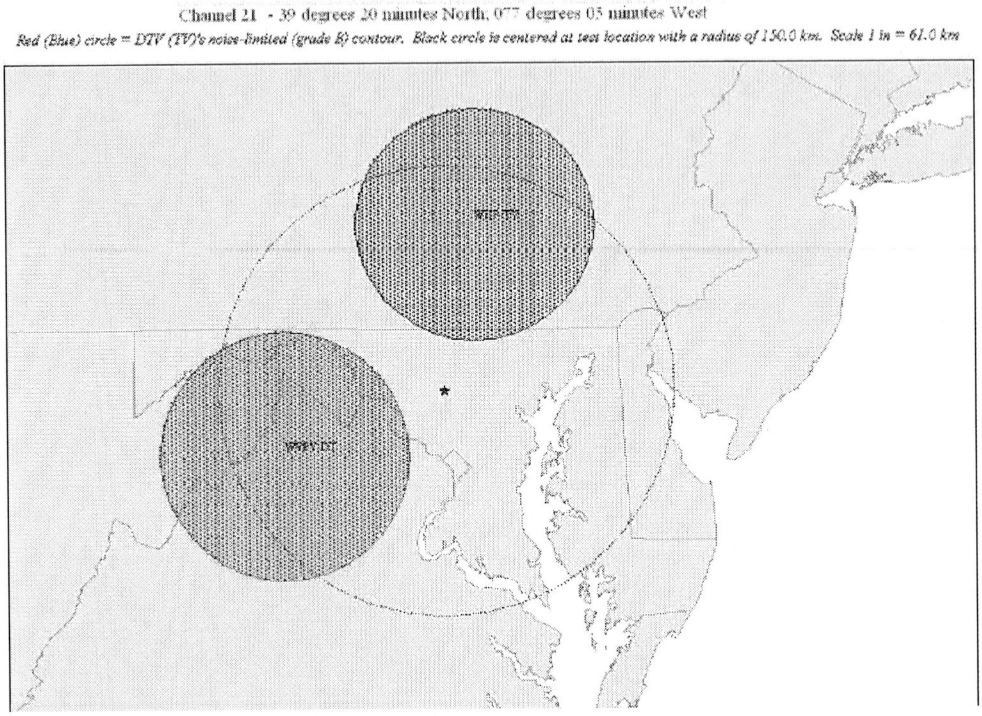

Figure 91. Channel 21 Service Contours Relative to Test Site 4

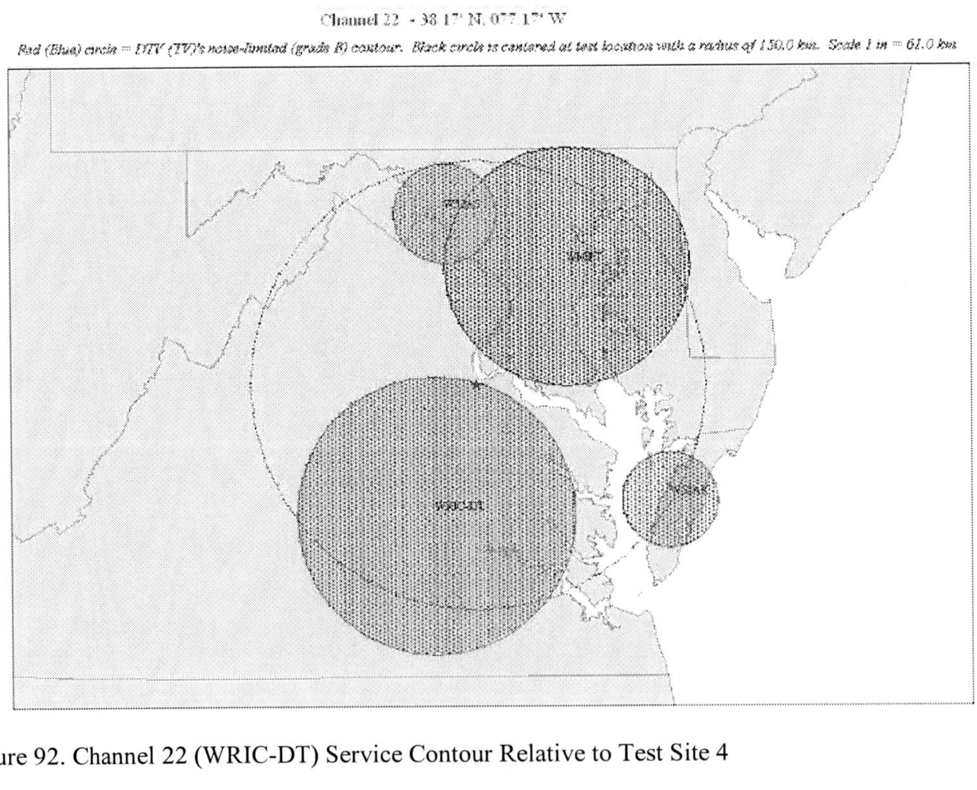

Figure 92. Channel 22 (WRIC-DT) Service Contour Relative to Test Site 4

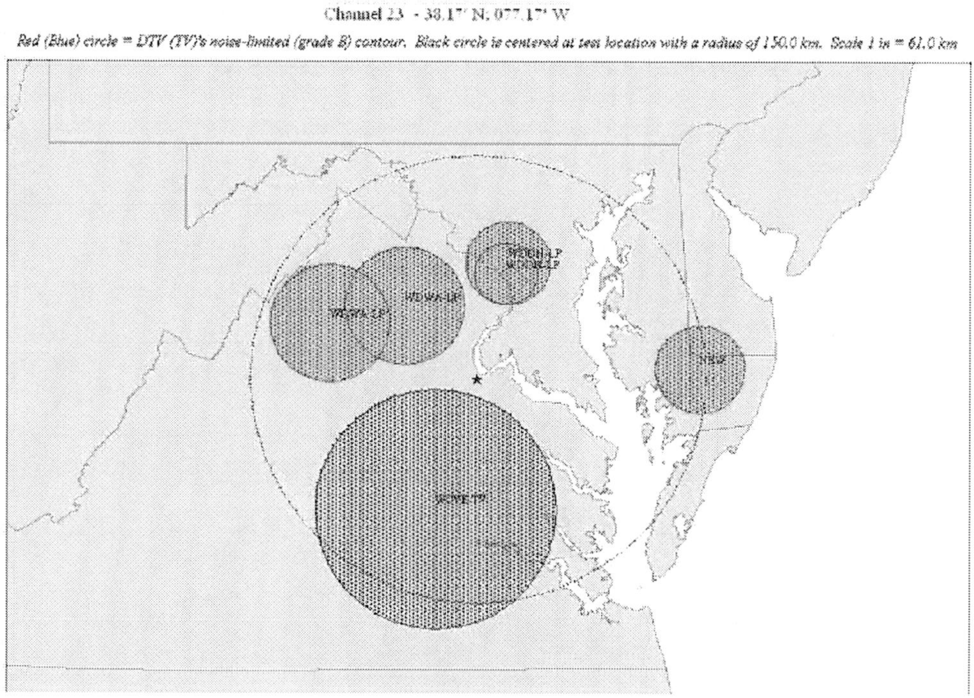

Figure 93. Channel 23 Service Contours Relative to Test Site 4

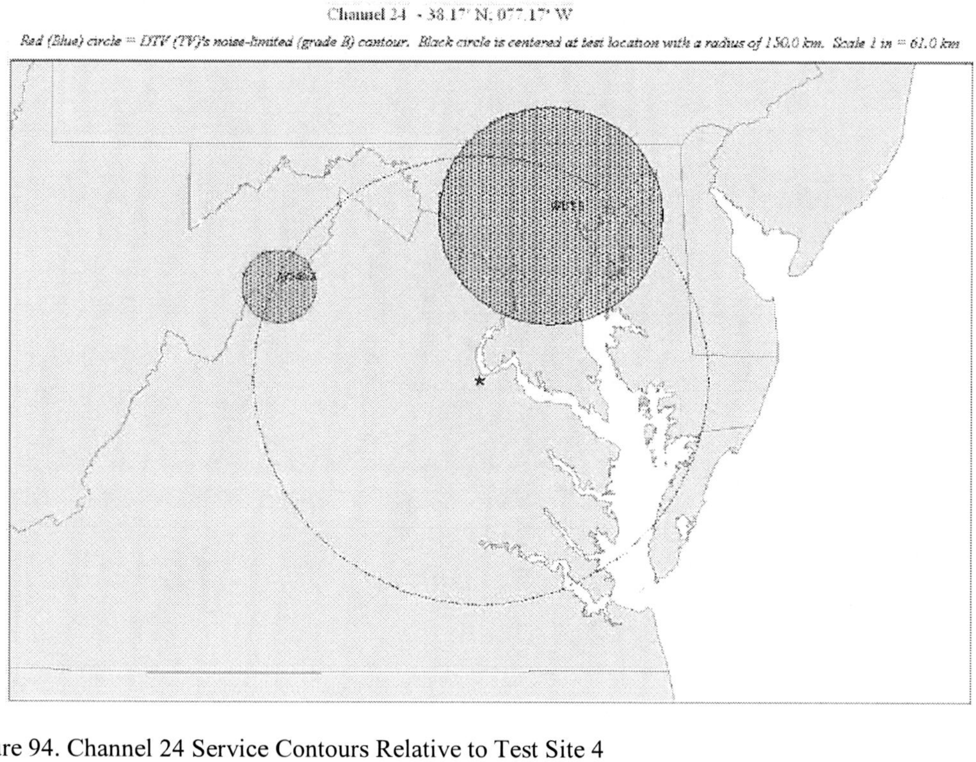

Figure 94. Channel 24 Service Contours Relative to Test Site 4

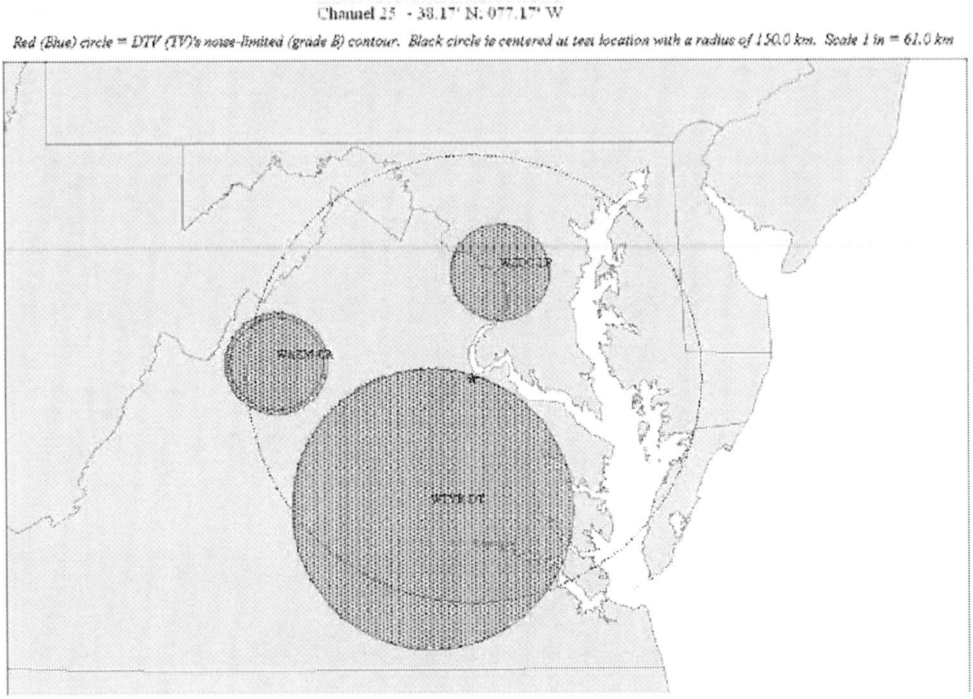

Figure 95. Channel 25 (WTVR-DT) Service Contour Relative to Test Site 4

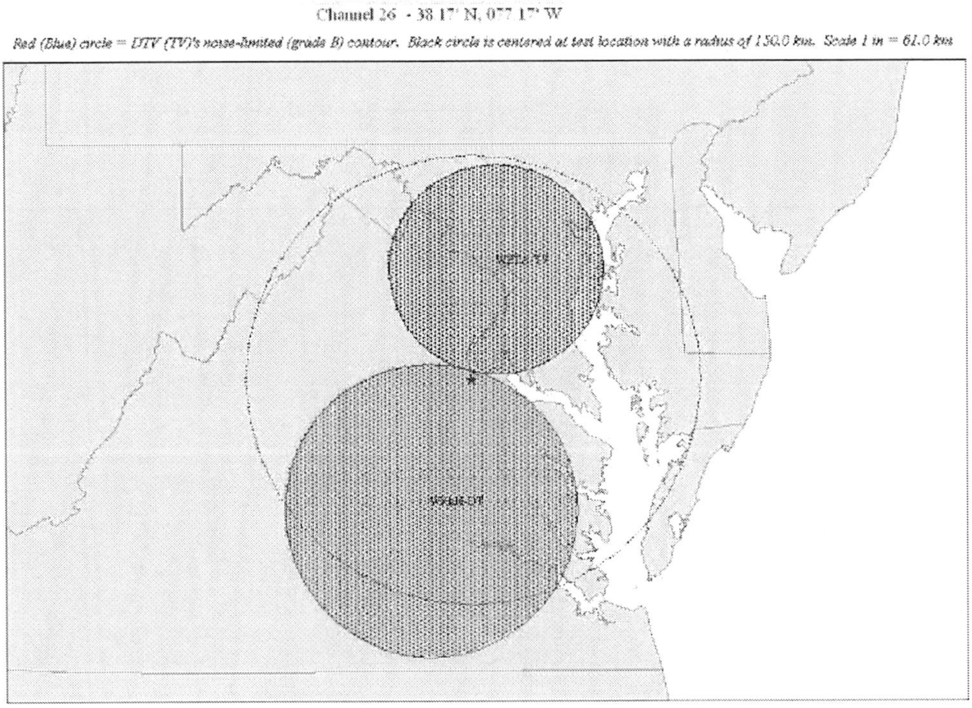

Figure 96. Channel 26 (WRLH-DT) Service Contour Relative to Test Site 4

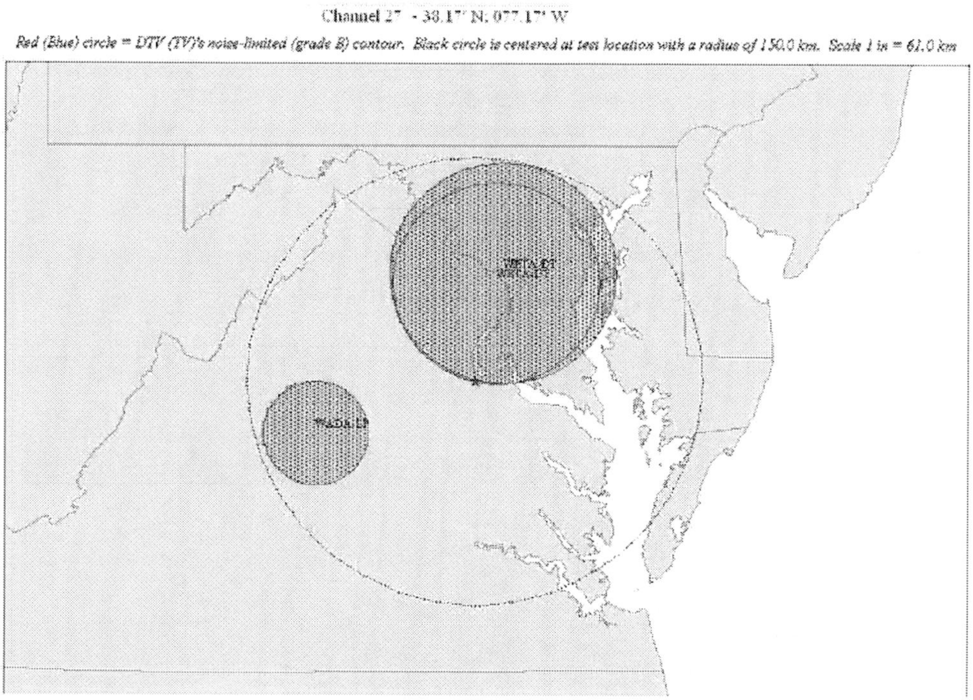

Figure 97. Channel 27 (WETA-DT) Service Contour Relative to Test Site 4

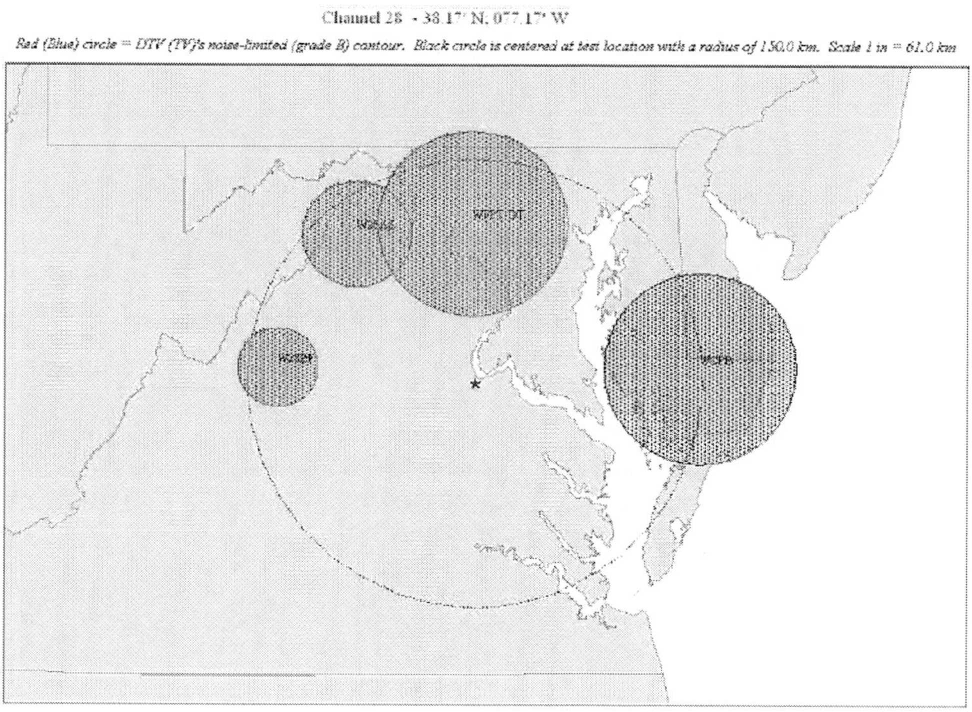

Figure 98. Channel 28 Service Contours Relative to Test Site 4

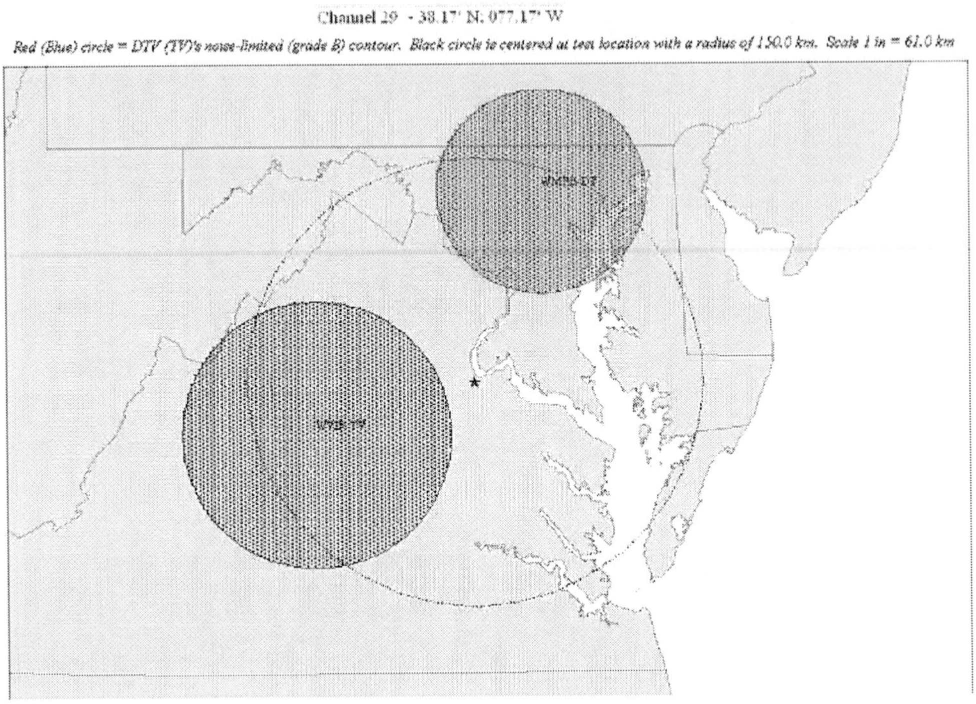

Figure 99. Channel 29 Service Contours Relative to Test Site 4

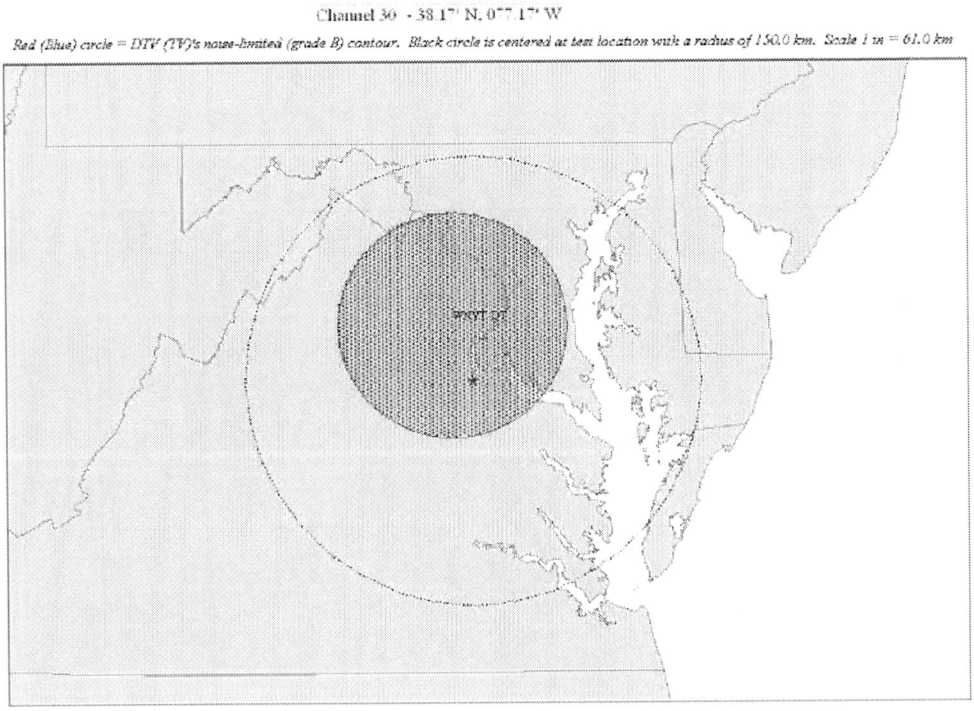

Figure 100. Channel 30 (WNVT-DT) Service Contour Relative to Test Site 4

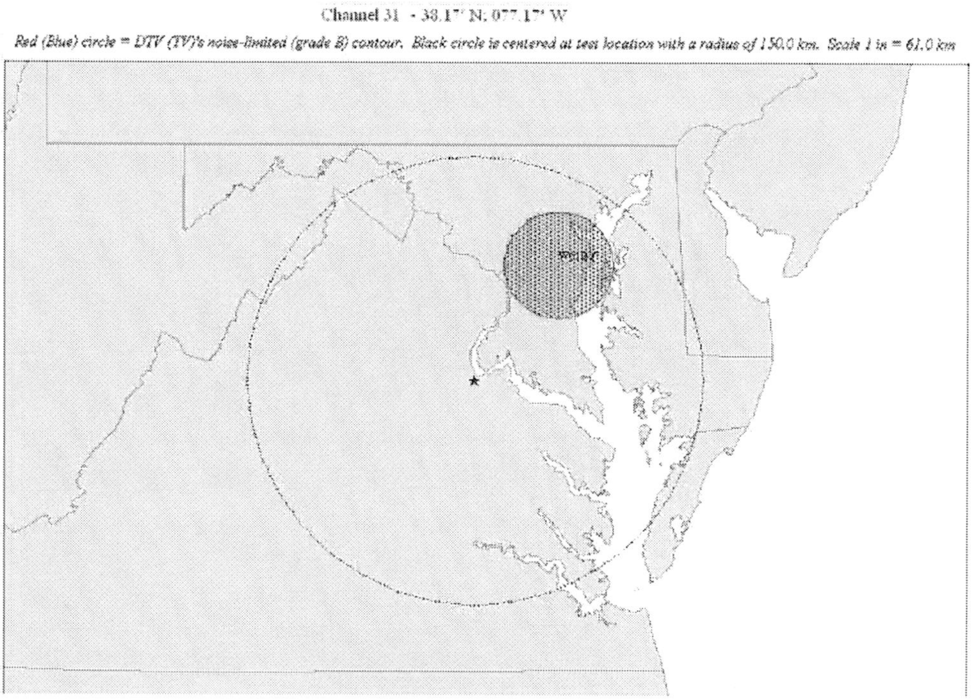

Figure 101. Channel 31 Service Contour Relative to Test Site 4

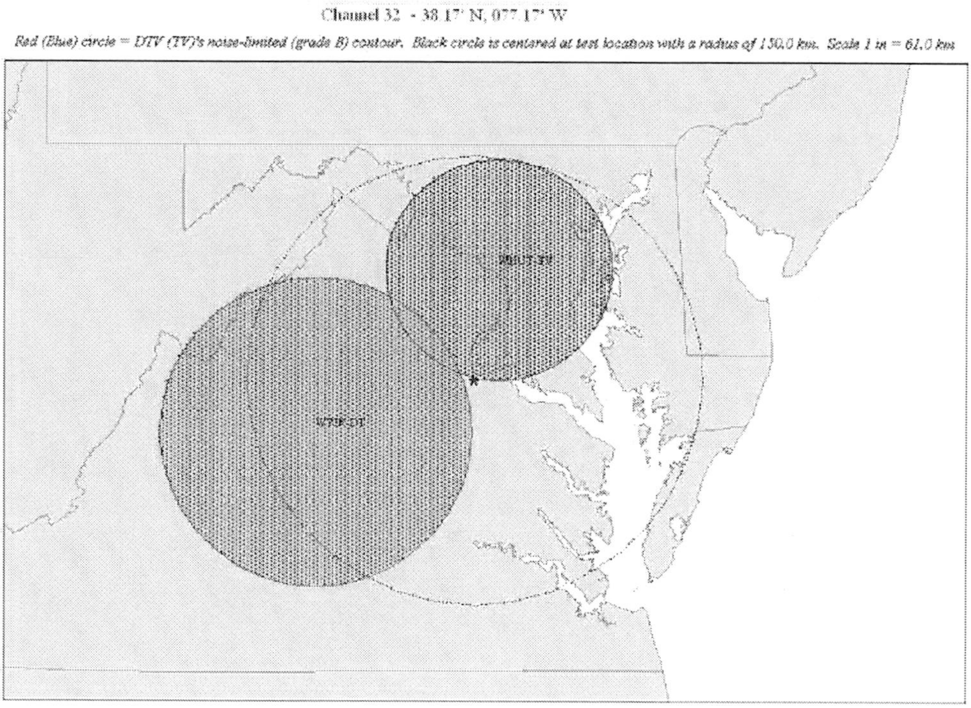

Figure 102. Channel 32 Service Contours Relative to Test Site 4

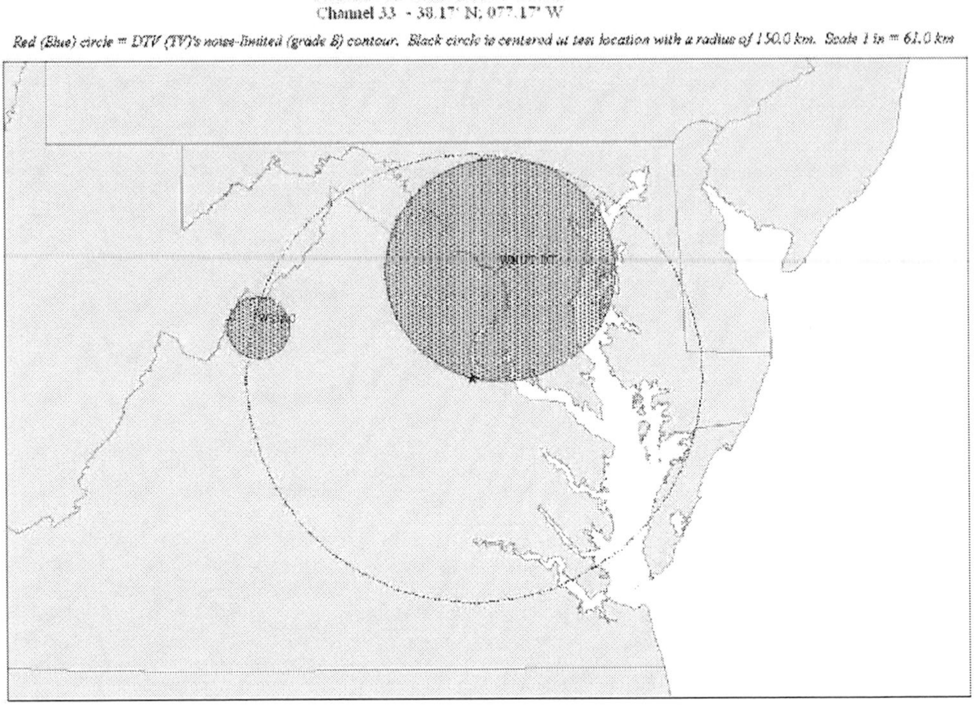

Figure 103. Channel 33 (WHUT-DT) Service Contour Relative to Test Site 4

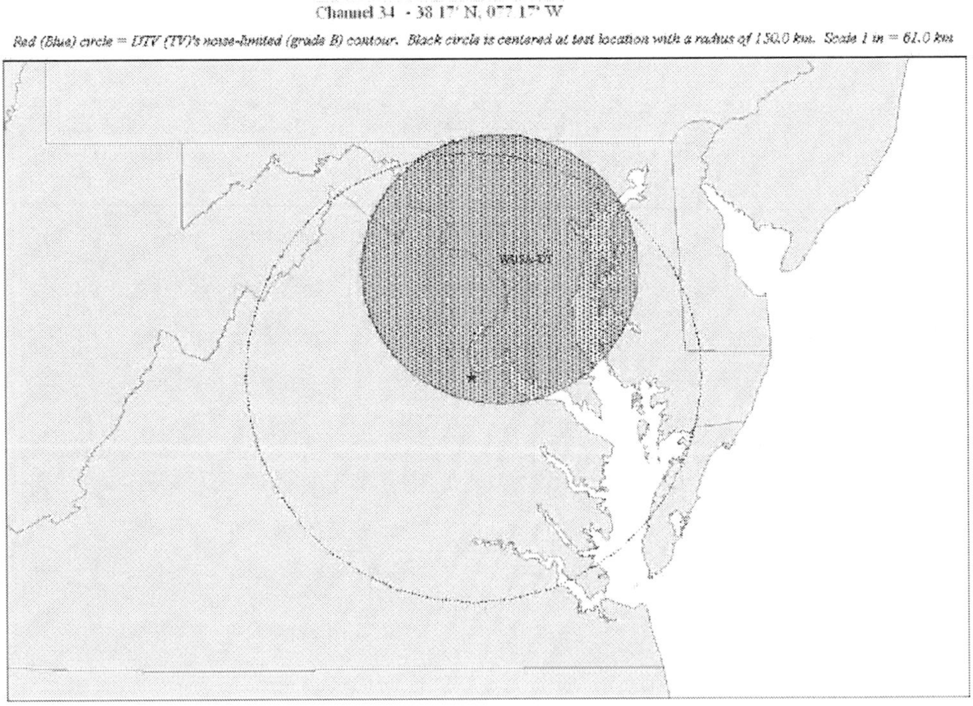

Figure 104. Channel 34 (WUSA-DT) Service Contour Relative to Test Site 4

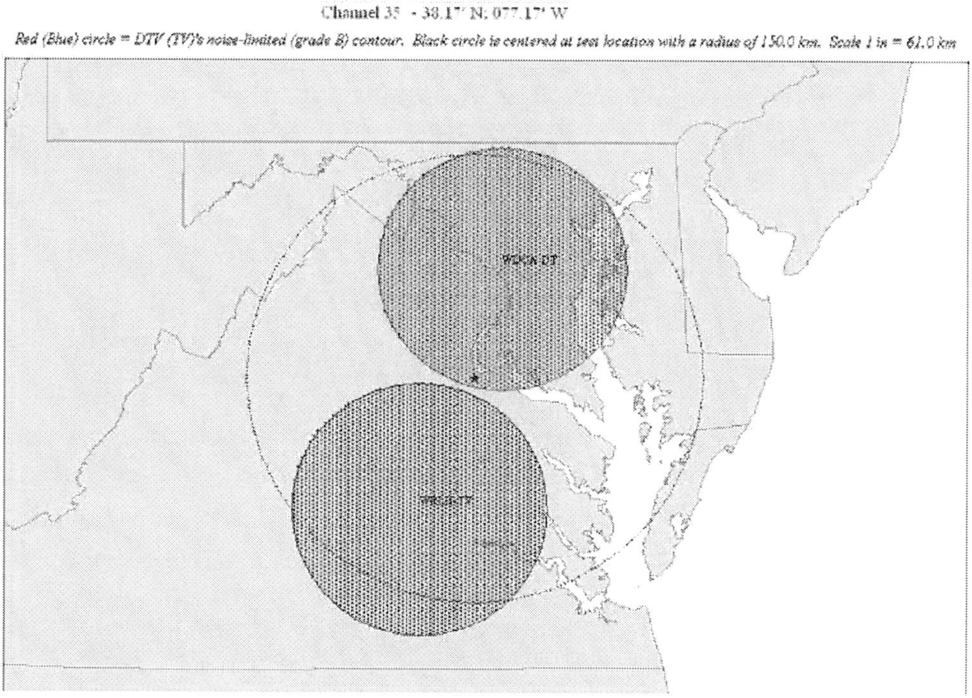

Figure 105. Channel 35 (WDCA-DT) Service Contour Relative to Test Site 4

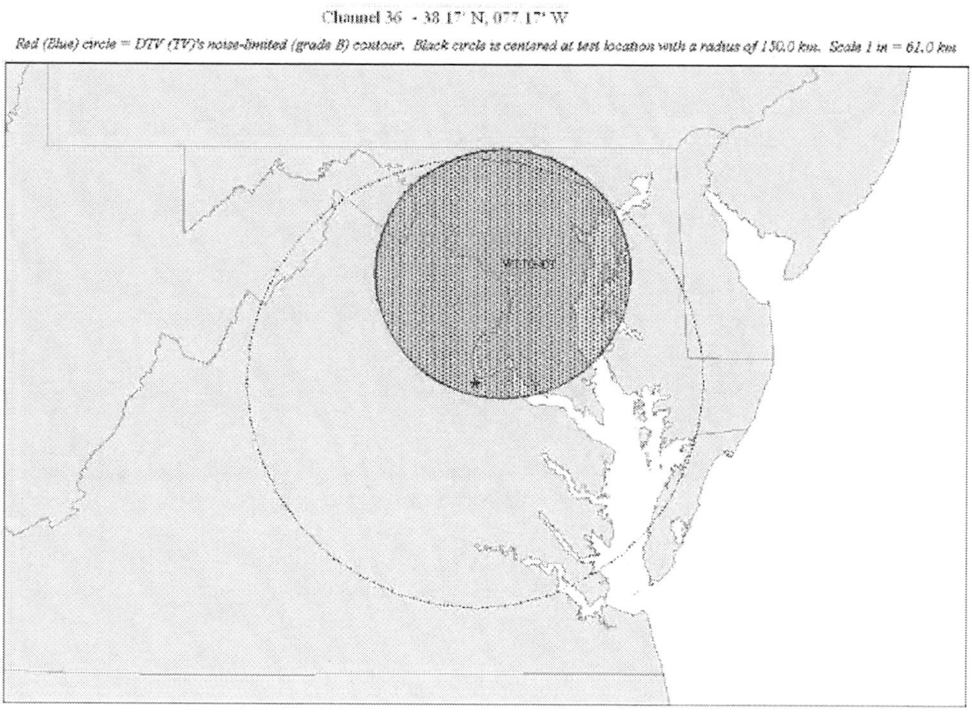

Figure 106. Channel 36 (WTTG-DT) Service Contour Relative to Test Site 4

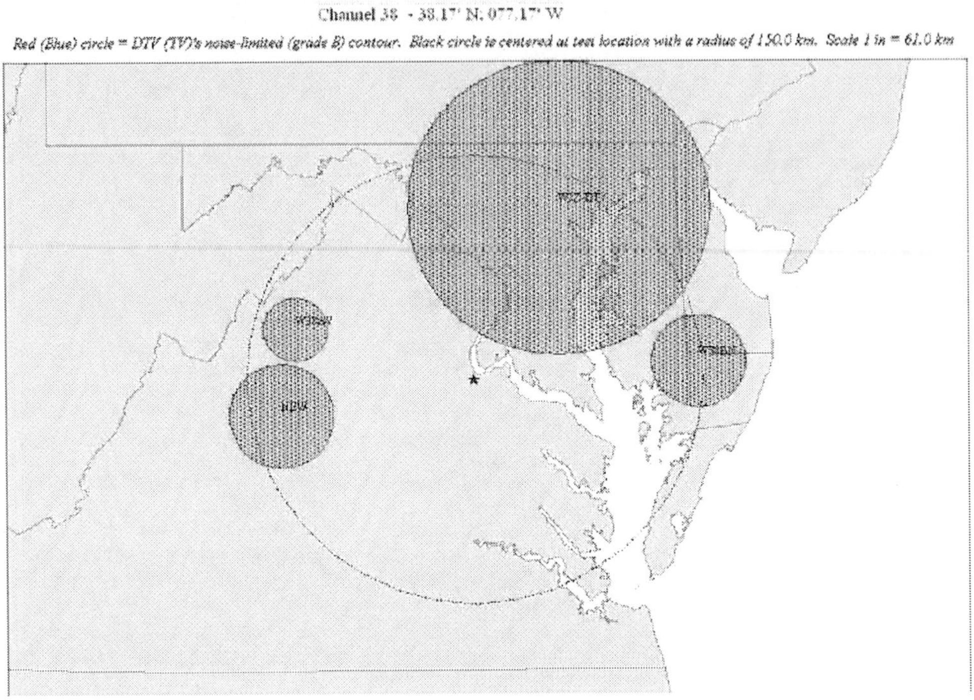

Figure 107. Channel 38 Service Contours Relative to Test Site 4

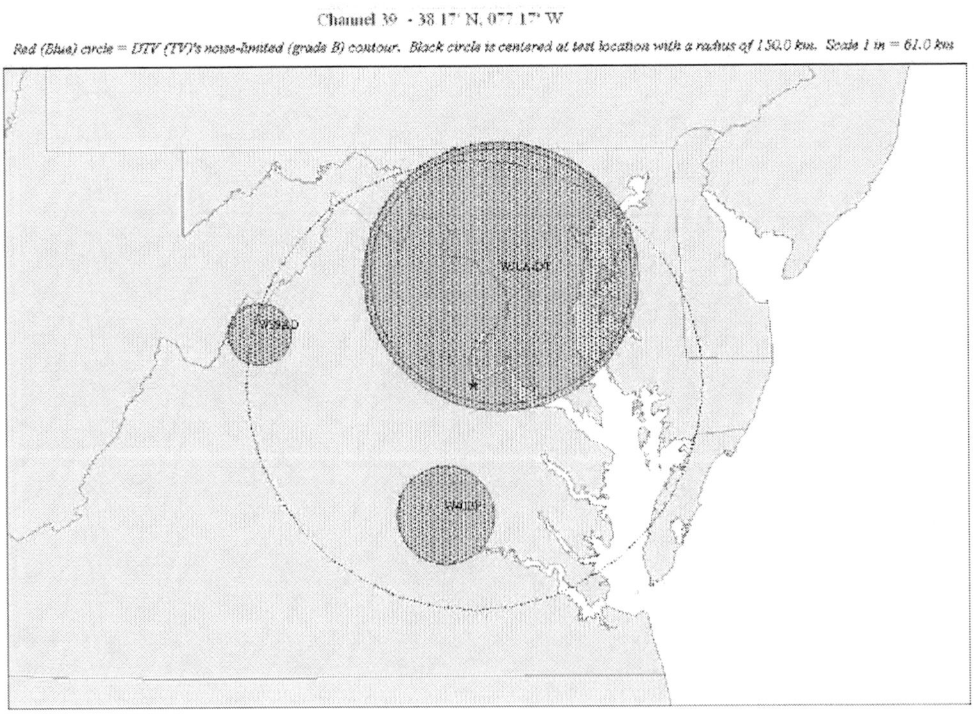

Figure 108. Channel 39 (WJLA-DT) Service Contour Relative to Test Site 4

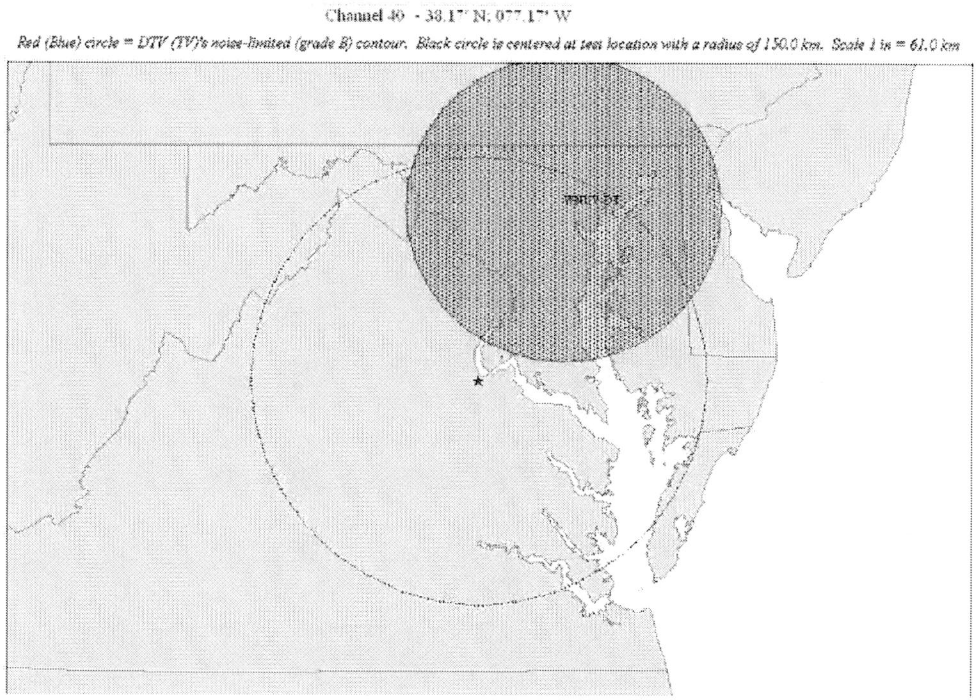

Figure 109. Channel 40 Service Contour Relative to Test Site 4

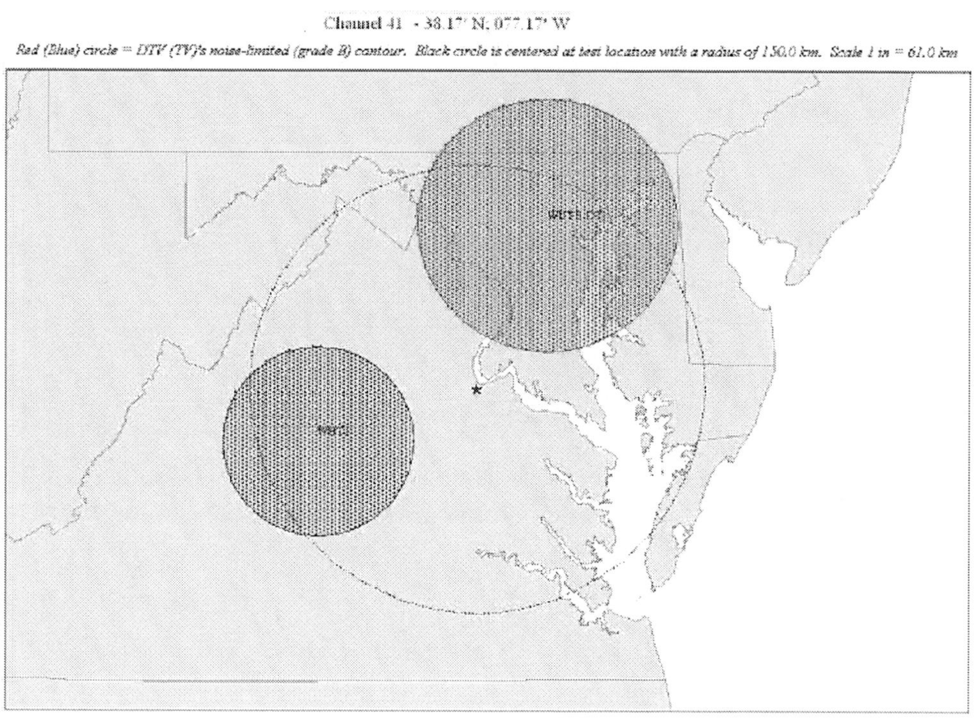

Figure 110. Channel 41 Service Contours Relative to Test Site 4

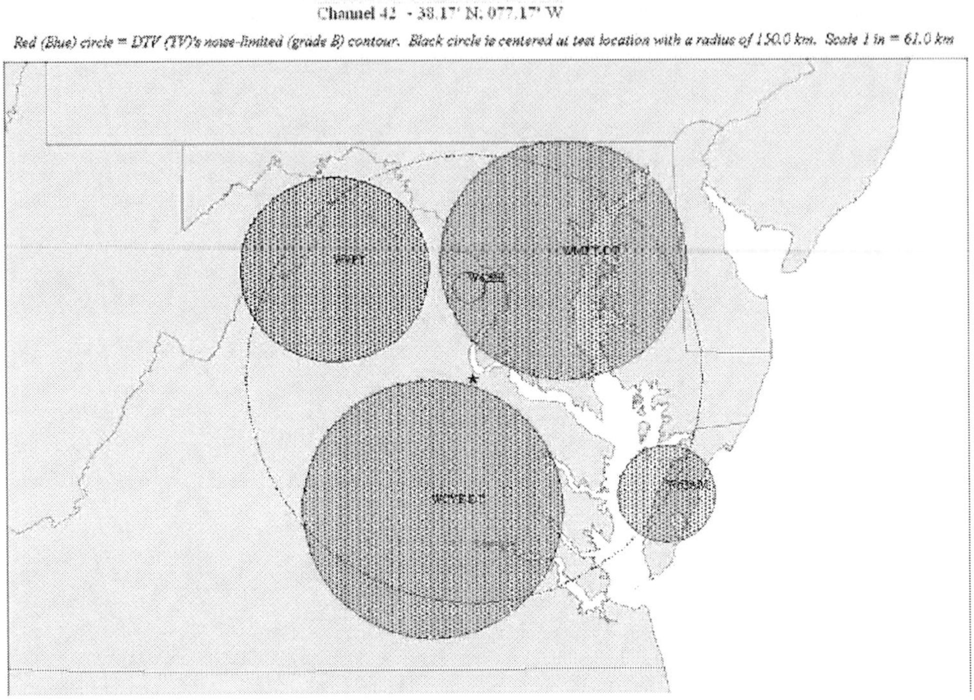

Figure 111. Channel 42 Service Contours Relative to Test Site 4

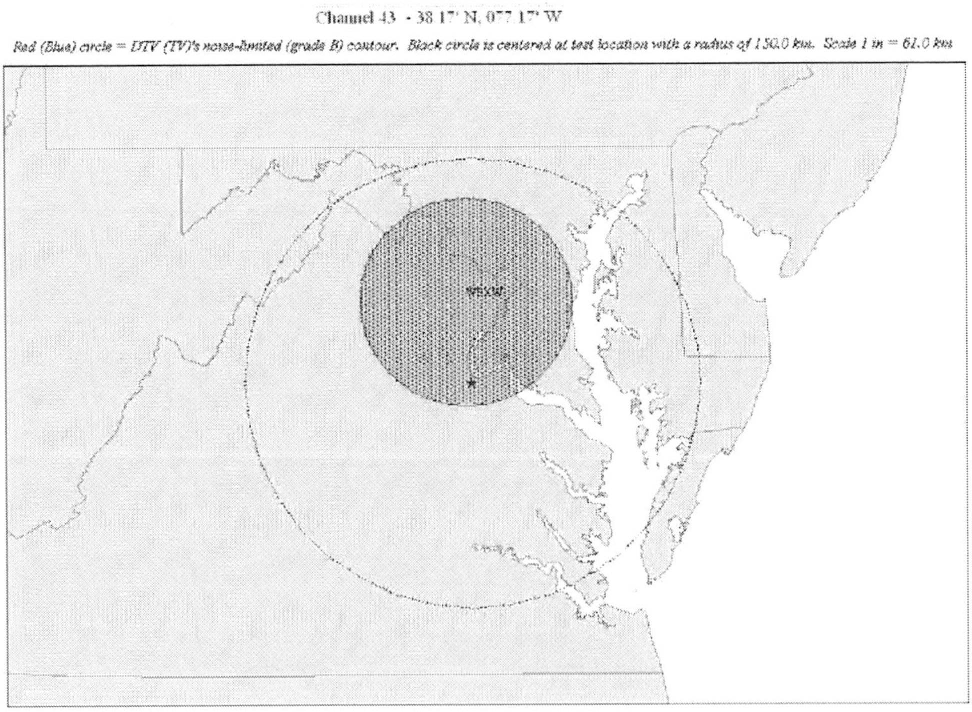

Figure 112. Channel 43 (WPXW-DT) Service Contour Relative to Test Site 4

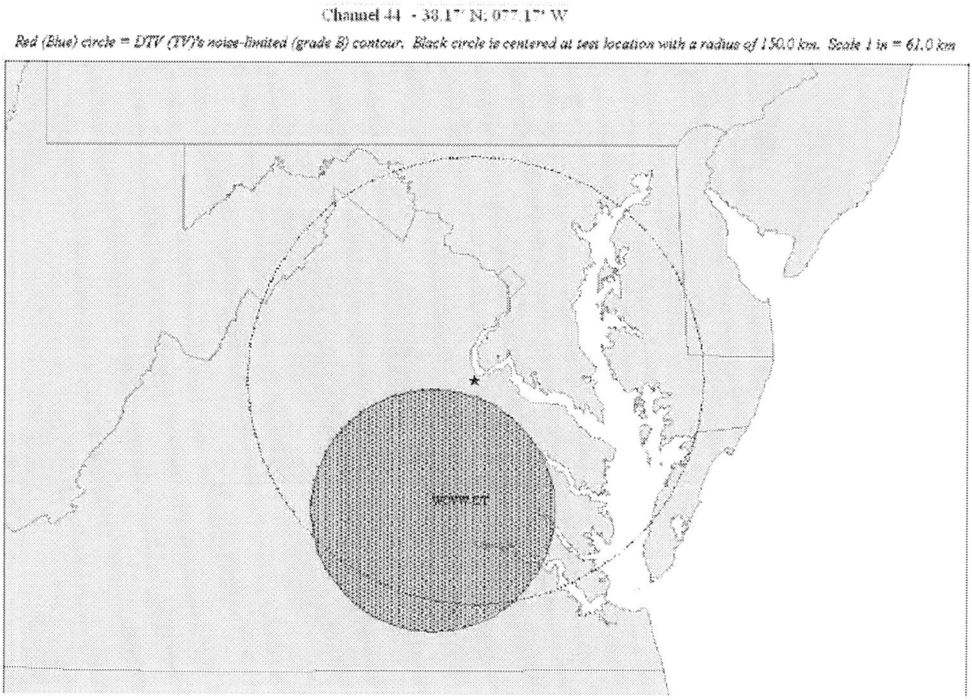

Figure 113. Channel 44 Service Contour Relative to Test Site 4

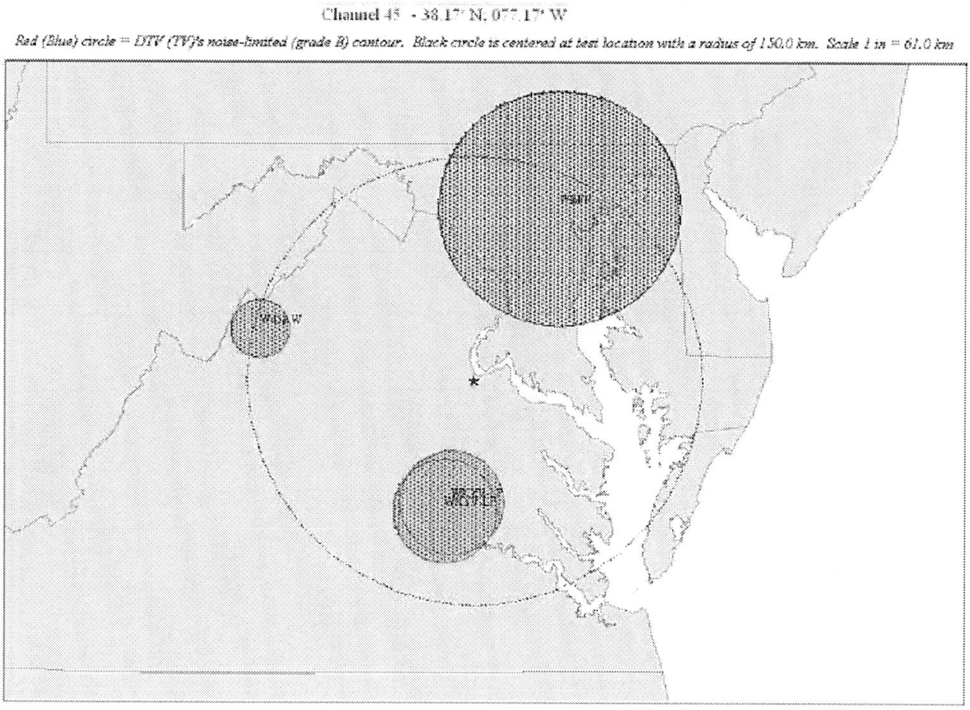

Figure 114. Channel 45 Service Contours Relative to Test Site 4

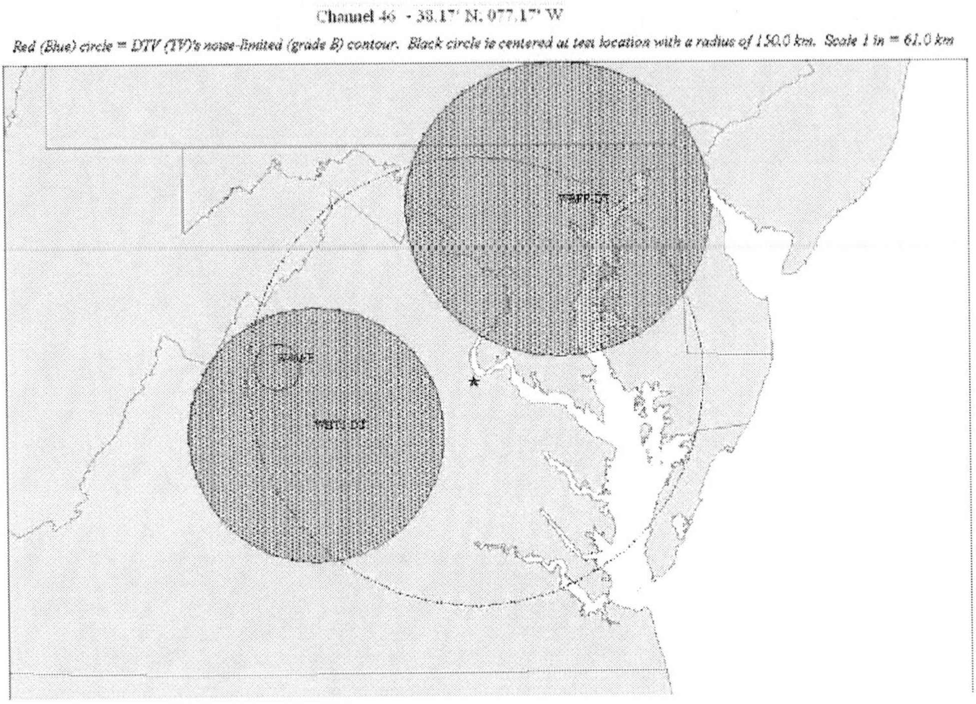

Figure 115. Channel 46 Service Contours Relative to Test Site 4

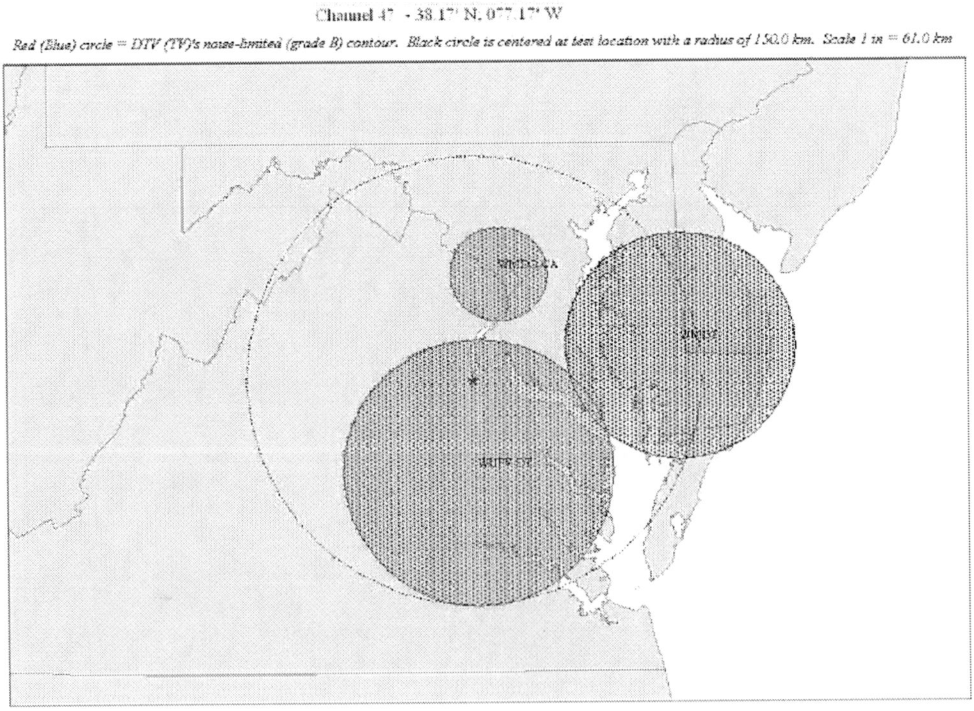

Figure 116. Channel 47 (WUPV-DT) Service Contour Relative to Test Site 4

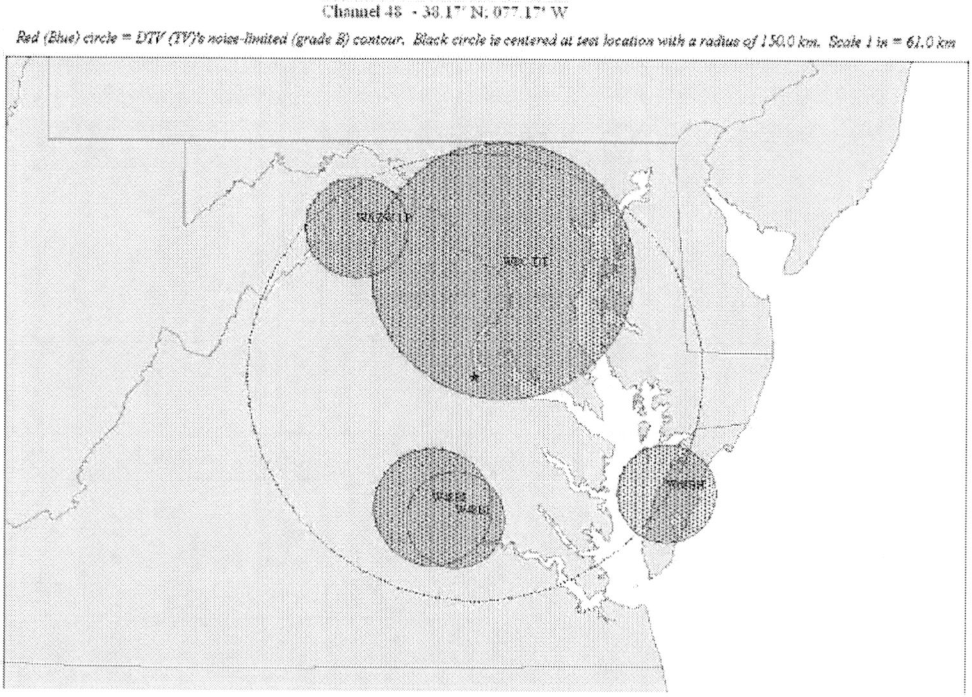

Figure 117. Channel 48 (WRC-DT) Service Contour Relative to Test Site 4

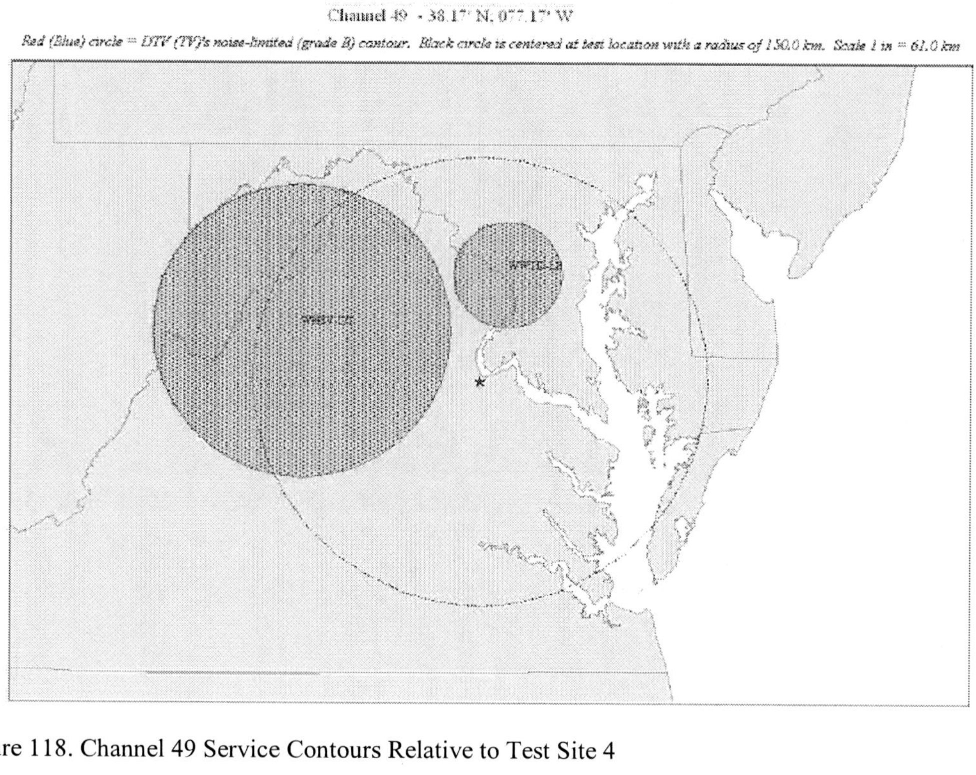

Figure 118. Channel 49 Service Contours Relative to Test Site 4

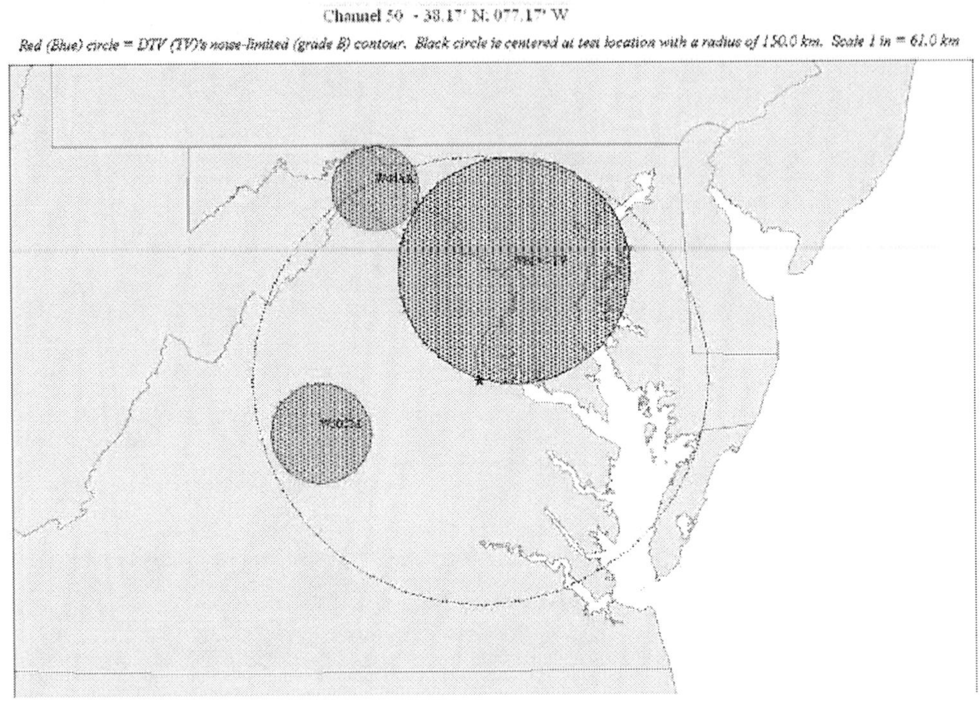

Figure 119. Channel 50 Service Contours Relative to Test Site 4

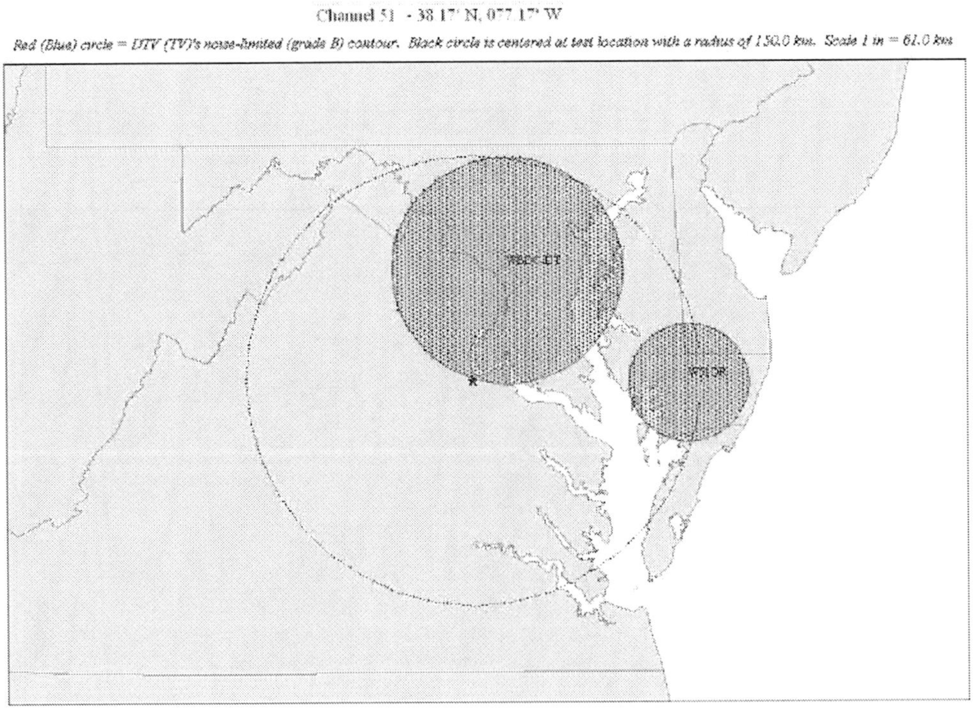

Figure 120. Channel 51 (WDCW-DT) Service Contour Relative to Test Site 4

In: Tapping the Television White Space Spectrum: A Revolution... ISBN: 978-1-60692-956-8
Editor: Julie E. Hendriks

Chapter 3

FEDERAL COMMUNICATIONS COMMISSION, PLAN FOR TESTS OF PROTOTYPE PERSONAL/PORTABLE TV WHITE SPACE DEVICES (PHASE II)

Technical Research Branch

INTRODUCTION

The Commission is conducting a proceeding in ET Docket No. 04-186 to consider authorizing the operation of new, low power devices in the television (TV) broadcast spectrum at locations where channels are not being used for authorized services.[1] This unused spectrum is often referred to as the "TV white spaces."

As part of this proceeding, the Office of Engineering and Technology (OET) is conducting a test program on TV white space prototype devices. The purpose of this testing program is to provide additional information for the record that will be considered in assessing the interference potential of such devices and appropriate requirements.

Initial tests (Phase I) were completed in July of 2007.[2] The Office of Engineering and Technology on October 5, 2007, issued a public notice inviting submittal of prototype devices for further tests (Phase II).[3] The Public Notice stated that further details on the testing would be released at a later time.

As part of this effort, several prototype devices were recently submitted for testing.[4] The devices are not finished consumer products. They are designed to demonstrate the feasibility of various concepts for sharing the TV broadcast spectrum without causing harmful interference. The Commission may ultimately establish requirements that the current prototype devices do not meet. However, all products would need to be certified as complying with the final rules before they may be marketed to consumers.

This document generally describes the tests that will be conducted in Phase II. The Phase II tests largely follow the procedures used for Phased I with some modifications. Comments and suggestions offered within the public record in the Commission's proceeding on the TV white spaces with respect to both the previous and current testing were considered and included where appropriate and practicable. The Phase II tests will be open to attendance and observation by any interested party.

The testing will include both laboratory (bench) tests and field tests. The laboratory tests will measure the performance capabilities of the prototype devices under controlled conditions. The field tests will be conducted at a variety of locations to provide information on the performance of the devices under real world conditions OET staff will adapt the test plan as appropriate based on the specific capabilities of each device and circumstances that may arise as the tests progress.

At the conclusion of the Phase II testing, a report will be issued and placed into the record in the Commission's TV white spaces proceeding. Any deviations from the test plan set forth herein will be described in this test report.

TEST PLAN

I. Television-Related Tests

A. Bench Test 1 (Single Laboratory-Grade DTV Input Signal)

1. These tests will be performed to establish the minimum DTV signal level that can be detected by the WSD scanner under "best-case" (*i.e.*, non-faded DTV signal) conditions.
2. Perform single signal scanning tests with a conducted "clean" laboratory signal using the same procedures as in the Phase I effort (*i.e.*, 30 trials over a range of input signal power levels encompassing at a minimum the thresholds proposed in the proceeding as well any different threshold claimed by the manufacturer).
3. Repeat the above procedure over 3 channels in the WSD tuning range (lower, middle and upper). If the results indicate consistent performance over the three test channels, then subsequent tests will be performed on a single detection channel in the tuning range (*e.g.*, the middle channel).

B. Bench Test 2 (Single RF-Capture DTV Input Signal)

1. Repeat the single signal test procedure described above using a single RF capture from those identified in the draft IEEE 802.22 standard as the input signal rather than a clean laboratory-generated signal.[5]
2. Depending on the outcome of the tests utilizing the single laboratory-grade DTV signal, these tests are envisioned to be performed using only one detection channel.
3. This test will be repeated until all 12 of the RF captures included in the draft IEEE 802.22 standard are utilized as input signals (minimum of 12 independent tests).

C. Bench Test 3 (Multiple DTV Signals)

1. This test is limited to two signal tests only.

2. The WSD will be subjected to a "clean" DTV signal input on a single detection channel (middle channel) that is stepped over the detection threshold range determined previously and another clean DTV signal at a high power level (-28 dBm) on a designated adjacent channel.
3. The WSD's detection capability will be tested under these conditions using the same procedures performed in the Phase I measurement effort but limited to a single detection channel.
4. These tests will be performed with the second DTV signal placed sequentially on each of the N±1 and N±2 channels relative to the test channel.

D. Bench Test 4 (Transmitter Emissions Characterization)

1. The transmitter emissions measurements are to be performed only on those prototypes submitted with transmit capability.
2. Spectral characteristics (*i.e.,* spectral signature, emissions bandwidth, out-ofband emissions and integrated channel power) will be measured and quantified.

E. Field Tests

1. The number of test sites and their locations will be decided as experience and practicality dictate. A diverse selection of sites will be chosen to represent realistic locations where personal/portable WSDs might operate in competition with over-the air (OTA) broadcast television reception. This will include a sample of personal residences (single family and multiple unit dwellings) and business structures located in urban, suburban and rural environs.
2. Preparation (performed for each test location identified)
   a. Identify test site and location (GPS coordinates).
   b. Run database search to identify and plot all licensed DTV stations with noise-limited contours that envelop the test site.
   c. Use the information to determine actual RF channel assignments and to predict the channels that should be receivable at the site and their estimated power levels.
   d. If cable television service is available at test site, gather information with regard to channel assignments and their correlation to RF broadcast channels.
3. On-Site Testing
   a. Verify the stations/channels that can be received on a DTV receiver (preferably an existing system).
   b. Use existing antenna downlead and spectrum analyzer to measure the power level associated with each DTV station identified from the database (including those where service could not be verified above). If an existing rooftop antenna is not available, other methods, such as the use of a telescoping mast, will be used to perform this measurement.
   c. Operate scanner function of WSD at various locations within (and outside of) the test site and record results (if possible, this procedure will be automated to electronically record information regarding channel occupancy/availability).

   i. These tests will utilize a common laboratory-grade calibrated receive antenna with all prototype WSDs
  d. Measure a sample of the available DTV signal levels with a spectrum analyzer and a calibrated test antenna that is placed as close as possible to the location where the scanner was operated.

F. Anecdotal TV Interference Investigations

1. Two separate types of anecdotal tests of WSD transmitter interference potential will be performed as a part of this effort. In addition, at sites where cable television service is available, a separate test to evaluate direct pick-up interference will be performed to the extent that it is feasible to do so.
2. The first anecdotal test will be performed at each site at which field tests are performed using the following procedure:
   a. Tune the WSD transmitter and the television to same channel as the lowest measured (but receivable) DTV station and activate the transmitter at each measurement location while observing picture quality for degradation.
   b. Repeat this test for the highest measured DTV signal location.
   c. Repeat for adjacent channel interactions if appropriate fixed-tuned transmission filters are available (this test may be limited to co-channel interactions because of external tuned-filter limitations and availability).
3. If cable service is available at the test site, the WSD transmitter will be tuned to a receivable cable channel (this will require preliminary access to cable providers channel mapping strategy) and activated at various locations while observing the television picture quality for signs of degradation.
4. A stand-alone anecdotal interference test, similar to that performed in the initial test effort, will be performed utilizing a similar procedures as follows:
   a. Set up a test range, likely on FCC laboratory property, where the WSD distance relative to a test DTV receiver set-up can be varied in quantifiable steps.
   b. Identify an OTA DTV signal that represents the lowest receivable signal level (as close as possible to the threshold of visibility (TOV)) relative to a typical outdoor TV antenna mounted on a rooftop, or alternatively, on a telescoping mast elevated to simulate a rooftop mounted antenna.
   c. A test DTV receiver connected to the test antenna will then be tuned to the selected weak signal channel.
   d. The WSD will be tuned to the same channel and activated to represent a co-channel interaction.
   e. The WSD will be moved relative to the DTV antenna in defined steps until the maximum distance at which observable degradation to the television picture quality is observed.
   f. Optimally these tests will be repeated to demonstrate the interference potential from WSD transmissions on each of the N±1 and N±2 adjacent channels (relative to the selected DTV channel); testing for adjacent channel will depend on whether appropriate filters are available.
   g. Preliminary scans indicate that the test channel most likely to be utilized in these tests is channel 28 (WFPT-DT with broadcast tower located in Frederick,

Maryland, approximately 26 miles WNW of the FCC laboratory). Testing on this channel will require external bandpass filters centered in channels 26, 27, 29 and 30 to enable testing interference interactions on all N±1 and N±2 adjacent channels.

The option to utilize band-limited additive white Gaussian noise (AWGN) will be retained as a potential method to simulate WSD emissions (in adherence to proposed emission mask); this could be used to avoid the problems associated with the external filter implementation limitations.

G. Investigation of Additional Interference Avoidance Features

We will also consider any additional features or functions that may be included in the device to facilitate interference avoidance, such as geolocation and data base access. Tests of such capabilities will be performed to the extent this may be feasible.

II. Wireless Microphone-Related Tests

The test program will use a variety of Part 74 wireless microphones from different manufacturers to examine the sensing capability and interference potential of WSDs to microphone systems. It is noted that wireless microphones employ different types of modulation and features such as tone keys. In the Phase II tests, the wireless microphone tests will include the microphones used in the initial tests and also microphones from additional providers. These new microphones will expand the range of modulation methods and features examined.

A. Bench Tests – Sensing

1. The microphones used in these tests will be modulated with a 1000 Hz tone at 24 kHz deviation and with no modulation.
2. Tests will be performed at operating frequencies in the center of the TV channel and near the channel edges.
3. The power level of the microphones will be varied to determine the highest and lowest levels at which the WSD will reliably detect the microphone signal.

B. Bench Tests – Interference

1. Interference to wireless microphones is defined as a reduction in SINAD to 30 dB for FM microphones and a specified BER for digital microphones at -80 dBm input signal level to the receiver caused by an undesired signal with the microphone modulated with a 1000 Hz tone at 24 kHz deviation.
2. Tests will be performed with the WSD signal co-channel and on the first and second adjacent channels to the wireless microphone signal.

C. Field Tests – Sensing

1. In order to prepare for field tests at events or other venues where Part 74 wireless microphones are in actual use, simulated tests will be conducted at the Laboratory facility using the available buildings and grounds. Tests will be performed in several different scenarios to obtain a sense of what to expect at "real world" sites and to resolve any problems that are revealed that may occur at a "real world" site.
2. Based on the experience gained from the simulated tests, field tests will be performed at as yet to be determined event locations where there is extensive use of wireless microphones. Possible sites include sports arenas and theaters where a number of wireless microphones are used and broadcasting events. A variety of challenging configurations will be explored at locations with different characteristics.

D. Field Tests – Interference

1. Provided a transmitter is included in the WSD, it will also be tested to examine the potential for causing interference to wireless microphones in simulated field scenarios at the Laboratory and the results compared to the results of tests and theoretical analyses submitted to the record in this proceeding. The methodology will be modified as necessary to yield useful and meaningful results and produce the highest confidence in the accuracy of the results. For example, assuming a typical scenario of a microphone receiver located 10 meters from the microphone, the degree of interference will be determined with the WSD at various distances from the receiver for co-channel and first and second adjacent channel operation.
2. The methodology derived from the simulated field tests will then be used to conduct actual field tests as in Section C.2 above.

TEST REPORT

At the conclusion of this test program, OET will prepare a report that sets forth the results of its Phase II tests of prototype personal/portable white space devices. This chapter will be inserted into the record of the Commission's proceeding in ET Docket No. 04-186 and an opportunity will be provided for interested parties to comment thereon.

End Notes

[1] FCC 06-156, *First Report and Order and Further Notice of Proposed Rulemaking in the Matter of Unlicensed Operation in the TV Broadcast Bands*, ET Docket No. 04-186, October 18, 2006.

[2] OET Report FCC/OET 07-TR-1006, ***Initial Evaluation of the Performance of Prototype TV-Band White Space Devices***, S. Jones and T. Phillips, July 31, 2007.

[3] See FCC Public Notice ***The Office of Engineering and Technology announces Additional Testing of TV White Space Devices***, DA 07-4179, released October 5, 2007.

[4] See ex-parte filings from Adaptrum, Motorola, Philips Electronics and Microsoft, Inc. in ET Docket No. 04-186.

[5] See ex parte submission from Qualcomm, dated August 24, 2007. Figures of the spectrum of each signal file can be obtained from:

http://grouper.ieee.org/groups

In: Tapping the Television White Space Spectrum: A Revolution... ISBN: 978-1-60692-956-8
Editor: Julie E. Hendriks

Chapter 4

PEER REVIEW PANEL REPORT ON OET REPORT FCC/OET 08-TR-1005 EVALUATION OF THE PERFORMANCE OF PROTOTYPE TV-BAND WHITE SPACE DEVICES PHASE II

Mary Bucher, George Dillon, James Higgins, Martin Liebman and Doug Miller

BACKGROUND

The following is the peer review of OET Report FCC/OET 08-TR-1005, entitled "Evaluation of the Performance of Prototype TV-Band White Spaces Devices Phase II." The report is a result of OET's second phase study of hardware development devices with capabilities for operating on an unlicensed basis in the TV bands as part of the "TV white spaces" proceeding, ET Docket No. 04-186. This second phase study examined four development devices and follows OET's initial study of TV white space development devices, which was released July 31, 2007, *see* "Initial Evaluation of the Performance of Prototype TV-Band White Space Devices," OET Report FCC/OET 07-TR-1006 (Phase I Report).[1] The devices examined in this second phase study were submitted by industry in response to an OET public notice inviting submittal of additional prototype white space devices and are product development platforms, rather than models of products that could actually be marketed. One of these devices has both transmitting and spectrum sensing capabilities; the other three have only spectrum sensing capability. Each of the devices has unique spectrum sensing capabilities and was examined in the laboratory and the field with respect to its ability to detect the presence of digital TV, analog TV, and/or Part 74 wireless microphone signals. The one device with a transmitter was also examined for its potential to cause interference to the reception of digital television signals and cable service.

Peer review of the OET report was performed as required under the OMD Information Improvement Act for influential scientific and engineering studies. The review panel was made up of five engineers, three from the Enforcement Bureau and two from the Wireless

Telecommunications Bureau. The review panel analyzed and discussed various subject areas in the OET report, both independently and jointly. Specifically, as requested in the OET memo, the review panel addressed the following:

1. Whether the scope of testing in terms of spectrum sensing abilities and signal conditions examined was appropriate and sufficient;
2. Whether the measurement methodologies used in the testing of the prototype devices spectrum sensing abilities was appropriate;
3. Whether the scope of testing of the device with transmitting capability for its potential to cause interference to digital TV, analog TV, and wireless microphone signals was appropriate
4. Whether the various tests performed were properly conducted consistent with the selected methodologies.

The response of the review panel is presented below for each question shown above:

1. Whether the Scope of Testing in Terms of Spectrum Sensing Abilities and Signal Conditions Examined Was Appropriate and Sufficient

In the opinion of the review panel, the overall scope of the spectrum sensing testing was appropriate. Tests were performed against a range of devices that utilized various signal sensing methods. Initial laboratory tests were performed under controlled conditions, followed up by field tests utilizing live over-the-air television signals.

For the laboratory tests, initial measurements were taken at multiple channels across the operating range of the prototype devices, which consisted of UHF channels 21-51, to determine if the devices responded consistently. The devices were tested against a clean digital TV signal, alone and in the presence of TV signals on both the first and second adjacent channels, as well as recorded RF captures of DTV signals to simulate various multi-path conditions. These tests were followed up by field tests at nine different locations representing a variety of topography including urban, suburban, rural, residential and business locations. Multiple measurements were taken at each location under varied indoor and outdoor conditions. The field tests were performed over the entire operating range of the devices and examined the ability of the devices to detect live over-the-air ATSC and NTSC signals. For these tests, channel occupancy based on TV contours was noted, but actual measurements were taken to determine the occupancy of each channel.

The wireless microphone portion of the testing was performed both in the laboratory and in the field. Laboratory tests were performed using signals generated by two different wireless microphones. The microphones were placed in an anechoic chamber to isolate the test signal and avoid undesired signals. White space device sensitivity measurements were taken for the wireless microphone signal alone and with simulated DTV signals at various first and second adjacent channels. The field tests were performed under actual real world conditions during a pre-season football game at FedEx Field in Landover, MD and before and during a performance of a Broadway play at the Majestic Theater in New York City. Measurements were taken at multiple locations and times at these locations.

The review panel believes that the approach taken and procedures followed by the staff were well thought out and provided an excellent analysis of the sensing capabilities of the WSDs. The review panel believes that if more time and testing equipment had been available for the laboratory testing, it would be more realistic to have tested multiple, adjacent DTV signals at the same time, rather than testing only a single adjacent channel at a time. In addition, if there had been no time constraints, performing more scans at the various field locations to obtain a larger set of results for determining detection reliability would have been useful.

2. Whether the Measurement Methodologies Used in the Testing of the Prototype Devices Spectrum Sensing Abilities was Appropriate

The review panel believes that the measurement methodologies used in the testing of the prototype devices were appropriate. OET developed a Phase II test plan which largely follows the procedures used for Phase I testing with some modifications. As in the Phase I test plan, guidance and procedures established to date by IEEE 802.22 for testing the spectrum sensing capability of fixed/access WSDs were considered where applicable. In the instances where the testing deviated from those methodologies, the review panel feels that the underlying objectives were met by the tests that were performed. OET indicated that comments and suggestions offered within the public record in the Commission's TV white spaces proceeding with respect to both previous and current testing were considered and included in the test plan where appropriate and practicable.

As stated above, both laboratory and field tests were performed to assess the scanning/sensing capability of the prototype WSDs for the television portion of the testing. Three separate bench tests were performed to determine the minimum DTV signal detection threshold for each of the prototype devices submitted for evaluation. The first laboratory test utilized a single, unimpaired, laboratory-grade DTV signal as the test input. The second laboratory test utilized recorded (captured) DTV signals that incorporated "real world" distortions resulting from reflections and multi-path fading. The third laboratory test utilized two unimpaired, laboratory-grade signals as the input, one on the detection channel and the other placed on the first or second adjacent channel. Measurements were taken to determine DTV power levels resulting in 100% detection reliability. Data was also collected for a few points above and below the breakpoint where the detection reliability degrades from 100%.

In the laboratory multiple measurements were taken in order to determine the percentage of successful detections with some statistical relevance, and initial tests were performed for multiple channels to investigate potential frequency related differences in performance. For the baseline detection signal test using a single DTV input, the input DTV signal was initially set to a measurable level and then further attenuated incrementally with the calibrated step attenuator bank while exercising the scanner over the occupied channel. At each attenuation step (input power level), thirty independent trials were performed. Tests were performed on three channels in the lower (channel 21), middle (channel 36 or 37) and upper (channel 51) portions of the WSD tuning range.

The review panel feels that the number of trials and channels tested was sufficient. Since the initial test results revealed that the sensing performance of the devices was fairly consistent over their tuning range, the remaining tests were performed on a single channel in the middle of the tuning range. The review panel agrees that testing on a single channel was appropriate given the results of the baseline testing.

The methodologies for the two DTV signal tests and the recorded DTV signal tests were similar to the methodology used in the single DTV input signal tests. For the two DTV signal test, a second DTV input signal was introduced at N±1 and N±2. Detection threshold measurements were performed with the adjacent channel signal power set to -28 dBm (high), -53dBm (moderate) and -68dBm (low) levels. The review panel feels that it would be helpful if the report explained the rational for the selection of these signal levels (i.e., an explanation as to why -28 dBm was considered a high signal level and why -68 dBm was considered a low signal level).

For both tests thirty independent trials were performed at each attenuation step (power level in detection channel) in order to determine the percentage of successful detections with some statistical relevance. Additional recorded DTV signal measurements utilizing one thousand independent trials were performed on the Motorola device to smooth out variations resulting from significant instantaneous temporal variations in the RF captures. Had time allowed, the panel feels that it would had been useful to perform these additional measurements on all of the devices.

Field tests were performed utilizing live OTA ATSC and NTSC television signals at nine different locations representing a variety of topography including urban, suburban, rural, residential and business locations. At each field test site, testing was conducted at two locations under varied indoor and outdoor conditions. At each of these locations, two scans were performed with each device over its entire channel range and the results were recorded. The review panel felt that the measurements taken at these locations provide a good initial assessment of the scanners/sensors performance under "real world" conditions. However, the review panel believes that it would have been useful to consider the potential effects of interfering structures on the results (i.e., by performing tests inside and outside at the same site and describing the nature of any interfering structures). The panel also believes that it may have been worthwhile to attempt to identify areas subject to fading and shadowing and to perform tests in and around these areas.

The review panel feels that the measurement methodologies used for the wireless microphone portion of the testing were appropriate to ascertain the ability of the devices to scan for and detect Part 74 wireless microphones. Two different types of microphones, FM modulated and digitally modulated, were used in these tests. Regarding these tests, the review panel did note that during most of the field testing scans were performed with the microphones either turned on or turned off. The panel believes that testing in a more dynamic environment with microphones being turned on and off intermittently during the scans would have been more beneficial. The report also indicates that there were a large number of false positive indications during the wireless microphone field testing. In relation to these results, the review panel believes that it would be helpful if the report could expand on possible explanations for these results and/or provide a rational for not performing additional tests to possibly identify their cause.

3). Whether the Scope of Testing of the Device with Transmitting Capability for its Potential to Cause Interference to Digital TV, Analog TV, and Wireless Microphone Signals Was Appropriate

It is the opinion of the review panel that the scope of the testing of the Adaptrum WSD prototype for its potential to cause interference to digital TV was appropriate, given the study's stated limitations. However, the review panel did not observe any discussion in the report describing tests that may have been done to analyze potential interference to analog TV signals or wireless microphone signals.

The project conducted limited, or "anecdotal," tests of the prototype WSD transmitter to provide information on its potential to interfere with digital TV reception. These tests were performed within the confines of the FCC laboratory facility with an unobstructed line-of-sight (LOS) propagation path between the WSD transmitter operating at approximately 150 mW EIRP and the DTV test receiver antenna. A test DTV receive antenna was mounted on an extendable mast to simulate a consumer installed roof-top antenna. All available channels were analyzed for the weakest receivable channel (i.e., the worst-case interference scenario) and channel 30 was ultimately used for that purpose. The WSD transmitter was then placed in the "main beam" of the receive antenna, tuned to each of the immediately adjacent channels, and activated at incremental distances from the DTV receive antenna while observing for interference effects to the picture quality. Tests were also performed with the WSD tuned to channel 30. The distance at which interference was observed was measured and recorded. The tests were repeated with the WSD transmitter located to the sides and behind the DTV receive antenna. Overall, the review panel feels that these tests provided valuable data on the effects of WSDs on DTV reception. The panel feels that additional analysis of the WSD transmitter operating at power levels other than an EIRP of 150 mW, e.g., the power level initially proposed by the Commission for mobile WSD devices, would be valuable.

The panel recognizes that use of OTA signals produces only anecdotal results and the ability to perform tests were limited by the area where the tests were being performed. Under different circumstances, the review panel believes that it would have been useful for a larger number of tests to have been conducted under more uniform conditions to provide a larger set of results that could be more easily correlated. The review panel also feels that normalizing the data in Table 4- 2 for a standard path length would perhaps make the results more meaningful. The review panel recognizes that the testing was limited by the fact that the Adaptrum WSD prototype was the only available working device with an active transmitter. Had more transmitting devices been available, the panel believes that simultaneous testing of multiple devices to determine the cumulative interference affects of those devices would have yielded valuable data.

4). Whether the Various Tests Performed Were Properly Conducted Consistent with the Selected Methodologies

In the opinion of the review panel, the tests were properly conducted consistent with the selected methodologies. OET developed a Phase II test plan which largely follows the procedures used for Phase I testing with some modifications. OET has indicated that comments and suggestions offered within the public record in the Commission's proceeding on the TV white spaces with respect to both previous and current testing were considered and included in the test plan where appropriate and practicable. For the most part, OET's testing was consistent with the established Phase II test plan and deviation from the test plan did not undermine the results. The review panel believes that the testing was well done and thorough. For more description of the testing methodology, see No 2 above.

End Notes

[1] Five devices were submitted for this phase of testing; however, one of the devices stopped working after a few tests were performed on it and the reports contain only partial data for that device.

In: Tapping the Television White Space Spectrum: A Revolution...ISBN: 978-1-60692-956-8
Editor: Julie E. Hendriks

Chapter 5

PEER REVIEW OF PROTOTYPE TV WHITE SPACE DEVICE PHASE II STUDY

Julius Knapp

DATE: October 28, 2008
TO: The Record
FROM: Julius Knapp
Chief, Office of Engineering and Technology
SUBJECT: Peer Review of Prototype TV White Space Device Phase II Study

The following information and comments are submitted in response to the October 1, 2008 memorandum of the peer review panel that provides their report on the review of the Office of Engineering and Technology's Phase II measurement study of prototype TV-band white space devices (WSDs). The study and its findings are set forth in a report entitled "Evaluation of the Performance of Prototype TV-Band White Space Devices, Phase II", OET Report FCC/OET 08-TR-1005. This study examined the performance of the "detect and avoid" (spectrum sensing) techniques used by several prototype WSDs to evaluate their ability to detect television channels occupied with incumbent signals (digital or analog TV and wireless microphones). The peer review of this study was performed by staff of the Commission's Wireless Telecommunications and Enforcement Bureaus pursuant to the Office of Management and Budget's requirement under the Information Quality Act that influential scientific assessments be subject to peer review to enhance the quality and credibility of the government's scientific information.

The peer review addressed the following subject areas: 1) whether the scope of testing in terms of spectrum sensing abilities and signal conditions examined was appropriate and sufficient; 2) whether the measurement methodologies used in the testing of the prototype devices spectrum sensing abilities was appropriate; 3) whether the scope of testing of the Adaptrum device's transmitting capability for its potential to cause interference to digital TV, analog TV, and wireless microphone signals was appropriate; and 4) whether the various tests performed were properly conducted consistent with the selected methodologies.

Overall the panel found that the testing was appropriate. Specifically, the panel found that (1) the overall scope of the spectrum sensing testing was appropriate; (2) that the measurement methodologies used in the testing of the prototype devices were appropriate; (3) that the scope of the testing of the Adaptrum prototype WSD (the only Phase II device with a transmitting capability) for its potential to cause interference to digital TV and wireless microphones was appropriate, given the study's stated limitations; and, (4) that the tests were properly conducted consistent with the selected methodologies. The review panel further found that the tests were properly performed consistent with the established test plan. In closing, the review panel concluded that it "believes the testing was well done and thorough."

The panel also offered additional comments regarding the subject project. In particular, the panel suggested that additional tests could have been performed in the laboratory for multiple signal types and that a larger number of scans could have been performed per data point. They also wondered if tests could have been performed to investigate potential causes of excessive false detections.

We agree that it may have been useful to have tested multiple, adjacent channel DTV signals at the same time in the adjacent channel measurements performed within the laboratory; however, there was a practical limitation on such tests due to the unavailability of the additional DTV signal generators necessary to simultaneously produce the multiple adjacent-channel signals. In addition, we believe that the field trial component of the test program served to evaluate the scanning/sensing performance of the prototype WSDs over a more diverse set of adjacent channel/amplitude combinations than could have been feasibly simulated on the bench. We also agree that additional data from the field trials, particularly from scans performed at additional locations at each test site, may have been useful. However, we believe that the data collected at all the locations taken together provide a reasonable assessment of the overall performance of the devices under a variety of conditions.

Increasing the number of independent trials from thirty to one thousand was not practical for many of the other devices. For example, the modified Adaptrum WSD has a 185-second scan time, requiring almost two hours to collect a single data point based on thirty independent scans. Increasing the number of independent trials to 1000 would have required more than 51 hours for each data point. Therefore, it was deemed impractical (and often unnecessary) to perform these tests on all of the prototype devices.

The limited measurements of incumbent channel signal strength performed at locations inside and outside at the same site in an effort to provide some data regarding the effects of intervening structures on the desired signal levels. Although we agree that operating the sensing function of the prototype devices at both indoor and outdoor locations at the same site might have been useful; the scan time limitations for the devices made it impractical to perform measurements at additional locations at each site.

The review panel suggests that turning the microphones on and off intermittently during scans would have been beneficial. It is not clear what benefit would derive from this test procedure which would have been difficult and time consuming. Also interpreting the results of such tests would be very difficult. The review panel's comment that an explanation of the cause of the large number of false positives would be helpful is well taken. Limited additional tests were performed to try to understand the test results and it was discovered that the devices produced radiated emissions that were a potential contributor to the reporting of false positives.

In response to the panel's comments regarding additional transmitter tests, we agree that such tests may have been useful; however, we were limited by the fact that only one of the prototype devices provided a transmit capability. We acknowledge that this single device could have been utilized in tests performed at multiple locations to produce a larger data set, but sensitivities regarding the potential for producing interference to the OTA reception of TV signals made us reluctant to activate the transmitter beyond the confines of the laboratory facility. We are not sure that such tests would have added significantly more information.

OET expresses gratitude to the reviewers for conducting a thoughtful review of a voluminous and complex report and for their earlier comments that contributed to development of an improved test plan for our study in Phase II of this project.

cc: Mary Bucher, WTB
George Dillon, EB
James Higgins, EB/Columbia
Martin Liebman, WTB
Doug Miller, EB/Atlanta

In: Tapping the Television White Space Spectrum: A Revolution... ISBN: 978-1-60692-956-8
Editor: Julie E. Hendriks

Chapter 6

FCC Adopts Rules for Unlicensed Use of Television White Spaces

Federal Communications Commission

In its continuing efforts to promote efficient use of spectrum and to extend the benefits of such use to the public, the Federal Communications Commission (FCC) today adopted a Second Report and Order (Second R&O) that establishes rules to allow new, sophisticated wireless devices to operate in broadcast television spectrum on a secondary basis at locations where that spectrum is open. (This unused TV spectrum is now commonly referred to as television "white spaces"). The rules adopted today will allow for the use of these new and innovative types of unlicensed devices in the unused spectrum to provide broadband data and other services for consumers and businesses.

The rules represent a careful first step to permit the operation of unlicensed devices in the TV white spaces and include numerous safeguards to protect incumbent services against harmful interference. The rules will allow for both fixed and personal/portable unlicensed devices. Such devices must include a geolocation capability and provisions to access over the Internet a data base of the incumbent services, such as full power and low power TV stations and cable system headends, in addition to spectrum-sensing technology. The data base will tell the white space device what spectrum may be used at that location.

Wireless microphones will be protected in a variety of ways. The locations where wireless microphones are used, such as sporting venues and event and production facilities, can be registered in the data base and will be protected in the same way as other services. The Commission also has required that devices include the ability to listen to the airwaves to sense wireless microphones as an additional measure of protection for these devices.

All white space devices are subject to equipment certification by the FCC Laboratory. The Laboratory will request samples of the devices for testing to ensure that they meet all the pertinent requirements.

The Commission also will permit certification of devices that do not include the geolocation and data base access capabilities, and instead rely solely on spectrum sensing to avoid causing harmful interference, subject to a much more rigorous approval process.

In a process that will be open to the public, applications will be released for public comment prior to agency action. Such devices will be tested by our Laboratory to a "Proof of Performance" standard both in the lab and in a variety of real-world environments to ensure they do not cause interference to licensed services when in use. The staff report and recommendation will also be released for public comment. For now, certification of any such device will require approval by the full Commission.

Manufacturers may continue to provide additional information to the Commission to support the use of higher power devices in adjacent channels. In addition, the Commission will explore in a separate Notice of Inquiry whether higher-powered unlicensed operations might be permitted in TV white spaces in rural areas.

The Commission will closely oversee and monitor the introduction of TV white space devices. The Commission will act promptly to remove from the market any equipment found to be causing harmful interference and will require the responsible parties to take appropriate actions to remedy any interference that may occur.

Action by the Commission November 4, 2008, by Second Report and Order (FCC 08-260). Chairman Martin, Commissioners Copps, Adelstein, and McDowell with Commissioner Tate approving and dissenting in part. Separate statements issued by Chairman Martin, Commissioners Copps, Adelstein, Tate and McDowell.

FCC

News and other information about the FCC is available at_www.fcc.gov

In: Tapping the Television White Space Spectrum: A Revolution... ISBN: 978-1-60692-956-8
Editor: Julie E. Hendriks

Chapter 7

Separate Statement of Michael J. Copps, Re: Unlicensed Operation in the TV Broadcast Bands, ET Docket No. 04-186, Second Report and Order and Memorandum Opinion and Order

Michael J. Copps

One of the great lessons of history I quickly learned here at the FCC is the power of technology to turn scarcity into abundance. Once upon a time, it seemed that certain swaths of the public airwaves could support only a few expensive, bulky "brick" telephones (so named because of their forbidding size and weight). These early phones permitted just 30 minutes of scratchy voice communications on a single battery charge. Today, thanks to the extraordinary work of America's great engineers and inventors, the same amount of spectrum can support millions of full-featured, relatively inexpensive, multimedia handsets—like the popular iPhone—that perform many of the functions of a desktop computer for days between charges.

Or, to take an example even closer to today's item, we have learned in the past decade that unlicensed bands—once derided as "junk spectrum" suitable only for garage door openers—can actually support Wi-Fi connectivity for tens and even hundreds of millions of users every day. Something that seemed of marginal value has given us broadband in homes, airports, hotels, coffee shops, and downtown areas—developments that are changing the ways in which we live. Just stop by a Starbucks and look at how many people are typing away on laptops. That, to me, is the most important lesson of the wireless revolution: as technology marches on, the real winners are American consumers. This process is as it should be—the airwaves, after all, are the ***people's*** airwaves. And it is the FCC's job to make sure that the American people continue to extract full value from their property.

So in setting responsible spectrum policy, the FCC must always recognize that technology cannot stand still, and that even as we protect existing users from harmful interference, we also have to promote innovation—because until someone finds a way to make more electromagnetic spectrum, it is only ***innovation*** that can improve the wireless choices available to consumers.

For four years, the FCC has been considering proposals to authorize unlicensed devices in the TV white spaces. The proponents have argued that we can enable a whole new generation of wireless devices—bringing new broadband connectivity to our rural and urban communities—without harming free, over-the-air TV. Does this seem almost too good to be true? Of course. But so did the modern cellular industry, the explosion of Wi-Fi devices, and so many other innovations at comparable stages in their development. Even the notion of transmitting high quality video through the air to millions of TV sets must have seemed pretty fantastical when it was first demonstrated decades ago. This is the history of wireless innovation in a nutshell—the nearly miraculous becomes commonplace.

My approach throughout this process has been to let the engineering analysis drive our decision-making. The good news is that the process—while far from perfect— has involved a great deal of serious study and public openness. In fact, the last round of testing was conducted with engineers and lawyers from all sides looking over the shoulders of the FCC's engineers as they performed their bench and field testing. While I do not envy the Commission's engineers for undergoing this level of scrutiny as they go about their daily tasks, I think it's fair to say that few other engineering analyses at the FCC have been as lengthy or open as this one.

The results of our testing have shown that there is merit in the initial positions of ***both*** sides in this debate—and I believe our conclusion today reflects this point. For example, the white spaces proponents initially argued that spectrum-sensing alone would be adequate. Our testing has shown that this approach, right now, is not ready for prime time. That is why our decision does not authorize devices today that rely exclusively on sensing—though we do create a process for evaluating the next generation of such devices.

Instead, today, we move towards a compromise technology—geo-location—that represents a simpler and more conservative approach to protecting existing licensed users of the television band. We also have built in several forms of belt-and-suspender protections. First, we still will have a chance to "kick the tires" of these devices at the certification stage—to make sure they operate as designed. Second, because we permit these devices to operate under our Part 15 unlicensed rules, they must protect broadcast licensees and shut off if they create harmful interference. Third, because we use geolocation technology, we retain the ability to require particular makes and models of white space devices to turn off if they have a manufacturing flaw. Fourth, to the extent that we consider certifying a next-generation sensing-only device, we have established a process to make sure that all parties have an opportunity to be present during the testing process. So, though there will never be metaphysical certainty when it comes to interference issues, I am comfortable that the FCC and its dedicated engineers have approached this issue with the necessary seriousness and developed an approach that will adequately protect over-the-air viewers.

In this lengthy process, some parties have come forward with arguments for licensed use of the TV white spaces, either in whole or in part. As I have stated before, I think that both licensed and unlicensed regimes benefit consumers. The key is to get the right mix. In the past few years, we have auctioned off almost 150 MHz of new ***licensed*** spectrum. The time has now come, I believe, to increase the amount of ***unlicensed*** spectrum—especially the amount of spectrum below 1 GHz. The order we vote on today will do exactly that.

Some have called this new technology "Wi-Fi on steroids"—and I hope they are right. Certainly, this new technology, taking advantage of the enhanced propagation characteristics of TV spectrum, should be of enormous benefit in solving the broadband deficit in many rural areas. Indeed, I agree with the view of several of my colleagues that it should be quite possible, at some point, to authorize higher-power devices in rural areas that will support backhaul and broadband infrastructure. That is why I support our commitment to issue a Notice of Inquiry on this topic—which should give the Commission more than enough time to build a record and act before the first commercially-ready devices enter the market.

Today's item represents the culmination of a lot of hard work by the Commission's skilled engineers and staff in our Office of Engineering and Technology and elsewhere. I want to thank them for their dedication to this task and their accomplishment in charting a course that accommodates both existing users and innovators in this spectrum. Ultimately, this careful threading of the needle between the path-breaking and the tried-and-true will reap huge benefits for the American people.

In: Tapping the Television White Space Spectrum: A Revolution... ISBN: 978-1-60692-956-8
Editor: Julie E. Hendriks

Chapter 8

SEPARATE STATEMENT OF DEBORAH TAYLOR TATE, APPROVING IN PART AND DISSENTING IN PART - RE: UNLICENSED OPERATION IN THE TV BROADCAST BANDS, ET DOCKET NO 04-186, SECOND REPORT AND ORDER AND MEMORANDUM OPINION AND ORDER

Deborah Taylor Tate

In the midst of the present economic downturn gripping our nation and the world, there is one bright spot on the horizon – American technology and the American entrepreneurial spirit, especially in the communications and technology sector, which hold the promise of future innovation, investment, jobs and economic growth. One of the more promising areas of innovation cannot even be seen – the "white spaces" that represent the unused portions of spectrum in the bands presently allocated to TV broadcast operations. Today's item advances us on a path to facilitate use of the white spaces for new wireless services, including broadband services, for all Americans.

These are first steps only, ones that hopefully will move us towards modernization and more effective use of the unused portions of this spectrum. Indeed, these are revolutionary times in an evolutionary industry, with the promise of even yet unseen innovative devices on the horizon. Many visions exist for how unlicensed use of the white spaces will evolve, but evolve they certainly will. Hopefully, this item will help facilitate the deployment of unlicensed devices and services that enable consumers to enjoy more television programming and video than ever before, and that will allow families to transfer and network information, photos, and much more across multiple devices, technologies and platforms in their homes. Similarly, communities of users may find they are able to communicate seamlessly through mesh networks rather than traditional phone lines. Finally, as someone who has championed rural America, I hope that this item will facilitate services, including broadband, to rural areas and thus help reduce the digital divide that is far too prevalent in rural communities across our nation.

In considering this order, I have listened to and weighed seriously the concerns of an amazingly broad array of interests. In the end, I have tried to reach a reasonable position that takes into account all interests, including, most especially, the interests of the American consumer. The order is not perfect – it precludes licensed services and lacks needed language regarding a specific and expedited complaint process for broadcasters, cable providers, wireless microphones and individual users in the case of interference. Nonetheless, the order ultimately may help promote the innovation and investment in advanced services that consumers have come to expect from the communications and technology sector. Accordingly, I would like to address several important public policy goals, as well as some of my concerns regarding how this item falls short of our mandate to advance the interests of the American public.

Measures to Address Interference

In crafting public policy, one must weigh the benefits and costs of various options. In this case, the potential costs of enabling unlicensed use of the white spaces are substantial. Specifically, this order enables a vision for entirely new services and devices but ones that also hold a real risk of causing interference with existing services, including broadcasting, cable television, and wireless microphones, among others.

Addressing interference in the use of the spectrum has always been one of the primary roles of the Federal Communications Commission since its creation by Congress in 1934. Our predecessor agency, the Federal Radio Commission, also assumed this important responsibility starting as early as 1927. Clearly, addressing interference is one of the crucial tasks this agency takes most seriously and performs most effectively.

Within the Commission, the Office of Engineering and Technology (OET) is the team responsible for conducting the research, analysis, and evaluation of devices that use the airwaves in this country. This group of talented staff has spent the past four years carefully considering whether and to what extent unlicensed devices may utilize the white spaces in the broadcast spectrum without causing undue harm to incumbent users. I have placed much faith in the Chief of OET, Julie Knapp, and his entire staff of engineers. Their recommendations, while always useful in any rulemaking, played an especially persuasive role in my final decision to support this item.

I have been assured by Chief Knapp and our top engineers that the risks of interference have been appropriately considered. Ultimately, as the Commission's most recent report states, the extensive testing by OET has provided "proof of concept" for some types of unlicensed devices to use the white spaces. Further, as this order makes very clear, *all devices* that in the future will be used in the U.S. market must first be tested and approved by the Commission, in effect demonstrating not only that there is "proof of concept" but also "proof in practice." While some commenters have criticized our testing process, both our 2007 and 2008 reports on the test results were peer reviewed by other expert staff within the Commission. In addition, these tests were open to the public, and at various points during this long process the work of our OET engineers was observed by members of the press,

Congressional staff, broadcasters and other industry representatives, as well as other interested parties. I am proud of this transparent and open process and encourage OET to continue this commitment going forward.

Based on the results of these tests, this order establishes a number of significant technical rules related to unlicensed use of the white spaces in the broadcast bands. Power limits are more restricted – to 40 milliwatts, or a fraction of what wireless microphones are authorized to employ in these frequencies – when operated in a channel immediately adjacent to a broadcast service, while a higher power limit applies in non-adjacent channels.

In addition, and of critical importance to incumbent providers, the item will create a database of existing operations in specific channels and entire geographic areas that will remain *unavailable* to all unlicensed operations. Incumbent providers may register their locations and unlicensed devices will be required to first verify that channels are available *prior to* accessing them. This requirement to use this database will ensure that broadcast operations, broadcast auxiliary services (BAS), cable head- ins, public safety operations, and venues such as sports stadiums and theatres may register their locations and receive complete protection.

The Commission, through our Enforcement Bureau and in consultation with the Office of Engineering and Technology, will investigate complaints of interference and take appropriate action, as we do with all cases of interference. I regret that my colleagues were unwilling to set forth in this item a more specific and swift process to deal with complaints of interference. I remain concerned that the item is too vague and does not provide necessary protections after the interference has occurred.

I also would like to acknowledge the legitimate concerns of the many industries that already provide valuable services using the core TV bands. Regarding the concerns of my friends in the broadcasting industry, I want to recognize the important service they provide, especially all they currently are doing to make the DTV transition as smooth as possible. Over 90 percent of Americans are now aware of the DTV transition, thanks in large part to the more than $1 billion invested by the broadcasters and the cable industry in PSAs and other public education efforts. The DTV transition is an extraordinary opportunity for consumers to receive revolutionary TV picture quality and additional programming, as well as new wireless services that will be available in the 700 MHz band when those channels are vacated. In addition, broadcasters play a key role in providing emergency alerts to the public. It is, therefore, imperative that TV broadcasting continue to be protected from interference. The power limits and other technical rules, proposed by OET and adopted herein, are designed to do just that.

Regarding similar concerns by cable operators, I also recognize the potential for interference, both to cable head ends as well as to cable connections inside the home. The item takes steps to protect cable head ends by restricting use of white spaces devices in the geographic areas in which head ends are located. With regard to cable systems in the home, OET has attempted to establish power limits that will lower the risk of interference between devices in the home, though this risk is not, in my mind, fully mitigated. However, the Commission does not generally focus on interference that users cause to themselves. Just as we have all learned to move devices away from each other in the office and we no longer place our cell phones next to the computer, so might consumers need to reduce interference inside their home by moving devices, at least until the next generation of cable equipment becomes more widely disseminated. I encourage device manufacturers and indeed the

Commission to provide information to consumers regarding these devices as they come online, as well as establishing 800 numbers, arming call centers with tech-specific solutions, email alerts, software "fixes" that lower power and other forms of outreach to consumers.

Regarding the concerns of facilities and events managers – including Broadway, sports stadiums, churches, my friends at the Grand Ole Opry, and many others – the geographic database provides a critical line of defense. In addition, the item will make spectrum available in two channels above Channel 20 in those markets that are particularly congested due to public safety operations in Channels 14 –20. Taken together, these measures should protect the vast majority of wireless microphones.

REMAINING CONCERNS

While the use of white spaces by unlicensed devices holds great promise and I am excited about their potential, I also have several significant concerns that I believe we should have addressed and encourage a future Commission to re-address.

With regard to the possibility of interference, I regret that the Order does not include language that would specifically state the legal responsibilities of those who provide these new unlicensed devices. In particular, I would like providers of these devices to have clear notice of what they must and must not do under the Commission's rules. Perhaps more important still, I wanted to ensure that our rules specify that, in the event of significant interference caused by an unlicensed device, the party responsible for this device will also be responsible for rectifying the problem and *assume the cost.* Some companies have assured us that this will be the case; that their business reputation requires it and it is indeed "good business" for them to correct the situation. However, the potential impact of millions of devices in the marketplace calls for a prudent process for recall, provisions for mitigating interference caused by faulty equipment, and a clear exposition of fines and penalties for violations. I felt it was imperative to deal with this on the front end, but today's item is not sufficiently clear on these matters. I hope the next Commission will address this *before,* rather than after, any harm occurs.

In addition, I am disappointed that this item does not take more specific steps to address higher-power fixed operations in rural areas, which could have been used to provide much needed backhaul, a key component of broadband service in rural communities. While I understand that the Commission will issue a Notice of Inquiry on this issue, this is too little and too late. Rural consumers need help now, and the record on the various options – including licensed approaches – is sufficiently developed for the Commission to take action today.

More fundamentally, and most troubling about today's order, the Commission's decision makes it difficult if not impossible to allow anything other than unlicensed use in the white spaces of the roughly 300 megahertz that comprise the TV broadcast spectrum. Other valuable uses, such as licensed operations, are precluded. I am not convinced that *all* of the white spaces in Channels 2 – 51 needed to be made available for unlicensed use. Indeed, many of the companies that have discussed with me their exciting new business models have focused only on the use of Channels 21 –51.

As with any policy decision, we must be cognizant of the opportunity cost – that is, we must look with a clear head at that which we are giving up. This is extremely valuable spectrum – beachfront property spectrum – with estimates of the value of white spaces under a licensed approach ranging from $8 billion to over $24 billion. Even if the low estimates in this range are the most accurate, we must consider this lost value to the U.S. Treasury and, ultimately, to taxpayers. Even more important than the lost auction revenues, the consumer benefits of spectrum use in a licensed regime are many times greater than what service providers pay to the U.S. Treasury. Of course, the benefits to consumers of services in unlicensed bands also may be significant. But if all white spaces spectrum in this band is set aside for unlicensed use – rather than appropriately balancing licensed and unlicensed approaches – we are likely to ignore the huge value of alternative uses. In short, we need to consider the tradeoffs, and the American consumer – in urban as well as rural areas – will suffer if we do not.

Based on my conversations with OET, independent engineers and various members of the White Spaces Coalition, it would appear that allowing unlicensed use of the white spaces only in Channels 31-51 would be sufficient to provide four channels – that is, 24 megahertz – in even the most congested markets and many, many more channels in suburban and rural markets. As an example, 24 megahertz is greater than the amount of spectrum made available in the largest block in the 700 MHz auction held this spring. In that auction, the C block sold for $3.7 billion, despite significant use restrictions that arguably lowered the final bid. While four channels of white spaces creating 24 megahertz would not represent contiguous spectrum, this nonetheless would be sufficient for broadband services in these highly congested markets. Other urban markets would have somewhat more spectrum available, while rural markets might have as much as 100 megahertz more spectrum available.

Accordingly, I am not convinced that making Channels 21 –51 available only for unlicensed use is necessary to create the types of exciting new services that have been predicted. I am even less convinced – and the record does not support – that we must make the entire core TV band, Channels 2 – 51, available for such use. This is more spectrum than was requested by most of the parties who argued that they could provide new and innovative services using Channels 21 – 51. Therefore, while I supported moving forward to allow a portion of the white spaces be made available for unlicensed use, I respectfully dissent from including all channels in the band plan in this order.

I thank Chief Julie Knapp, the fine staff at OET, and all those who have worked so hard on this item, including the many hours spent testing and analyzing devices. However, this was only a step on this incredible technology journey. Testing on individual devices will now commence and must be subject to the same stringent, thorough, and transparent procedures and reporting we have seen thus far in this process, consistent with the Commission's procedures for all device certifications. In addition, it is of critical importance that the Commission establishes and ensures a process that will immediately respond to and mitigate any interference experienced by incumbent users and individual consumers alike.

In: Tapping the Television White Space Spectrum: A Revolution... ISBN: 978-1-60692-956-8
Editor: Julie E. Hendriks

Chapter 9

STATEMENT OF KEVIN J. MARTIN, RE: UNLICENSED OPERATION IN THE TV BROADCAST BANDS; ADDITIONAL SPECTRUM FOR UNLICENSED DEVICES BELOW 900 MHZ AND IN THE 3 GHZ BAND, SECOND REPORT AND ORDER AND MEMORANDUM OPINION AND ORDER

Kevin J. Martin

Today's approval of rules authorizing the use of television (TV) "white spaces" spectrum is a significant victory for consumers.

Today's action was advocated by a diverse coalition of consumer groups, technology leaders and Internet pioneers. Opening the white spaces will allow for the creation of a WiFi on steroids. It has the potential to improve wireless broadband connectivity and inspire an ever-widening array of new Internet based products and services for consumers. Consumers across the country will have access to devices and services that they may have only dreamed about before. I fully expect that everything from enhanced home broadband networks, to intelligent peer-to-peer devices, and even small communications networks will come into being in TV "white spaces."

More specifically, the rules adopted by the Commission establish a framework for enhanced wireless communications by building on a proven concept: the safe deployment of new, intelligent devices in the unused spectrum that exists between television channels.

We do so without disrupting TV services, wireless microphones, or other permitted uses of this spectrum. I have always said that opening the white spaces must be conditioned on protecting primary spectrum users from interference.

We have been cautious in our approach. Significantly, the Commission embarked on extraordinary testing. Not only did the Commission consider thousands of comments, ideas and recommendations submitted by the public, we opened up the testing of the prototype white spaces devices that had been submitted to the FCC. For months, both proponents and opponents of opening the white spaces participated in laboratory and field testing conducted by our engineers. OET also released a lengthy and detailed report of its findings to the public.

Normally, the Commission adopts prospective rules about interference and then certifies devices to ensure they are in compliance. Here, we took the extraordinary step of first conducting this extensive interference testing in order to prove the concept that white space devices could be safely deployed.

Prior to going to market any white space device will undergo a rigorous certification process. Today's item protects broadcasters' operations. It also protects entertainment, sports and other significant venues, including the unlicensed operation of many wireless microphones and other wireless devices in those areas. Additionally, channels have even been set aside to protect wireless microphones in major markets.

In order to protect broadcasters, all white space devices must initially use geo-location in conjunction with database consultation to ensure they can operate in a particular location without interference. Manufacturers remain free to demonstrate that sensing-only devices can operate safely and efficiently, without causing interference to television viewers. Such devices will only be deployed after they have demonstrated their effectiveness through a vigorous, public, proof of performance process.

Similarly, manufacturers are challenged to show ways that devices using higher power levels that authorized today can be deployed on adjacent channels.

In conjunction with our other actions today regarding Verizon-Alltel and Sprint-Clearwire, opening up the TV white spaces furthers the goal of making wireless a powerful, competitive broadband platform and offers significant benefits to consumers across the country.

In: Tapping the Television White Space Spectrum: A Revolution... ISBN: 978-1-60692-956-8
Editor: Julie E. Hendriks

Chapter 10

STATEMENT OF COMMISSIONER ROBERT M. MCDOWELL, RE: UNLICENSED OPERATION IN THE TV BROADCAST BANDS, ET DOCKET NO. 04-186 AND ADDITIONAL SPECTRUM FOR UNLICENSED DEVICES BELOW 900 MHZ AND IN THE 3 GHZ BAND, ET DOCKET NO. 02-380, SECOND REPORT AND ORDER AND MEMORANDUM OPINION AND ORDER, FCC 08-260

Robert M. McDowell

To paraphrase astronaut Neil Armstrong as he became the first person to step on the moon, today the FCC is making both a small step and a giant leap. It will be a giant leap for American consumers to be able to use the untapped television "white spaces." At the same time, we enter this new frontier with a small step in the form of a prudent and cautious order that sets up safeguards to ensure that new unlicensed devices do not cause harmful interference with licensees and other users of these frequencies.

While new broadband technologies are the most likely uses of these channels, the most exciting part about our action today is that we are creating the opportunity for an explosion of entrepreneurial brilliance. Our de-regulatory order will allow the market place to produce new devices and new applications that we can't even imagine today. Not only will the lives of millions of Americans be enriched by these new technologies, but I am confident that imaginative use of the TV white spaces could actually improve our safety as well.

We owe this historic moment to several white spaces pioneers including former FCC Chairman Michael Powell who initiated this proceeding four years ago, many Members of Congress, countless entrepreneurs and, of course, my Commission colleagues. I also acknowledge our chief engineer and technologist, Julie Knapp, and his entire OET team. Thank you for your patience, openness, fortitude and counsel.

To those who have expressed concern regarding the results of this proceeding, I would like to offer them comfort in the form of the text of today's order. As a preliminary matter, I note that the empirical data we have studied suggests a very real potential for deployment of new personal/portable devices, and it would be premature to either wholly endorse or close the door entirely on future developments. Accordingly, it is appropriate that the steps today are limited in scope, define an outline for our future approach toward device certification and allow for any and all changes that may be required by future circumstances.

We have listened carefully to the arguments of broadcasters, cable TV operators, wireless microphone companies and entertainers, and today's order addresses their concerns. In short, all avenues of modification and improvement remain open. And I thank my colleagues for working collectively to strengthen and clarify these protections. In fact, as technology improves, I expect that one day we will likely look back on this order and think of it as quaint; but today it is state-of-the-art.

Our decision today also obviates the need for artificial government mandates, such as those imposed on the C Block of our 700 MHz auction. Make no mistake, I have long advocated application and device portability as well as free and open networks. Consumers want these features, and the market started working on delivering them years before unnecessary, counterproductive and after-the-fact Commission mandates. Therefore, if such mandates were not necessary then, they are even more unnecessary after our action today. Robust unlicensed use of white spaces will give nimble entrepreneurs the freedom to disrupt the market in positive and constructive ways that will force incumbents to keep pace with this new revolution. As a result, the pressure created by dynamic competition will knock down barriers created by walled gardens and pry open closed networks. This liberation will come about not through increased regulation, but through increased competition. Our ultimate shareholders, American consumers, will reap the dividends.

I also appreciate my colleagues' support for a notice of inquiry to commence an examination of possible other limited uses of white spaces, such as point-to-point backhaul in rural areas as a substitute for special access. This is an idea that was raised in our original notice, is thoroughly discussed on the record, and is already ripe for decision. I still believe that a further notice of proposed rulemaking would have presented us with a clearer path toward a closer examination. Nonetheless, I hope my colleagues in the next Commission will move forward expeditiously on what I believe to be a win-win solution.

Here is the challenge: all wireless services have to be backhauled to the PSTN and the Internet via a network of some kind. Over the years, some people in the tech industry, as well as some of my colleagues, have complained about a lack of competition in the special access market which, they allege, artificially drives up backhaul costs. Competitive local exchange carriers (CLECs) and competitive wireless carriers presented us with a possible solution to this challenge for rural areas which enjoy more unused spectrum than urban areas and suffer from the least amount of special access competition. Some form of limited white spaces point-to-point licensing may allow entrepreneurs to find more efficient paths for their backhaul needs while leaving the lion's share of white spaces spectrum on the table for unlicensed users.

I am disappointed by some in the tech industry and public interest community who seem to oppose the Commission even discussing this issue while also complaining about the condition of the special access market. All of us should remember that CLECs, wireless companies and their supplier partners are also part of the tech industry and their voices should be heard as well. Hopefully, this notice of inquiry can plant the seeds of progress to find a workable solution that inures to the benefit of all parties, but mainly, American consumers.

Our decision today is revolutionary. It does more than simply foster the development of exciting new wireless services. It also has the potential to sustain the continuing good health of an "old" technology: television broadcasting, by protecting the interests of the millions of consumers who continue to rely on broadcasting for their news, entertainment and public safety information.

Accordingly, I strongly support today's order.

In: Tapping the Television White Space Spectrum: A Revolution... ISBN: 978-1-60692-956-8
Editor: Julie E. Hendriks

Chapter 11

STATEMENT OF COMMISSIONER JONATHAN S. ADELSTEIN, RE: UNLICENSED OPERATION IN THE TV BROADCAST BANDS; SECOND REPORT AND ORDER AND MEMORANDUM OPINION AND ORDER, ET DOCKET NO. 04-186

Jonathan S. Adelstein

Today the Commission takes a critically important step towards managing the public's spectrum to promote efficiency, and to encourage the development and availability of innovative devices and services.

White spaces are the blank pages on which we will write our broadband future. I have long called for not only a "third pipe" into the home but a "third channel" directly to consumers, wherever they may roam. Today, we make a key advance toward a third channel for broadband competition and deployment. This Order is a central element of the national broadband strategy that I have advocated for quite some time.

Today's decision is consequential to our nation's future because wireless broadband has the potential to improve our economy and quality of life in even the remotest areas. One of the best options for promoting broadband and competition across the country, particularly in rural areas, is maximizing the potential of spectrum-based services. Because we are a nation of innovators and entrepreneurs, the Commission's decision to open fallow spectrum to new uses will give our country an opportunity to reclaim its place as a world leader in broadband deployment.

Wireless devices have become a central part of many of our lives. As the market continues to expand, so too does our need for spectrum and infrastructure that can meet the rising demands. Our job as a Commission is to pack as much data as possible over the public spectrum without causing harmful interference. Unlicensed spectrum holds by far the most promise for maximizing the use of white spaces. Our balanced approach in this order provides the flexibility and low barriers to entry needed to provide an opportunity for everyone to make the best use of this under-used spectrum. It also implements safeguards to protect those that already make valuable use of the spectrum.

The results of the Office of Engineering and Technology's study show that geolocation technology offers an accurate and dependable means of protecting spectrum incumbents from interference. I am pleased with our decision to require both master- client based and independent based devices to utilize geo-location technology and database access for identifying vacant channels. Because broadcasters have such a vital role in local communities, preservation of consumers' television reception and other services is critical to acceptance of new white spaces devices. Although spectrum-sensing technology has not yet been proven to offer the same level of protection, I applaud our support for its continuing development and look forward to the innovation it may bring.

I am pleased that my colleagues have agreed to provide details regarding the certification process for white space devices. These additional details will provide an avenue for public comment on this process and ensure that the resulting process is fully vetted by interested parties.

Many have raised the concern that broadcasters and users of wireless microphones will suffer harmful interference as a result of this Order. The Office of Engineering and Technology assures us that there are adequate protections here to prevent harmful interference to licensed users, and I accept their assurances.

It is small consolation to those that feel justifiably aggrieved by the process that there was overwhelming support for acting now, and that more time was unlikely to change the outcome. This process is not a model of transparency. When the Commission puts expediency ahead of an open process, it creates unnecessary resentment from parties that believe they were not given a fair hearing. This is not the process I would have undertaken, but since our independent engineering staff justifies the decision on the basis of their best expert analysis, I approve this item.

An issue of great concern remains, though, that many unwary consumers purchased wireless microphones that may not comport with Commission rules. An investigation is underway, and if we find that Commission rules were violated, we must act to protect the many musicians, performers and others who will need to ensure their equipment continues to work. While I am hopeful the protections we adopt today are sufficient, we will need to closely follow the impact on users of wireless microphones, and devise solutions, considering all spectrum at our disposal, so they can continue to use them with confidence they will function as they have come to expect.

I am pleased that the Chairman and my colleagues have agreed to launch a separate Notice of Inquiry (NOI) into the best use of the white space spectrum for higher powered unlicensed operations in rural areas. In order for the white spaces to achieve maximum utilization in rural areas, rural wireless Internet service providers will need cheap, available and reliable backhaul. We need to explore all ways of achieving this. Variable power limits deserve our consideration as one possible means. I would have preferred that these questions were raised in a Notice of Proposed Rulemaking rather than an NOI, and that they covered all spectrum, but getting this discussion underway is a positive development.

Historians looking at American telecommunications policy may well view today as a day that heralded change not only with the nationwide vote that is taking place at the polls, but also the vote that takes place here to open the white spaces for broadband deployment. Both portend great progress for our future technological development. Just as I enthusiastically voted at the polls earlier this morning, I vote again enthusiastically for this item here today.

CHAPTER SOURCES

The following chapters have been previously published:

Chapter 1 – This is an edited, excerpted and augmented edition of a Federal Communications Division, Office of Engineering and Technology, Technical Research Branch Laboratory Division Report, OET Report FCC/OET 08-TR-1005, FCC-DA-08-2243A3, dated October 15, 2008.

Chapter 2 – This is an edited, excerpted and augmented edition of a Federal Communications Division, Office of Engineering and Technology, Technical Research Branch Laboratory Division Report, OET Report FCC/OET 07-TR-1006, FCC-DOC-275666A2, dated July 31, 2007.

Chapter 3 – This is an edited, excerpted and augmented edition of a Federal Communications Division, Office of Engineering and Technology, Technical Research Branch Laboratory Division Report, FCC-DA-08-118A2, dated January 17, 2008.

Chapter 4 – This is an edited, excerpted and augmented edition of a Federal Communications Division, Office of Engineering and Technology, Technical Research Branch Laboratory Division Report, OET Report FCC/OET 08-TR-1005, FCC-DA-08-2390A3, dated October 2008.

Chapter 5 – This is an edited, excerpted and augmented edition of a United States Government Memorandum, dated October 28, 2008.

Chapter 6 – This is an edited, excerpted and augmented edition of a Federal Communications Commission News Release, dated November 4, 2008.

Chapter 7 – This chapter is edited and excerpted testimony by Michael J. Copps for the Second Report and Order and Memorandum Opinion and Order, dated November 2008.

Chapter 8 – This chapter is edited and excerpted testimony by Deborah Taylor Tate for the Second Report and Order and Memorandum Opinion and Order, dated November 2008.

Chapter 9 – This chapter is edited and excerpted testimony by Kevin J. Martin for the Second Report and Order and Memorandum Opinion and Order, dated November 2008.

Chapter 10 – This chapter is edited and excerpted testimony by Robert M. McDowell for the Second Report and Order and Memorandum Opinion and Order, dated November 2008.

Chapter 11 – This chapter is edited and excerpted testimony by Jonathan S. Adelstein for the Second Report and Order and Memorandum Opinion and Order, dated November 2008.

INDEX

/

E

F

G

H

I

J

L

M

N